CITIES IN ACTION

Society and the Environment

SOCIETY AND THE ENVIRONMENT

The impact of humans on the natural environment is one of the most pressing issues of the twenty-first century. Key topics of concern include mounting natural resource pressures, accelerating environmental degradation, and the rising frequency and intensity of disasters. Governmental and nongovernmental actors have responded to these challenges through increasing environmental action and advocacy, expanding the scope of environmental policy and governance, and encouraging the development of the so-called green economy. Society and the Environment encompasses a range of social science research, aiming to unify perspectives and advance scholarship. Books in the series focus on cutting-edge global issues at the nexus of society and the environment.

Series Editors
Dana R. Fisher
Evan Schofer

Saving Ourselves: From Climate Shocks to Climate Action, Dana R. Fisher
Reforesting the Earth: The Human Drivers of Forest Conservation, Restoration, and Expansion, Thomas K. Rudel
Underwater: Loss, Flood Insurance, and the Moral Economy of Climate Change in the United States, Rebecca Elliott
Super Polluters: Tackling the World's Largest Sources of Climate-Disrupting Emissions, Don Grant, Andrew Jorgenson, and Wesley Longhofer

Cities in Action

Organizations, Institutions,
and Urban Climate Strategies

Christof Brandtner

Columbia University Press   New York

Columbia University Press
Publishers Since 1893
New York   Chichester, West Sussex
cup.columbia.edu

Cataloging-in-Publication data is available from the Library of Congress.

ISBN 9780231202381 (hardback)
ISBN 9780231202398 (paper)
ISBN 9780231554534 (epub)
ISBN 9780231564991 (PDF)

Cover design: Chang Jae Lee
Cover image: Oscar M Caballero

GPSR Authorized Representative: Easy Access System Europe,
Mustamäe tee 50, 10621 Tallinn, Estonia, gpsr.requests@easproject.com

For Zenelia and all in our footsteps

Contents

Preface   ix

CHAPTER ONE
Acting   1

CHAPTER TWO
Planning   25

CHAPTER THREE
Learning   62

CHAPTER FOUR
Leading   101

CHAPTER FIVE
Scaling   146

Conclusion   185

Acknowledgments   211
Methodological Appendix   217
Notes   229
Bibliography   279
Index   305

Preface

From Ebenezer Howard's 1898 *Garden City of To-Morrow* to Ernest Callenbach's 1975 *Ecotopia*, grand visions have accompanied the rise of cities.[1] In 1964, the English architect Ron Herron's *Walking City* imagined people living in giant, mobile pods that roamed the world adapting to the environment, connecting and disconnecting from one another, and making and unmaking cities as needed.[2] Although twenty-first-century cities are a far cry from Herron's vision, the urgency of adapting to a changing environment resonates with his fantasy. Whereas critics misunderstood the robotic aesthetic as reminiscent of a weapon, Herron had thought of the design as a survival pod for dire times. This book, in contrast to the evocative proposals of avant-garde architects, is not an exercise in radical philosophy. Rather, my quest is to understand how cities—as they are—act in response to climate change.

Even if they are not literally on telescopic legs, many of today's cities respond directly to a changing environment. They choose their own destinations, walk their own paths, and build connections with each other. Cities craft sophisticated strategies, compete to be the "greenest," learn from each other, and even engage in collective action.[3] In short, some cities act against climate change. But for a social scientist, this simple observation leads straight to a conceptual conundrum. Cities are complex, emergent phenomena with many moving parts and interests.

The idea that cities can be *actors* contrasts with the conventional portrayal of cities as stages of what the great urbanist Lewis Mumford called the "urban drama."[4] What does it mean to say that cities act? Can we seriously understand cities not just as the stages of urban drama but also as actors in it? If cities can, in fact, act to make the world a better place, why do so many fail to do so?

Sociologists faced an analogous conundrum in early studies of organizations. James Thompson's trailblazing 1967 book *Organizations in Action* was among the first to theorize organizational behavior—in contrast to the individual behavior located inside organizations.[5] Germinal work in organizational sociology responded to Thompson's effort to understand what determines the action of organizations in their own right, rather than simply as a product of managers and workers. Early organizational studies could well have been written with some of the hot topics of contemporary urban studies in mind: bureaucratic efficiency, permeable boundaries, network governance, cooptation, and collective action. Some of these trailblazing contributions to organizational theory were, in fact, written about cities and how they are organized—the diffusion of civil service reforms in the early twentieth century, different patterns of municipal management and their relation to effective governance, and the cooptation of a reformist public utility company by local elites.[6] The analogy to Thompson's work motivates the title of this book, *Cities in Action*, and its agenda: to examine what it means for cities to be *in action* and to explain why some cities are restricted to inaction.

Studying cities through an organizational lens provides a fantastic opportunity to put two lines of research into conversation—organizational and urban studies—that, to my ongoing bewilderment, rarely speak to each other. For one, cities are an incredibly important setting for students of organizations—a role you apparently never grow out of, regardless of academic rank. Local governments are the largest employers in many parts of the world. Planners and architects shape the spatial dimensions of cities based on their training, philosophies, and sense of aesthetics. City managers are professionals who know perfectly well how to navigate the hybridity of having to be both effective and politically savvy. As I will argue, cities have taken on some features of organizational actors—such as strategic goals, purposive practices, and concern for legitimacy—that make them fascinating objects of study.

Yet organizational scholars have given scant consideration to cities as of late.[7] At the same time, urbanists have paid far too little attention to what past decades of work on organizations, both in sociology departments and business schools, have revealed about organizations and their propensity to change in response to a shifting social structure.

One of the greatest revelations in organization studies was the insight that organizations never act in isolation, but that they are open systems. Dominant theories of cities and city government have perhaps noted but certainly not fully internalized the reality that urban politics and economic development occur not in a vacuum but rather in the context of cultural beliefs, power structures, and social networks that exceed municipal boundaries. Opening up urban theory requires middle-range theories that consider how organizations change in the face of broader shifts like climate change, the swinging pendulum of democracy and autocracy, or technological transformations, including the internet and artificial intelligence. In this book, I show the generative potential of connecting urban and organizational perspectives to understand how broader institutions and local organizations shape cities.

My organizational perspective on cities requires a more expansive understanding of a city than the encyclopedia definition of a "large, urban settlement," but it stops shy of examining "the urban" writ large.[8] I understand the city as a place-based community of people and organizations whose interactions shape urban practices and outcomes. Taking a page out of the book of open systems theories of organizations suggests that cities are embedded in and act in response to social structures—what I will refer to as local organizational infrastructures specific to each city and global institutional superstructures shared among many cities. The view of cities as embedded in a global institutional environment *and* made up of local organizations promises not only a fresh look at cities but also illuminates the power and limits of institutions where the rubber hits the road.

City climate action is a fertile setting for making the first steps toward an integrative organizational theory of how social structure shapes cities. Rather than taking a strong a priori stance about what it means for cities to act in response to climate change, I bring an inquisitive, social constructivist lens to take a fresh look at the problem of cities and organizations—peeking over the shoulder of many organizational scholars of companies, schools, states, and associations before me while scrutinizing databases and interrogating those on the ground of urban sustainability. Although I want to stress my humility about the ability to derive practical insight from following the path of the sociologically curious, I offer some pointers about what makes some cities more proactive and capable than others.

Thinking through and spelling out the theoretical and practical implications of an integrative organizational account of cities will require the lifetime of a scholar, and perhaps a generation of scholars, to do—and you are welcome to join in this effort. My view of cities as open systems composed of interorganizational networks is not meant to undermine the excellent existing scholarship on the political economy of cities, the ecology of urban economies, or the culture of place in an increasingly globalized world, but rather to complement this canon. If anything, I hope the image of the *Walking City*, depicted in Oscar M Caballero's *Automaton* in Figure 0.1, inspires our collective sociological imagination: curiosity about how contemporary cities are bound or emboldened by wider social structures, and optimism that shaping these structures is not outside the realm of the possible.

Lyon, France
October 28, 2025

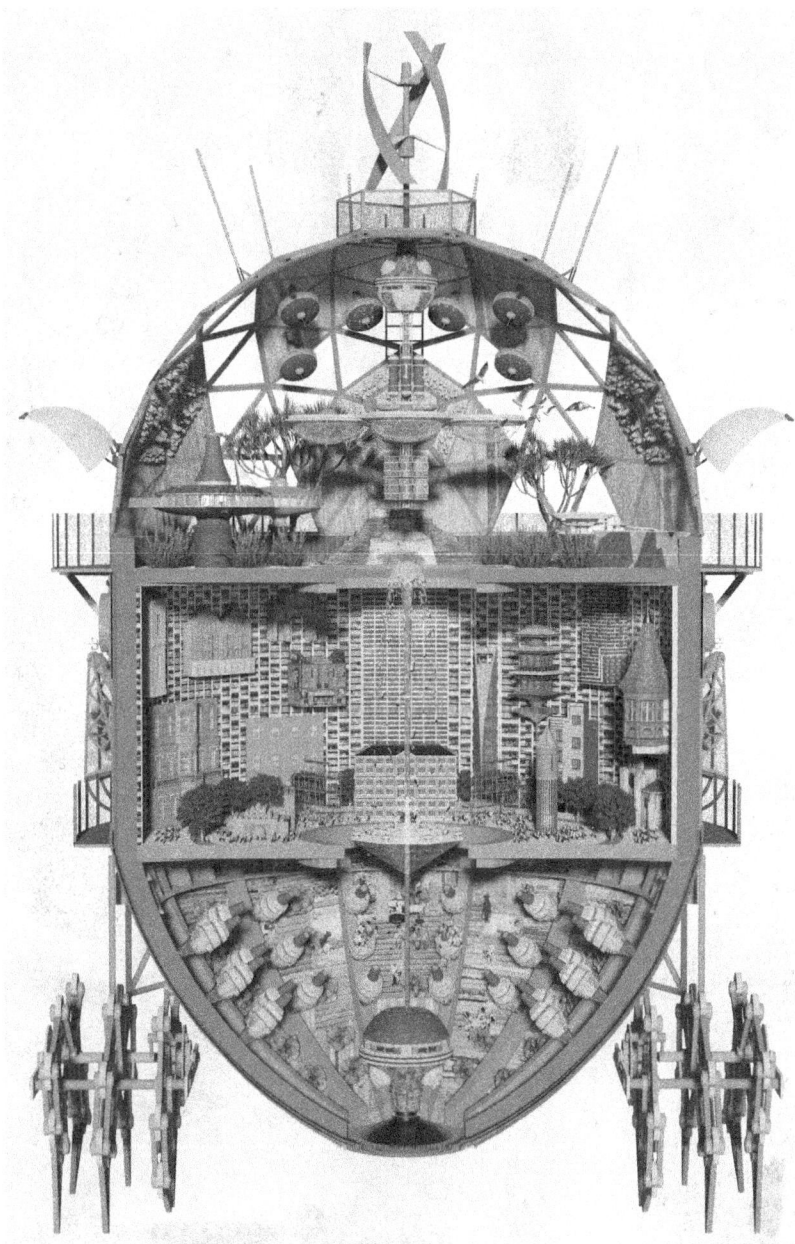

FIGURE 0.1 *Automaton* city pod by Oscar M Caballero (2025), inspired by Ron Herron's *Walking City on the Ocean* (1966).

CITIES IN ACTION

Acting

The Feds are completely abandoning climate issues in general and cities
like [ours] need to step up and keep doing more. . . . The mayor is keenly
aware of that role for [our city] as a national leader, as a global leader.

—CHIEF RESILIENCE OFFICER OF A MAJOR US CITY

In June 2017, the president of the United States stepped into the White
House Rose Garden to announce his intention to withdraw from the
Paris Agreement on climate change. He declared it was "time to put
Youngstown, Ohio; Detroit, Michigan; and Pittsburgh, Pennsylvania;
along with many, many other locations within our great country, before
Paris, France." With these words, the world's second-largest carbon
emitter started walking away from the most promising international
agreement to reduce carbon emissions since the 1992 Kyoto Protocol.
But the president had reckoned without his mayors. Pittsburgh Mayor
Bill Peduto promptly promised to "follow the guidelines of the Paris
Agreement." Greg Stanton, mayor of Phoenix, scathingly diagnosed the
federal government with a "refusal to lead" that he and other mayors
were prepared to compensate for. Chicago Mayor Rahm Emanuel, in the
British *Guardian*, promised that if the federal government "won't tackle
climate change, then Chicago will." Within a couple of days, more than
250 mayors had signed an open letter reaffirming the Paris Agreement's
goal to keep global temperatures from rising 1.5°C above preindustrial
levels. The local leaders supporting what became known as the "We're
Still In" campaign represented half the US population, and many came

from the president's own party. When the United States withdrew from the Paris Agreement for the second time, in January 2025, the mayors of Chicago, Phoenix, Seattle, and Boston, alongside those of Paris and London, once again instantly decried the executive order as an "act of climate vandalism" and, in a statement issued by the City Climate Leadership Group (known as the C40 cities), reaffirmed their commitment to "working tirelessly to cut emissions, build a green economy through a just transition, protect residents from the impacts of the climate crisis, and continue international collaboration." A well-orchestrated group of cities representing more than 20 percent of the global economy asserted their role as "bastions of climate progress," building on over a decade of work to position cities as key actors on climate change.[1]

This growing movement of cities taking climate action was not limited to cities responding to national-level politics in the United States but in fact involved cities from Accra, Ghana, to Zapopan, Mexico. City managers, urban planners, administrators, and mayors, with help from their publics, had quietly prepared plans for and taken action toward a more sustainable future long before the dramatic showdowns of 2017 and 2025. They exchanged new ideas, participated in conferences and webinars to share best practices, and prepared guidelines and ordinances to compel laggards to take sustainability seriously. Since 2014, more than thirteen thousand cities from six continents have signed the Global Covenant of Mayors, a pledge to dramatically reduce carbon emissions. Yet this number, while heartening, also implies that roughly three in four cities around the globe have *not* signed the multilateral agreement and have made no promises or plans to mitigate their emissions of greenhouse gases.[2] This disparity illustrates that the biggest concern about city climate action is not with the cities already proactive about climate change; it is that some cities are holding back.[3] All the talk about sustainability and the climate may be too little, too late if only some cities participate.[4] The heterogeneity of cities' responses also raises the question: *What does it mean for cities to act on climate change, and why are some not acting?*

I argue that the answer lies in how organizations and institutions shape the emergence, diffusion, and implementation of practices and policies among city administrations (*administrative action*) and city's inhabitants (*distributed action*). Applying what sociologists might call

Dual embeddedness		City action		

Dual embeddedness
Institutional superstructure &
Organizational infrastructure

Attribution

Aggregation

City action
Administrative &
distributed

Timely
decarbonization
Emissions

Just
transitions
Equity

Responsiveness
to shocks
Resilience

Social structure   **Outcomes**   Consequences

FIGURE 1.1 Visual summary of the arguments introduced in chapter 1.

an *organizational lens* shows that cities are embedded in two types of social structures, each of which shapes city climate action. On the one hand, a city's institutional superstructure—defined as the wider institutional environment it shares with many other cities—enables it to act as an entity in and of itself ("Jakarta designates X billion dollars for sea walls."). On the other hand, a city's organizational infrastructure—defined as its distinct community of local organizations—enables it to act as a sum of its parts ("Johannesburg's nonprofits lead the nation in environmental protection."). Figure 1.1 offers a visual summary of my overarching argument and points to its significance in terms of timely decarbonization, a just transition, and resilience to external shocks, with an emphasis on understanding the disparities in city climate action.

HOW CITIES ACT ON CLIMATE CHANGE

Let's begin with what it means for cities to act. City climate action involves but goes beyond signing treaties and making bold claims; it also entails setting goals, such as carbon neutrality by a certain date, crafting climate action plans, and working with other cities in pursuit of these goals. That is, city climate action is discretionary (setting goals that nobody forces them to adopt), purposive (laying out a clear relationship between these goals and a series of measures), and relational (learning from other cities and organizations). To understand the disparities among cities in terms of their discretionary, purposive, and relational city climate action, I began my research in the offices of people who had written the plans that the mayors then pulled out of their drawers.

Two weeks after the US withdrawal from the Paris Agreement was announced, four Metro stops away from the White House, I met one of the unlikely leaders of this move against the federal government's parochialism. A senior sustainability planner with a figurative pencil behind his ear shrugged off the fatalism about his country's climate future. In a small, windowless room, he rattled off the measures his city was taking to meet, or exceed, the ambitious climate goals set by the international community: green roofs, electric vehicles, energy-efficient buildings, growing the tree canopy. According to the sustainability planner, the media were eager to learn that D.C. was prepared: "I don't know that it changes anything. Everything you sign onto has to have some teeth to it. We feel empowered to continue our Sustainable DC work [and are] still committed to working with other cities and states that are concerned about the climate." Sustainable DC, the capital's strategic plan, understands this commitment to sustainability as a collective effort: "The District must plan ahead . . . and everyone must pitch in." Constrained by the federal government like no other place in the United States, Washington, D.C. exemplifies how cities have positioned themselves as key actors responsible for protecting future generations and the natural environment—abstract entities that can hardly object to being targets of political ambition.

My trip to D.C. showed that, by the time the mayoral campaign to save the Paris Agreement gained traction, solid strategies and networks for climate collaboration had already emerged. Some cases, such as New York's desire to compensate for national climate action, bore the mark of the mayor. Federal and state governments also played a role, like when President Obama's State Department encouraged municipal involvement in the 2015 United Nations Climate Change Conference (COP21) negotiations to ramp up its bargaining power in Paris or when California began convening climate policy leaders in Sacramento. A new generation of young protestors and even corporate activists demanding that leaders take their future more seriously has increased the pressure being placed on town squares. But the key to city climate action was inside city halls all along.

Over the course of almost two decades of research, I saw myriad plans like D.C.'s. Cities from Palo Alto to Hong Kong and from Copenhagen to Cape Town have been taking action related to climate change. Their plans have included regulating energy consumption and waste, revamping transportation and food systems, and promoting green architecture, greenspace, and sustainable finance. Although the right mix of practices is a matter of the city's specific geography, demography, and politics, there is ample potential to act. Many cities have come to be "in action": they *plan*, *learn*, *lead*, and *scale* sustainable practices that supplement and often exceed the ambitions of the nation-states in which they are located.[5]

This observation raised the question of what enables some cities but not others to engage in climate action. What I found is that, in the years leading up to their public pronouncements, city managers and urban planners, with support from their publics, had quietly prepared plans for a more sustainable future—long before the mayoral moves and movements, and climate apathy and disbelief spread among both national leaders and voters. Not from ostentatious staterooms nor the social media crowd, but from stuffy conference rooms and messy design studios, civil servants and civil society leaders got their cities moving. They exchanged new ideas, participated in conferences and webinars to share best practices, and prepared guidelines and ordinances to compel laggards to take sustainability seriously. Although enterprising mayors officially put climate action on cities' agenda, their initiatives were embedded in a broader social structure—which is the focus of this book.

In short, what cities have done and are doing to protect the climate shows that they are not just *sites* of action but actors in their own right. In this book, I explain what it means for a city to act, turn the spotlight on cities that act and show what distinguishes them from those that do not, and use an organizational sociologist's perspective to argue that whether and how cities act is a function of their *dual embeddedness* in a superstructure of global associations and peer cities and in an infrastructure of local organizations.

WHY CITY CLIMATE ACTION MATTERS

Some cities act. So what? Why cities take their role seriously in an overheating climate or remain inactive is not only important to study because cities shape climate change outcomes. How cities address climate change also offers an analytical window into the fundamental question of what it means for a city to take leadership in regard to social and environmental problems in general.

Cities' actions matter because cities play a pivotal role in both causing and addressing climate change. On the one hand, what cities do—or don't do—is a primary *cause* of climate change. The soaring significance of cities around the world has created several stylized facts about urbanization that you have probably heard nearly as often as they deserve to be repeated. Globally, the percentage of people living in cities climbed from 3 percent in 1800 to 30 percent in 1950 and then propelled past 50 percent in 2007.[6] According to the United Nations World Urbanization Prospects, roughly two-thirds of the world's population will live in cities by 2050. Cities' shares in production, innovation, and pollution are also disproportionately high. In the United States, the one hundred largest cities cover just 12 percent of the land mass but shelter two of every three US Americans, create 75 percent of the nation's gross domestic product (GDP), and are the source of 92 percent of new patents annually.[7] Cities also have a major environmental footprint: Urban dwellers worldwide consume around 75 percent of the world's energy and emit up to 70 percent of all greenhouse gases. Environmental concerns are profoundly local.[8] Therefore, the question of city climate action is not just a theoretical exercise but also an existential question. Each city's efforts are interconnected with those of neighboring communities and beyond. Even if New York City and San Francisco make significant strides in reducing carbon emissions, their progress can be undermined if nearby cities like Jersey City and Fremont do not join in.

On the other hand, cities are at the forefront of *addressing* climate change, both in terms of adaptation and mitigation. Cities also have an urgent need to develop strategies to prevent and respond to climate disasters. As urban sociologist Eric Klinenberg plainly states, "climate change virtually guarantees that, in the next century, major cities all over the world will endure longer, more frequent, and more intense heat waves—along with frankenstorms, hurricanes, blizzards, and rising seas."[9] Beyond their local needs, cities and other subnational entities like counties, regions, and states are also assuming more responsibility for climate-related planning and governance.[10] Cities are politically nimbler than nation-states, which has earned them the label of "first responders to climate change."[11] They are catalysts in the diffusion of already existing technologies and also can, at times, experiment with new practices more easily than higher-level governments. Additionally, the presence and absence of city action related to climate change and sustainability is the source of drastic disparities related to both equity and emissions. When climate action is voluntary rather than determined from the top, it is inevitably skewed—which makes the question of how to achieve not only a transition toward a more sustainable future but also a *just* transition imperative. Although some cities do a lot to further sustainability, others fall behind—leading to unequal access to sustainable neighborhoods that are clean, walkable, and healthy. This is especially true for minorities and working-class people, who often live near hazardous industrial sites and in neighborhoods that are downright toxic. Knowing which city administrations replace lead pipes, expand their public transportation, or adopt fracking bans is part of the answer to the puzzle.[12] But deciding what to do and what not to do is a difficult and often political process.

Most contemporary research on what enables cities to activate "pathways" toward sustainability transitions works under the assumption that cities can scale and share what works and ultimately can provide myriad ingenious solutions that move us closer to carbon neutrality.[13] Environmental sociologists often criticize this line of work for its ecomodernist bent, meaning that it emphasizes technological solutions, obscuring the social and political processes that shape the uptake of and resistance against new technologies.[14] *City climate action* refers to the social and political processes that shape whether and how cities address climate change, whatever they deem the appropriate technology to do so.

The case of city climate action also provides fertile ground for understanding city action in general, as it applies to various contemporary issues, such as sanctuary for undocumented immigrants, protection of civil rights, and public health crises, as with the COVID-19 shelter-in-place policies.[15] Many cities have pursued economic and social agendas that go well beyond the professionally, constitutionally, and even politically defined requirements of city governments and their administrations. Whether cities make a dent in global emissions is of utmost importance, and the case of city climate action also opens a window into a fundamental question: How and why do cities take on responsibilities and provide innovative solutions to important issues in the first place, *acting* at the scale of the city? Why do some cities act and others do not?

CITY CLIMATE ACTION
AS AN ORGANIZATIONAL PROBLEM

The most fruitful way of looking at variation in city climate action—why some cities act and others hold back—is through an *organizational* lens. An organizational lens brings into focus how sustainability solutions emerge, diffuse, and are implemented in and by cities and their organizations. It emphasizes the antecedents to actions and seeks to reveal the origins and spread of practices and policies. Not all of these practices and policies will do what they promise, or enough of it, and despite unintended consequences and contradictions—which are worth exploring—doing something to address climate change is usually preferable to doing nothing. Better options and bolder plans surely exist, but while major national and global action—banning fossil fuels, a Green New Deal, or a global carbon tax— is stalling, the contributions of individual cities add up to something remarkable in the aggregate.

Cities are both complex organizations *and* also are made up of myriad organizations. Cities do not have a simple hierarchy that allows mayors to run the city via fiat, and so the actions of cities are not limited to the upper echelons of city hall but require the participation of many actors.[16] Moreover, city climate action involves not just local governments but also thousands of private organizations, businesses, and residents, each with their interests and motivations. The distributed nature of climate action makes it challenging to assess the true impact of climate policies solely by adding up official carbon reduction pledges. Just as proactive city policies affect a city's greenhouse gas output, so do the sustainable—or unsustainable—practices of manufacturers, utilities, and service providers, within that city, which in turn trickle down to individuals employed or served by such firms.[17] Understanding how this cast of diverse people and entities acts together—or fails to—can help us identify ways to encourage more cities to take meaningful action against climate change and ensure that the benefits are shared more equitably. Like the city itself, sustainability sprawls. But exactly how this sprawl occurs in cities that are in themselves communities of interacting private and public organizations with different motives, incentives, and constraints is poorly understood.

Because my organizational lens focuses on the question of city action, certain questions remain outside the purview of this book. The emphasis on when and how cities act comes at a cost—because every lens that brings certain things into focus makes others blurry. The trade-off is that the intended and unintended consequences of these actions are in the background. Although consequences are of utmost importance and how practices and policies are chosen, contested, and can go awry is in fact the subject of "implementation science," questions about the best policies and their consequences are out of this book's focus. For instance, whether a city should incentivize sustainable transport by expanding bike paths, incentivizing public transportation, or building electric vehicle charging infrastructure is a question for another book. And whether green construction brings gentrification and displacement of marginalized communities is an important question but is not one that I examine here.

Organization studies and urban studies have chemistry—but their reactive potential is mostly unrealized. Where an organization is located is a constituting aspect of its institutional environment and its identity; place shapes an organization's people, practices, and partnerships.[18] In turn, organizations give cities much of their attraction: Museums provide access to the cultural riches of modern civilizations, public agencies at various governmental levels coordinate and finance public infrastructure and social security, nonprofits and neighborhood organizations facilitate the meaningful social relationships that structure our everyday lives, businesses—from multinational corporations to mom-and-pop stores—employ and feed a majority of urban dwellers, and schools and universities produce the human resources that underlie the gravitational force of cities as creative centers.[19]

In urban studies, card-carrying sociologists have investigated the role of organizations in constituting community. For instance, Mario Small has argued that childcare centers act as "organizational brokers" that create ties between parents that are otherwise unlikely to be made and has showed that a neighborhood's dominant race shapes whether a bank or a pawn shop is closer to your home. Robert Sampson's work on neighborhood effects understands nonprofit and social movement organizations as a source of "civic action" that is required for collective efficacy. Jeremy Levine has noted that community is constructed through nonprofits, which—channeling the earlier work of Nicole Marwell on Brooklyn—operate as unelected neighborhood representatives.[20] Organizations are the fabric that, once woven together, makes cities colorful, resourceful, and flawed.[21]

In contrast, in organization studies, some scholars have examined how the urban context shapes organizations. Christopher Marquis and Julie Battilana have argued that communities shape organizations through standard institutional channels—market mechanisms, regulative influences, social-normative influences, and cultural-cognitive influences. For instance, companies in the same metropolitan area tend to have similar levels of corporate social responsibility because they are all subject to the same cultural pressures.[22] But organization theorists could never quite settle on the boundary of their investigations, examining not only place-based communities but also digital (like collaborative online platforms) and epistemic ones (like the professions). Some organizational sociologists have documented the pervasive influence of communities over the long term through their institutional legacies, showing that companies tend to support their communities in the aftermath of disasters and mega-events and that shared experiences in the past "echo" decades later in the creation of community-oriented organizations.[23]

Although both scholarly traditions recognize that cities are inhabited by organizations, they have not produced an integrative organizational theory that accounts for the relationship between place-based communities and organizations. My historian colleague Claire Dunning and I have argued that this relationship is bidirectional and coevolving, meaning that just as much as cities are sites of organizational action, they are also drivers of it. How a community, such as a city, appears and acts is then a result of the organizations that constitute it. Take Renaissance Florence as an example. Sociologist John Padgett and his colleagues describe the city-state as a complex system of multiple overlapping networks, which means that individuals can be tied together at various levels, such as politically, socially, and commercially, or through kinship.[24] Interactions across these "multiplex networks" have contributed to the city-state's formation by giving robust power to those who speak to more than one audience at once, such as the Sphinx-like, multivocal Cosimo di Medici. Such multiplexity also gave rise to the modern partnership system, which was transposed from family to commerce and became one of the organizing principles of capitalism. Arguably, the institutions that made Florence what it once was are the outcome of these networked interactions.[25] The city shapes its organizations, and organizations make the city.

CITY ACTION AND SOCIAL STRUCTURE

City action is a special form of *social action*, and as such, it can only be understood in the context of social structure. Social action typically refers to individuals' capacity to make independent decisions about what to do—for instance, about where they live and work and whom they choose as friends and partners.[26] But what does it mean to apply the concept of social action to a *city*? This is a tricky question. A person has a central nervous system and can speak with one voice. A company may rely on a public relations office to appear as having a consistent corporate identity despite the "unresolved political problems" that it likely features.[27] Regardless of how much identity, culture, and even strategy are ascribed to a city, messy interest politics and complex, multilevel governance systems mean that cities as entities are not freely making decisions.[28] Rather, city action is possible through two complementary processes: aggregation and attribution.

Aggregation refers to how individual micro-level interactions add up to macro-level phenomena, which in turn further shape individual behaviors.[29] For instance, people who oppose animal cruelty may form an activist group that influences public opinion about animal rights, or stockbrokers reacting to economic developments may influence markets, which then shapes individual payouts. *Attribution*, in contrast, refers to how macro-level forces shape the perception of certain types of organizations as possessing "actorhood."[30] The broad concept of actorhood suggests that collectives are not so much made up of individual agents, but rather that such collectives are perceived as having a sense of self, boundaries, and responsibility for their actions.[31] For instance, once a group of people has a name and logo, puts a mission statement on a website, and registers its tax-exempt status with the treasury department, it is considered a nonprofit organization. Or in the case of this book, once a place has a city planner's office and a metro, has a museum with recognized art, and attracts company headquarters, it can be considered a "real" city. Nothing else may have changed about the group or place. The constructivist view of social action, then, allows us to understand cities as having the potential to respond, as an entity, to climate change.[32]

City action is constituted through a mix of aggregation and attribution; the city as a whole and the city as its parts are intertwined— not mutually exclusive—loci of action. Attribution places the emphasis on *administrative city action*, which includes the adoption of practices and policies by city administrators—when a city adopts a recycling policy, it is perceived as "greener" than a city without such a policy. Being "sustainable" is attributed to the city; "sustainability" is understood as an attribute of the city. But individual components of the city's organizational infrastructure also add up to collective perceptions—such as when the majority of city dwellers and local businesses recycle, their aggregated actions also make the city "sustainable," which I call *distributed city action*. In other words, attribution and aggregation lead to administrative and distributed city action, as shown in figure 1.1.

Neither aggregation nor attribution occur in a vacuum; they are shaped by social structure, by which I mean persistent, macro-level systems that cannot be easily manipulated by any one individual.[33] Social structure constrains and enables action in many ways, including systems of power, cultural norms, and social networks.[34] Individuals are shaped by their social positions by way of ethnicity, gender, race, nationality, class, and the neighborhoods in which they live. Organizations are shaped by laws, market conditions, social connections, and reputations. As organizational sociologists Patricia Bromley and Amanda Sharkey put it, organizations are not "mere contexts for action or tools for achieving the goals of owners" but rather are shaped by cultural norms and expectations.[35] To understand what determines city action, the same logic applies: Cities can be understood as acting—and as actors— when the people and organizations that constitute them behave in a way that aggregates to a coherent set of behaviors, and when others attribute action to them.

Before I use my organizational-institutional lens to examine the evidence, I will spend a moment on its intellectual underpinnings—the vantage point from which I look through it. Two well-established perspectives on the structural influences on cities rely on fundamentally different views of what drives city development: the imperatives of economic interests and social norms. A third perspective emphasizes the fact that cities are not merely determined by social structure but rather are dually embedded in it—through organizations and institutions.

The first perspective is at home in Karl Marx and Friedrich Engel's Manchester. The political economy view emphasizes competition between places. Cities try to outperform each other to attract capital and people.[36] As a result, cities are urged to become entrepreneurial by boosting their creative industries, investing in hip neighborhoods, or backing the financial sector. Urban sociologists John Logan and Harvey Molotch argue that cities are growth machines.[37] Real estate investors develop the urban environment to keep the land market afloat. Capital owners band together in growth coalitions that pursue constant development. Growth coalitions are an important part of the power elite that rule the city, as Floyd Hunter found in an influential study of Atlanta.[38] The political-economic argument offers an appealing explanation for cities to pursue sustainable development. On one hand, private investors and landowners are among the key players controlling sustainable development, and the greening process can drive land prices up. On the other, mayors see themselves as competing for prestige and economic resources. As emphasized by several local officials with whom I spoke, there is money at stake in the form of competitive grants from the federal government and private foundations, and for decision-makers in cities, the perceived rat race among cities is a motivating factor. With coastal flooding threatening business assets and homeowners caring about green space and pollution-free air, big bucks are at stake in keeping elites local.[39]

The second perspective traces its intellectual heritage to Émile Durkheim's Paris. The sociocultural view emphasizes that cities, like other organizational forms, are subject to social norms that influence their practices. Building on Durkheim's *Elementary Forms*, social constructivists Peter Berger and Thomas Luckmann provide an influential description of the process by which certain realities become constructed as durable social facts, cultural understandings, rules, scripts, and constraints that shape the activities of actors.[40] Sociologists John Meyer and Brian Rowan further developed the idea that such institutions determine the policies and formal structures that organizations perceive as normal and that they expect to make them appear legitimate. One of the first empirical studies of the process of institutionalization was Pam Tolbert and Lynne Zucker's 1983 study of the adoption of civil service reform in corruption-ridden US municipalities around the turn of the twentieth century.[41] After early adoption among middle-class cities, civil service reform became the new default of municipal administration—even when it was not mandated by state requirements—and spread indiscriminately to an increasing number of US cities. It is exactly these social processes that, for Durkheim, "produce our mental categories—our ideas—and those mental categories have real effects in the world."[42] The application of the sociocultural perspective to city climate action is equally appealing, as it explains how individual nation-states, as members of an organizational field in the form of a "world society," have responded to institutional influences in their adoption of national environmental protection policies.[43] Cities are likewise highly aware of each other when choosing their goals and the means to reach them, as evident in the remarkable homogeneity in the tools cities use to respond to different problems. This homogeneity is driven in part by some of the typical protagonists in institutional analysis: Professional urban planners carry ideas about state-of-the-art planning from universities into cities, they meet at regular conferences to discuss what's new, and an entire industry of urban design consultants is geared toward delivering strategic plans.

Both political economy and institutional perspectives are good candidates for explaining differences in city action. On their own, however, neither gets to the core of the issue: *Why do some cities act whereas others do not?* Political economic theories are clear about the interests at stake, but they are vague about how such interests are formed and expressed. Institutionalists adhering to the sociocultural perspective highlight that organizations are susceptible to common influences but fail to account for variation in local actions.[44] These broad research programs make appearances in the subsequent chapters—in different manifestations, such as growth coalition theory, urban regime theory, the global cities thesis, and world society theory. My perspective builds on these established frameworks but enriches them with a third, meso-level lens emphasizing agentic potential within institutional constraints.[45]

My perspective originates in Vienna, where Max Weber lectured around the time he finished his essay on *The Nature of the City*, later published as part of his magnum opus *Economy and Society*. Weber described a city as an organization of people living in proximity and one that has fortified boundaries, a market for economic exchange, an autonomous court, and an administration autonomous from the state and the church.[46] Weber's version of the city brings to mind the fact that the different groups living in cities are vying for status and resources and cannot be reduced to either public *or* private actors. It would be remiss to reduce the city to the sum of the public sector entities that act on its behalf—the councils, administrations, and courts that pass and enforce legislation. Private entities trade in the marketplace, protect the city from intruders, and promote policies that bind them together. The private sector certainly includes profit-oriented manufacturers and greedy developers, but nonprofit service providers, neighborhood advocacy groups, and other civil society organizations also play their parts in the realm of civil society but without a public mandate. Ultimately, a city is both an autonomous collective and an aggregation of individual social actions.

To understand how cities as complex networks of organizations reflect society, it makes sense to look to one of Weber's contemporaries across the Atlantic. In Chicago, eminent urban sociologist Robert Park developed his *human ecology* of the city.[47] Inspired by the biological sciences, his view likened the city to a "web of life." Like nature, this complex network is characterized by interdependence, competition, and coordination.[48] This complex system of creatures does not operate in isolation but is also exposed to nature at large. For urban society, the "symbiotic substructure" of competition for scarce resources is complemented by a "cultural superstructure" that provides moral guidance through communication and consensus.[49] Park's characterization of cities as microscopes that magnify general societal developments is also consistent with Weber's view of the city as a site reflecting general economic and political changes in the world. Symbiotic substructure and cultural superstructure are thus not independent but rather are interdependent and intertwined, shaping and reflecting each other.[50] Organizational interactions—both cooperative and competitive—are critical. Cities do not exist in isolation from the social world and are shaped by social structures, power relations, and norms. For instance, municipal governance is not isolated from the city's private sector (its companies and elites), which shapes its economy, politics, and social norms.[51]

This line of reasoning pioneered by Weber and Park inspires my argument: Whether or not cities act is influenced by the groups that constitute the city (its organizational infrastructure) as well as its wider context (its institutional superstructure), which determines what practices and policies are seen as desirable. That is, cities are *dually embedded* within an organizational infrastructure and an institutional superstructure—the primary explanatory concepts of this book.

THE DUAL-EMBEDDEDNESS MODEL OF CITY ACTION

The dual-embeddedness model of city action suggests that cities act when they are empowered to do so by a local organizational infrastructure that is distinct to them and a global institutional superstructure that they share with other cities and organizations. Figure 1.2 illustrates the dual embeddedness of cities from the perspective of a single city (right) and a comparative perspective (left).

Each city has a distinct *organizational infrastructure* that includes local governments and their administrations and other formal and informal organizations within the city, such as corporations and nonprofit organizations. First and foremost, city administrations are the custodians of cities and enact a significant share of public policies, so their presence is a scope condition for explaining city action—meaning that any prediction only holds if there is some sort of city administration.[52] Corporations are also often deeply involved in city politics and the development of real estate and utilities, and, despite varying levels of interest in and responsibility for their local communities, they provide employment and broker connections.[53] Among the most important local influences are civil society organizations, which often take the nonprofit

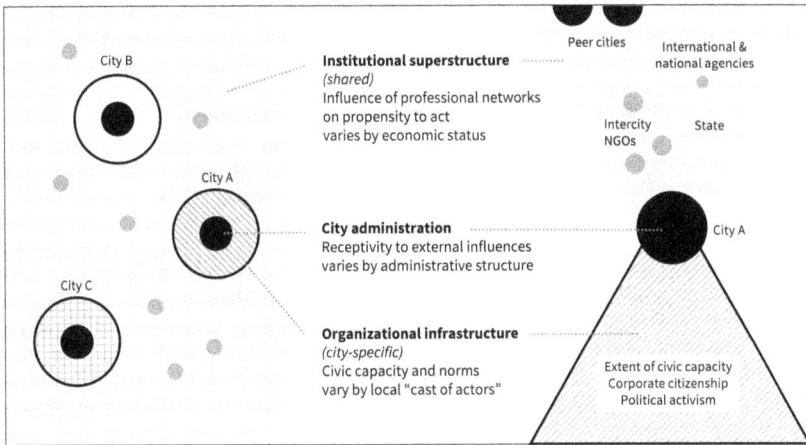

FIGURE 1.2 Conceptual representation of the dual embeddedness of cities. *Note*: *Right panel*: City administration (black dot) is dually embedded in its city-specific organizational infrastructure (patterned, bottom pyramid) and a shared institutional superstructure (top, gray pyramid). *Left panel*: Each city administration (black dot) is dually embedded in its city-specific organizational infrastructure (patterned ring) and a shared institutional superstructure (gray background). Interorganizational associations and agencies (gray dots) are connecting cities to each other.

form (although they can also take other forms like informal neighborhood groups, social enterprises, or corporations under public law). Civil society organizations are important sources of civic capacity to recognize problems, learn about solutions, and gather a critical mass to advocate that these solutions be enacted. Among the many ways in which civil society organizations underpin cities, the most important are their role in urban governance, their ability to forge social capital, and their role as brokers in economic networks. Urban social movements also shape political discourses and priorities through agitation and mobilization, even when they are not formally incorporated as nonprofits. Cities embedded in communities that expect their city to act are more likely to assume an enterprising and proactive role in society than those that are comparatively isolated or whose capacity to act is diminished by a single powerful interest group.[54]

Cities also share with each other an *institutional superstructure* that shapes the social actions of people and organizations. The institutional superstructure—the wider cultural, social, and political context a city shares with other cities—determines what ideas are considered legitimate: what a city administration or a city's inhabitants might consider desirable or at least within the realm of possibilities. The institutional superstructure is actively shaped by national and international professional associations that provide technical skills, diagnoses, and solutions to the city's problems. In addition to providing technical assistance, these associations connect cities with their institutional environment, involving peer cities and the political context emanating from regional and national governments and the economic influence of international corporations and changing global markets. Importantly, local organizations may also interpret and translate the knowledge available in these networks, pulling the information down to the local level and reconfiguring it for the local context. In sum, this environment provides role models with cautionary tales and represents institutional expectations for what a contemporary city needs to do to be perceived as a legitimate player on the national or world stage.

The framework of dual embeddedness has the power to answer our big question, namely why certain cities engage in city action—understood as discretionary, purposive, and relational practices to advance the city's prosperity—whereas others hang back. Cities with an enabling local organizational infrastructure that are also well connected to the global institutional superstructure are more likely to act. What is *enabling* is an open question that I tackle in the following chapters in greater detail, but in short, I argue that these enabling factors stem from organizations that help city administrations plan, lead, learn, and scale social innovations. When talking about city action in response to climate change, special consideration must be given to aspects of the infra- and superstructure that are specific to climate change. The content, or substantive orientation, of these practices depends on how the distinct organizational infrastructure interacts ($\times$) with the shared institutional superstructure. These structural conditions enable and constrain city action, shaping the context in which city administrators, mayors, patrons, and civil society leaders also exhibit agency, which is essential to help some cities punch above their weight, here captured by the error term (ε). This is my argument in a nutshell:

city action = institutional superstructure × organizational infrastructure + ε

In the subsequent chapters, I link organizational infrastructure and institutional superstructure to varying levels of city action. First, city administrations strive to appear legitimate and professional, which creates expectations to act. City leaders tap new sources of support by providing comprehensive strategies that highlight their commitment to critical issues of our time, engaging in branding, staging large-scale citizen involvement and stakeholder engagement, and working with outside experts. That is, city leaders *plan* for climate change—with consequences that are sometimes materially but always symbolically meaningful. Cities are also important nodes in the diffusion of novel practices and policies—they *learn* from each other and adopt practices in response to external expectations. This expectation for cities to take a *lead* on climate action is commonplace. But not all city administrators and mayors act on it; some cities are more receptive to concrete demands than others. One reason for this differential receptivity is that cities build different forms of capacity that are crucial in both administrative and distributed city action. First, city administrations with capable bureaucracies and high state capacity are more receptive to ideas from the institutional environment, but many ideas are experimented with and first introduced by actors who are willing to take risks to prove a point. A second important source of the ability to *scale* these initial ideas is civic capacity: Prosocial organizations such as nonprofits contribute to a population's ability to recognize social problems, organize solutions, and mobilize support for these solutions.

In the following chapters, I consider how organizational mechanisms such as catalysis, diffusion, adoption, and implementation link the causes and outcomes of city climate action. My organizational lens does not invalidate political economy explanations of city action that begin their analyses with political party preferences or coalitions of interest groups, or cultural ones that suggest that cities have come to embrace new roles because of broader institutional shifts. Rather, this book operationalizes contemporary theories from political sociology and institutional theory to explain the actions of cities in both form and content—the presence of a plan and how it is specified and enacted. Power struggles and political negotiations that shape city action can be quite nuanced, as ethnographers of urban governance and politics have described in detail. But what cities say in strategic plans, which problems they focus on, and which solutions they champion to a large degree depend on the kinds of interested parties that inhabit and govern cities. The dual-embeddedness view suggests that who participates in a city's governance is more important than how these interests are pursued and defended.

HOW WE KNOW

As one of the chief sustainability officers I spoke to emphasized: "In god we trust; all others need to bring data." To explain why some cities engage in climate action while others do not, I draw on multiple novel quantitative datasets and interviews with sixty leaders of cities and city associations in the United States, Europe, Australia, and Asia.[55] Broad data from around the world and deep data from one country has allowed me to model different aspects of organizational and institutional influences on cities. The data include a unique corpus of strategic plans from the world's 360 largest cities (how cities *plan*), these cities' membership in collaborative and economic intercity networks (how cities *learn*), the sustainability practices of city administrations in 1,540 small to midsize US cities from the International City/County Management Association and the National Center on Charitable Statistics (how cities *lead*), and green construction policies and construction data from the US Green Building Council and the Census Bureau across more than five thousand US places (how cities *scale* their actions). The data span many places—including cities, towns, and villages with more than twenty-five thousand inhabitants—and a broad spectrum of climate-related practices, from policy and practice adoption to actual implementation.

By including not only the largest global cities and but also all cities, towns, and other place-based communities in the United States with more than twenty-five thousand residents in the sampling frame, I have avoided common problems in city research such as selection bias of self-reported data and the overrepresentation of successful cases. Surveying a wide range of cities, including those that typically lag in climate action or do not self-report emissions or practices, has minimized sampling biases. Additionally, the long-term horizon of the study, spanning nearly two decades from 2000 to 2023, allows for examining the evolution and temporal dynamics of cities' practices and policies.

These aspects of the data—longitudinal and comparative—give this organizational lens the sharpness needed to examine the diffusion of practices in the private and public realms and the main drivers among their adoption.

Although quantitative and comparative data are well-suited for capturing general patterns related to the emergence and diffusion of practice and policies, it is also crucial to speak to people who put plans into action to understand the underlying mechanisms. Interviews alone should not be taken at face value as evidence of cities doing a good job—why would they undermine their own work?. Nevertheless, they gave me a sense of what I was looking at when I later dug through archives, websites, and survey responses of thousands of cities, providing insight into the strategic planning processes and the dynamics of climate action initiatives. I have sought to provide sufficient context for what they shared about their remarkable efforts while masking their identities to protect them from potential professional repercussions. This blend of quantitative breadth and qualitative depth allows the research to capture both the macroscopic patterns of city behavior and the microscopic details of administrative and strategic decision-making that give rise to these patterns.

As such, in this book, I provide the most empirically comprehensive and theoretically grounded work focused on understanding city action to date. My research designs are laid out in each chapter, and I direct readers seeking more methodological detail to the methodological appendix and several peer-reviewed academic papers underlying the results cited there.

WHAT WE WILL LOOK AT

In the following chapters, I will explain why and how cities take climate action, by examining how *cities plan* their bold goals related to climate change and sustainable development more broadly, how *cities learn* through relationships with each other and external experts to implement these goals, how *cities lead* climate action by adopting exemplary sustainable practices that inspire others, and how *cities scale* proofs of concept to achieve meaningful transitions to sustainability. Across these four instantiations of city action, city administrations are the primary drivers behind city climate action, but their success is critically contingent on support from various organizations in the public and private spheres—from intercity associations to states to local nonprofit organizations. Together, I use the case of climate change to show, first, how cities act discretionarily, purposively, and relationally and, second, why some cities are *in action* whereas others show *inaction*.

Two separate datasets form the study's evidentiary basis. Chapters 2 and 3 feature data on the creation and adoption, and substantive thrust, of city strategic plans to demonstrate that some global cities are more proactive about climate change than others because they are integrated in global networks that facilitate learning, social networking among city administrators, and sharing of cultural norms. That is, the institutional superstructure in which cities are embedded shapes whether and how cities plan and learn. Chapters 4 and 5 draw on survey data of US city administrations as well as data on the uptake of buildings and policies related to green construction from the US Green Building Council. I explain how local organizational infrastructures create civic and state capacity and thus shape the implementation and governance of sustainability programs. The chapters are outlined as follows.

In chapter 2, I trace the remarkable rise of city action among the largest cities worldwide. The practice of strategic planning showcases the purposive nature of city action. To craft strategic plans, city administrations work with stakeholders to make long-term, holistic, and comprehensive plans that formulate their vision for the city and set out their goals and the means for achieving them. What leading cities propose to do is not only relevant for their functional response to climate change but also is also an indication of the extent to which cities act by attribution—much like firms or other organizations manage their identity and legitimacy. Examining the adoption and content of public-facing city strategies, I show that the best way to understand city climate action is as a *rationalized practice*—an action that satisfies expectations emanating from the wider institutional superstructure but with open-ended outcomes.

In chapter 3, I examine how cities' emerging institutional superstructure has shaped the adoption and content of strategic plans. Which cities first adopted strategic planning as a response to climate change, and why, is directly related to those cities' positions in the broader social fabric, or institutional superstructure. I focus on the learning that occurs among cities to understand how the development of a network of professional associations has affected the form and content of city strategies. Learning from best practices and participating in global debates about climate change reflects the fundamentally relational, collaborative aspect of city action. At the same time, being left out of the institutional superstructure—because countries curb local autonomy or because competition takes precedence over collaboration—holds some cities back.

In chapter 4, I probe the constraining role of higher-level governments to understand how discretionary city action really is. Examining the uptake of sustainability practices among city administrations in the United States and barriers to doing so reveals that planning is part of a suite of actions through which cities lead. I provide evidence of how the local organizational infrastructure adds to and amplifies influences from the institutional superstructure. I also show how cities in places where climate action is less legitimate overcome the institutional barriers to urban sustainability. Holding the political context constant to a single country shows how even smaller cities are activated by global discourses and expertise from intercity associations. Cities, however, can be stifled by inactive and preemptive states and a lack of both material and symbolic resources to overcome these barriers.

In chapter 5, I take a step back from attributing action to city administrations and ask how the entire urban ecosystem works together to aggregate city action. Drawing on detailed data of the diffusion of an urban innovation, I show how certifications for energy-efficient construction spread in and among cities, and what sped up this distributed diffusion process. Realizing the idea of cities as networks of public *and* private organizations, I show under what conditions cities' sustainability strategies translate into meaningful action. Most sharply, I show the vital role of civic capacity in activating cities—both by being early adopters of sustainable practices and by providing proofs of concept that city administrations can act on.

In the conclusion, I review the book's main themes and findings regarding the drivers of city climate action and develop an outlook for how cities may effectively address social and environmental problems in the future. I summarize what the organizational lens showed us about emissions, equality, and innovation. I close with suggestions for understanding and enacting sustainable development as a case of urban innovation for scholarship and policymaking, offer a bottom-up view of the important role of cities in contemporary social transformations, and revisit the idea that local organizations can shape communities and society in important ways.

WHAT WE WILL SEE

Cities are not only one of the major causes of climate change; they also hold promise in proposing and implementing solutions. Cities can do a lot to lessen the blow of climate change, decrease climate impact, and prepare for unexpected crises—adaptation, mitigation, and resilience are different aspects of what I refer to as city climate action. Cities' strategies cannot be viewed in isolation; no city with a solid mitigation strategy cares little about resilience or adaptation. Some cities are exemplary on all fronts, while others lack the capacity or will to care about resilience or adaptation. The specific shape city climate action takes is contingent—it depends on what malleable terms like climate action or sustainability have come to mean at a given point in time and space.[56] Understanding why some contemporary cities are more proactive about climate change sooner than others—and how this has come to be—is my primary puzzle. My goal, with respect to city climate action, is to explain that it is not just what cities *can* do but also what enables them to do it and why. I show how city climate action—understood as discretionary, purposive, and relational attempts by cities to tackle climate change—is fashioned by an interplay of cities' organizational infrastructure and their wider institutional superstructure. The framework I develop and apply can also be useful to understand other urban frontiers, like digitalization or social equality, and may hold lessons for the governance of other entities, like nation-states or firms. Most of all, this framework crafts an organizational understanding of cities in an era in which cities, where most humans now live, are a major influence on the climate.

CHAPTER TWO

Planning

> Hurricane Sandy absolutely changed the conversation on climate,
> because [Mayor Michael] Bloomberg stood up and said we are going to
> not just respond to Hurricane Sandy. We are going to use this moment
> to prepare our city and adapt for the risks of climate change. And that
> opened the floodgates of people being willing to talk about it. . . .
> PlaNYC alone launched a whole revolution among cities that were
> doing sustainability planning.
>
> —NEW YORK CITY'S CHIEF RESILIENCE OFFICER (CRO)

A decade before I wrote this book, I worked with colleagues trying to comprehend a curious new practice among the city administrations of this world that my friends had noticed working with city managers in Sydney, Australia, and Vienna, Austria. Cities had turned to thinking about their futures not in terms of something that could be planned but something that should be strategized.[1] Sydney had learned from consultants who had an eye toward the corporate world. Vienna's city administration was shaped by a trend toward new public management, trying to breathe fresh air into the magnificent offices of the nineteenth-century city hall. Both were now thinking and talking about key performance indicators, milestones, and a strategic mindset to gain a competitive edge over other cities with whom they were in fierce competition for company headquarters, capital investments, and creative talent. Richard Florida, who claims that cities benefit from fostering the creative class, was a well-recognized name among the town halls of this world.[2] I had become familiar with both Vienna's and Sydney's strategic plans, which talked about different projects underway to make the place both competitive and livable. But then I got my hands on a plan that felt different.

In 2011, Copenhagen announced its bold promise to become the world's "Green and Blue Capital" by being the first completely carbon-neutral city. The city explained this goal in a detailed document, CPH 2025. The strategic plan included a variety of actions in the context of energy consumption, energy production, mobility, and city administration initiatives. What surprised me was not just the boldness of their plan—becoming entirely carbon neutral by 2025 while also expecting to grow by 20 percent seemed next to impossible. This vision was also vivid—the city not only wanted to become the greenest city in the world but also wanted its citizens and visitors to be able to swim in the harbor. The second thing that surprised me was an accompanying plan that still sits in my office, titled *Copenhagen: Solutions for Sustainable Cities*, which was published in collaboration with private partners. The document was odd because it was not in Danish, and its audience wasn't the general public. Because Copenhagen wanted other cities to copy them, they laid out several strategies that were working for them: bike highways, integrated public transit, swimming in the harbor, drinking water from the tap, a skyline of wind turbines, biomass heating, and an underground water-cooling system. It felt like I had a piece of the future in my hands—and the city wanted everyone to know about it.

Is there a measurable strategic turn that—even when it does not equate to urban sustainability—indicates that cities are becoming active about climate change? Was Copenhagen's plan indicative of a broader shift, or was it an outlier? This is the key question of this chapter, as it was my question then. I first took a closer look at New York's 2007 PlaNYC, a significant document produced during Mayor Michael Bloomberg's tenure of running a town hall like a business and possibly the first of its kind of this new generation of strategic plans. PlaNYC was hailed as one of a prototypical instance of a strategic plan that was emulated by many cities around the world, as the CRO proudly claimed in the introductory quote. The plan for "a greener, greater New York" laid out strategies for protecting New Yorkers from extreme heat, flooding, and energy shortages. It also strove to improve their quality of life through the construction of green spaces, waterways, public transportation, and sustainable food sources and to build a green economic engine with a focus on both green and circular economy. The plan depicted a future intended to be a drastic break from its past: "Thirty years ago, a plan for New York's future would have seemed futile. The city was focused entirely on solving immediate crises. Government flirted with bankruptcy. Businesses pulled up stakes. Homes were abandoned. Parks were neglected. Neighborhoods collapsed. Subways broke down. Crime spiraled out of control. New York seemed unsafe, undesirable, ungovernable, unsolvable. Today, the city is stronger than ever."[3]

Although PlaNYC described the newfound concern with a greener future as a big shift, its emphasis on sustainable development did not seem distinct. By the time I had moved to the San Francisco Bay Area to study sociology in 2012, the city of Vienna had published a new strategy outlining how it was going to become a smart city that keeps a tight grip on being the world's most livable city, according to a ranking by the US consulting firm Mercer.[4] This trend also did not appear to be confined to North America, Europe, and Australia, as I discovered after just returning from studying abroad at the University of Hong Kong (HKU). Hong Kong, too, had published a strategic plan—Hong Kong 2030—and the urban planning department at HKU was abuzz with debates about the role of public consultation to ensure that planned megaprojects like the Central–Wan Chai bypass would be sustainable.[5]

I also noticed something curious about Vienna's earlier STEP plan from 2004. Vienna is globally renowned for its history of public housing, which is a deeply ideological, socialist project. Deep in its strategy plan, however, a note appeared as to *why* the city needed to continue its important focus on social housing. The city pointed out that the construction of public housing projects would be more energy efficient than single-occupancy housing, and as a result, would be more suitable to reach the Kyoto goals—the pre–Paris Agreement to lower human-made greenhouse gas (GHG) emissions. Climate change was becoming what institutional theorists call a "master frame" used to justify all sorts of activities at the organization's core.[6]

Fellow researchers saw this proliferation of strategic plans as evidence of a general trend from city government to network governance—a presumed process in which cross-sector partnerships take over from town halls.[7] Having earned my academic spurs not far from Vienna's city hall, which employs more people than the entire administrative apparatus of the European Union, I could not quite believe that governments were on their way out.[8] Nevertheless, I wanted to learn more about the institutional changes underway in the world of cities. I was also intrigued that sustainability and climate change were such popular themes of strategies ostensibly meant to boost business. Of course, Copenhagen, Sydney, and Vienna not only were among the most livable cities in the world but also were some of the richest and most international. As I was wrapping up our work on understanding what these strategic plans can tell us about urban governance, I became interested in how many cities have these plans, and what they say. I decided to find out.

In this chapter, I will show that city climate action emerged as a global phenomenon, as viewed through cities' proclamations. Looking at any one city at a time, it is tempting to discard these proclamations as uninformative or even phony. From a comparative perspective, however, the global diffusion of strategic planning points to larger transformations in cities' social structure rather than any one city's idiosyncratic motives. I start by examining plans not because all the projects listed in them will necessarily be implemented, despite a plausible association between talking about decarbonization and moving toward it. Nor is it because plans mobilize a consensus about the city's future among stakeholders with drastically differing interests, although that may well be why city administrations are prone to involving the public in the planning process. The reason to take strategic plans seriously is what their presence, timing, and content tell us about those cities' purposive action for the climate relative to those that do not plan, plan late, or plan about things other than sustainability. Figure 2.1 summarizes this point visually, and subsequent chapters will add complexity to this picture by explaining when and how cities plan and act.

As I will argue, strategic plans are ideal indicators of two important features of city climate action. First, they explicitly state climate change as an important goal of cities. If and how these goals translate into sufficient outputs and outcomes is a more complex question that we will begin to

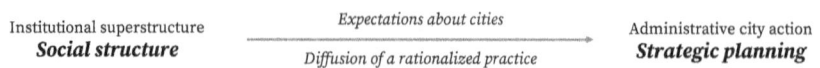

Institutional superstructure	*Expectations about cities*	Administrative city action
Social structure	*Diffusion of a rationalized practice*	**Strategic planning**

FIGURE 2.1 Visual summary of chapter 2.

grapple with and return to in the following chapters.[9] For now, however, it is helpful to know that even though they may rarely come to full fruition, the presence of bold plans tends to be associated with resources and a process in place that leads to subsequent actions. This is due to the second important feature of strategic plans: They are *purposive*, meaning that they lay out a series of interventions to meet a particular goal and indicators that can verify when that goal has been reached. As such, strategic plans are a first, albeit incomplete, indication of whether cities are positioning themselves to tackle climate change. If cities had not even expressed an intention to tackle climate change, the situation would be dire—but as it turns out, they have, and increasingly so.

Analyzing all publicly available plans of the 360 largest cities in the world, I found that not all cities have strategic plans, and not all strategic plans acknowledge climate change and the need to do something about it. This disparity underscores the problem of variation in city climate action that I laid out in chapter 1: societal problems—climate change being a prime example—require the coordination of several, if not all, local governments to be resolved. Examining the practice of creating strategies and what they say about the climate gives us a sense of the phenomenon of city climate action and shows us why it is so crucial to understand this action through an organizational lens.

I will first explain how I collected and analyzed the plans, provide examples of what these plans look like, and summarize their content showing that, indeed, climate change is among their primary goals. I will then draw on interviews with the authors and funders of the plans to scrutinize why these documents are so important for understanding city climate action: not because they offer proof of lower carbon emissions but instead because they show that many cities have adopted strategic plans as "rationalized practices" (i.e., a suite of purposive actions intended to achieve the goal of mitigating and adapting to climate change that make the city look legitimate in the eye of observers, whether or not they work.)

PILES OF PLANS

Is city strategy, and in particular cities' strategic positioning about climate change, really on the rise? It ended up taking half a dozen undergraduates and me a few months, spread out over several years, to answer this question in the form of several piles of plans (both virtual and physical) published by cities. We pulled together a distinct archive of city strategic plans and tried to identify when each city started publishing its plans, and with what goals in mind. Three methodological decisions at this stage deserve consideration; for others, please see the methodological appendix.

The first decision we made determined what we would look for. Although we initially drew a wide net, city strategies had to meet three exclusion criteria before we considered them evidence of a profusion of city action: They had to be *comprehensive*, city-authored, and public. First, documents had to be comprehensive in that they related to the entire city rather than to a specific project or task. Some plans had a narrow focus on specific challenges that the city encountered, such as trying to improve its tourism profile or addressing the renewal of a particular neighborhood. Many Chinese cities have produced specific development plans for the national government's five-year plans, but have failed to create a broader, independent vision for the city with an eye on the local and international communities. Second, the plans had to be *authored by the city* rather than a funder or another external stakeholder. Several sophisticated plans were written by prestigious research institutes at universities, think tanks, or national planning agencies that assisted local planners. Many African and Latin American plans, for instance, were produced either by international organizations like the World Bank or nonprofit research institutes; many Indian plans were written with the help and quill of the national planning commission or state development agencies.[10] We had to exclude those documents because they did not reflect discretionary goal setting in the context of city action. Third, documents had to reflect a *public mandate* rather than just offer a technical plan for internal use. Several documents did offer a long-term plan for the city, but they did so in strictly technical or physical terms without painting a broader picture of the city's vision, without the widespread participation of the public in producing the plan, and without articulating any strategic goals. The act of long-term planning is hardly a new development; it is rather the public audience and articulation of people-centered goals that makes this new suite of plans noteworthy.

The second decision we made directed where we would look. We decided to examine city websites—current ones as well as those archived by the Internet Archive, a nonprofit digital library that has been making snapshots of webpages since their inception. This, of course, does not mean that no urban planning whatsoever takes place in cities without a strategy document. It is the public-facing, ambitious positioning of a plan like Copenhagen's or New York's that we were interested in.

The third decision we made identified which cities we would search. Many contemporary studies of cities cherry-pick those that you have most likely heard of London, Paris, Berlin, Chicago, or Vienna. The landscape of today's cities looks a little different than it did in 1900, which is when these six cities dominated the urban landscape. The largest cities in the world are no longer London and Paris. They are Tokyo, Delhi, Shanghai, Sao Paulo, Mexico City, and Cairo. Among the top one hundred largest cities, twenty are in China—including Dongguan, Guangdong; Changchun, Jilin; and Hangzhou, Zhejiang. Forty-six Indian cities exceed the population of Amsterdam—including Durg-Bhilainagar, Chhattisgarh; Bhopal, Madhya Pradesh; and Lucknow, Uttar Pradesh. In other words, the sample—and therefore the urban landscape—becomes highly skewed when one only focuses on cities that are historically highly integrated into the world system, regardless of their population.

We therefore examined all 360 cities above 1.25 million inhabitants worldwide—not an indisputable cutoff, but one that meant that we included the majority of capitals and had a manageable workload. Our primary goal was to capture the state of the world in 2016 before the fallout of the US withdrawal from the Paris Agreement, but how far did we have to go back? It is difficult to state exactly when the first formal documents outlining a city's future in a comprehensive and strategic manner appeared, but it was probably sometime in the early 2000s. Some would argue that New York's PlaNYC in 2007 was the first real plan, but other reasonable contenders include Paris's 2007 Climate Protection Plan, Ottawa's 2003 environmental strategy, and Vienna's STEP 2000. Some plans in London, Seoul, Ottawa, Sapporo, Montréal, and Calgary had strategic features before New York's watershed moment chronicled at the beginning of this chapter. None of these plans met all the criteria of a full-fledged strategic plan, but they were nonetheless unusual documents in the grand scheme of historical urban planning.

This collection of strategic plans alone, illustrated in figure 2.2, reveals a familiar story about their publication. The number of new plans has

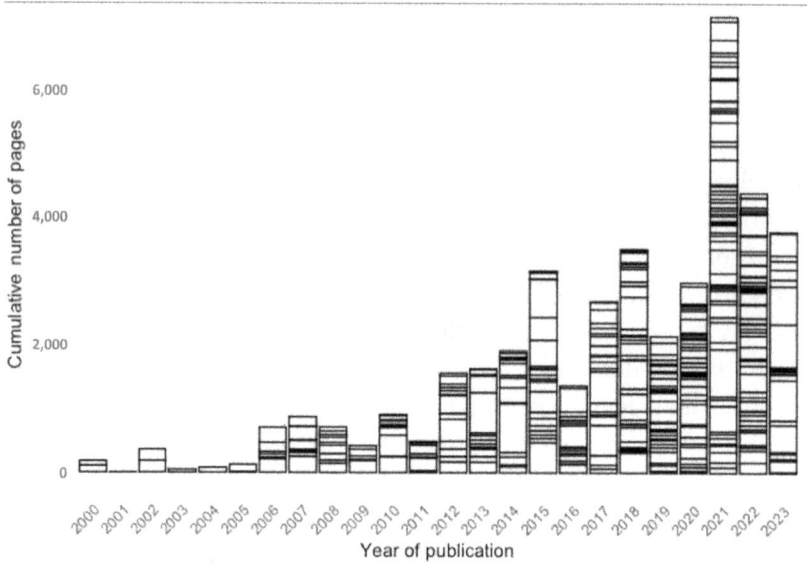

FIGURE 2.2 Strategic plans published by the world's 360 largest cities as of 2023.
Note: Increasing page count illustrates the steep increase in city-level strategic planning since 2000 and the data underlying the textual analysis.

steadily increased since the first plans were produced in the early 2000s but has slowly decelerated since peaking in 2021. This progression is typical of the diffusion pattern of innovative practices—whether that practice is changing the fridge in your kitchen to be HFC-free or giving disgraced upper management golden parachutes—and we will see it a couple of times in this book.[11] This pattern was first discussed by Everett Rogers, the pioneer of the study of how innovative technologies and practices diffuse. Initially, only a few innovators pioneer a practice.[12] After a minority follows suit and potentially perfects the practice, a majority shift gears. At some point, those that are likely to ever pursue this kind of practice is reached, and the number of new adoptions plateaus. Just like most households have a fridge and a television, by 2023, most cities had adopted a strategic plan. That does not mean that strategic planning and strategic plans do not continue to be a hot topic among planners, but rather that new plans are revisions, updates, and reports instead of entirely new documents for a city that had never before had a strategic plan.

By mapping the strategic plans over time, it is possible to reveal the global dimension of this diffusion process. Only twenty-six cities have a plan on record that dates before 2006, as shown in the top panel of figure 2.3. These early plans, however, were not the last of their kind. In 2016, 158 of the 360 largest cities already had a strategic plan in place, shown in the bottom panel of figure 2.3. The map shows the adoption of strategic plans among cities in the sample by economic status, indicating that city strategies are not limited to the economically most prosperous cities or those in the Global North. As I had realized in Hong Kong, these plans were not necessarily clustered in North America and Europe but spread from Curitiba to Kyoto and Brisbane to Seoul—but in the big picture, they were scarce.

We conducted another search in 2023 and found that many cities were still active in updating, reporting on, and producing new strategies, but a few cities that didn't already have a strategy plan in place by 2016 had produced a new one by then. This means that around the time many mayors started proclaiming their intention to act strategically on behalf of the natural environment in the mid-2010s, the majority of cities that were ever going to publish a strategic plan had already done so. By then, some 190 of the 360 largest cities had published plans to position themselves strategically and to imagine a better future.[13] Thus, more than half of the largest cities worldwide have published such documents. That is a remarkable level of dissemination, considering that most practices in the world of business or even the arts are not widely disseminated.[14]

In a way, this result spells good news: Many of the world's largest and most influential cities seem prepared to take climate action. At the same time, however, almost half the cities still have *no* plans. That is, although strategic planning has become an ostensibly global and universal practice, even some of the largest cities in the world may lack the capacity or willingness to articulate a strategic position in the global system, and they certainly do not position themselves as first responders to climate change. To bring these numbers to life, by the time the Paris Agreement was rolled out in 2016, cities with a strategic plan roughly accounted for 770 million people, whereas 570 million people called a city home that lacked a strategic plan. More than half a billion

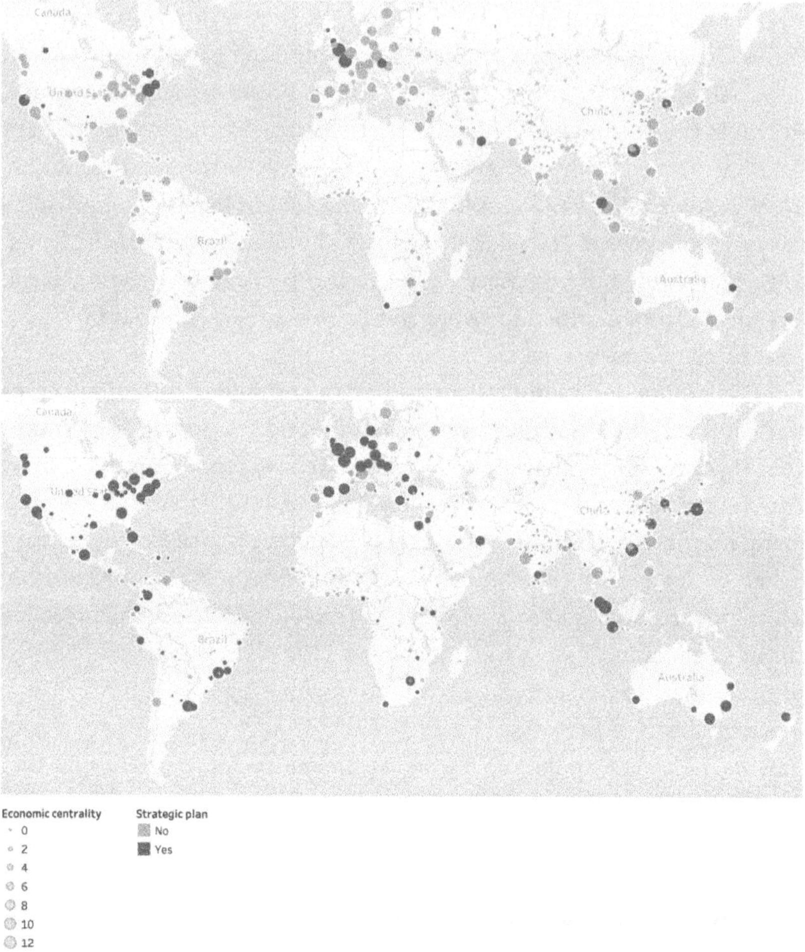
Canada
Brazil
Australia
Economic centrality
0
2
4
6
8
10
12
Strategic plan
No
Yes

FIGURE 2.3 Adoption of strategic plans over time by economic status of city.
Note: Figure shows that few high-status cities had adopted a strategy plan in 2006 but more than half had done so by 2016. Economic status measured as rank in Globalization and World Cities (see chapter 3 and methodological appendix).
Source: Map uses map data from Mapbox, https://www.mapbox.com/about/maps; OpenStreetMap made available under the Open Database License, https://www.openstreetmap.org/copyright.

people were living in cities whose planners were not participating in the global discourse about greener, smarter, better cities at all. And many of these planners had joined the discussion late! The absence of data on these cities is meaningful because it illustrates once again the great disparity in the global discourse about city climate action—with potentially grave consequences that we have yet to explore. (Explaining which of these buckets—in action or not—a city falls into is the focus of chapter 3.)

Most of the nonadopters are clustered in China and India, both of which have strong national planning regimes and a history of internal rather than public planning. In other regions, the reasons for nonadoption are more idiosyncratic—like in Leo Tolstoy's *Anna Karenina*: "All happy families are alike; each unhappy family is unhappy in its own way." Nonetheless, any city that could muster the resources, capacity, and political will was highly likely to have adopted a strategic plan at some point. If we understand the adoption of city strategy over the past two decades as a case of the diffusion of innovation, we have now reached the point at which few new cities are taking on this practice. Among these plans, however, are truly remarkable statements of cities trying to stay ahead of the curve.

STRATEGIZING SUSTAINABILITY PURPOSEFULLY

What do these strategies look like? The strategic plans had a lot in common even beyond the exclusion criteria of our search (i.e., comprehensive, authored by the city, publicly accessible). For one, they all sought to come across as *visionary*—drawing up imagery of a desirable long-term future that justifies certain interventions and policies. Many plans were professionally produced, with images of citizens, buildings, and trees. Once distant dates often divisible by five featured prominently in the titles of the plans: Abu Dhabi, Berlin, Bogotá, and Warshaw 2030; Zurich 2035; Wellington 2040; Cairo and Nairobi 2050. In a signed preface, mayors would tell the reader that their city faces grave challenges and also terrific opportunities. Most claimed a *holistic* goal—one that covered the entire city and all its issues instead of just a specific subset, such as transportation planning or urban renewal—as the primary focus of their plan: Sustainable, Resilient,

Smart, Overall Terrific! Cities stopped shy of saying that they wanted to be the best place to live in the world but that was the sentiment they expressed.

The reason these strategies are remarkable—and immediately relevant to documenting both the uptick and divergence of city climate action—is not just that they are visionary and holistic. It is also *how* they seek to accomplish these goals. City strategies provide for urgent action and implementation through milestones and key performance indicators of the long-term vision; they emphasize collaboration with other strategic actors to incentivize private actors and leverage private sector investments, and they position the city as a leader by example while acknowledging the complexity of multilevel governance by making reference to federal, state, and local levels.

New York's 2007 PlaNYC perfectly illustrates this purposive nature of plans. To put all these goals and indicators into context, a benchmark page compares New York's performance to that of other global cities from Rio de Janeiro to Johannesburg. PlaNYC was eventually replaced by a new document, "OneNYC: Build a strong and fair city," when Bill de Blasio took over the key to the city from his predecessor. Strategic plans bear the heavy imprint of the city's mayors; nevertheless, the practice of strategic planning persists beyond the power of any individual leader or politician. Any new call for an inclusive economy, a vibrant democracy, and a livable climate is a readjustment of areas where the city had already planned to go. The NYC Office for Climate and Environmental Justice called climate change a "monumental challenge" that also "presents an opportunity for us to reimagine our communities and to create a more equitable, healthy, and resilient future."

Looking back to the successes of PlaNYC after nearly two decades of work on sustainability, in a 2023 plan titled "PlaNYC: Getting Sustainability Done," a letter attributed to Mayor Eric Adams emphasizes that the plan will "produce results that New Yorkers will feel all year long, in blue skies and gray skies alike. This plan will help ensure that every New Yorker is protected from climate threats, has improved quality of life, and can take advantage of some of the more than 230,000 green economy jobs that this city will host by 2030." The report for PlaNYC is available in Spanish, Chinese, and English and proudly announces:

New York City has been a leader on climate action since the release of the first PlaNYC 16 years ago, and today our leadership is more necessary than ever. Climate change has shifted from a threat on the horizon to a recurring aspect of our weather, with impacts felt disproportionately in vulnerable communities—leading to hundreds of preventable deaths every year. At the same time, the benefits of climate action are increasingly clear, such as cleaner air, better mobility, safer homes, and growing green jobs and businesses. Responding to and preparing for climate change means improvements to our daily lives today, and a future that is more equitable, healthy, and resilient.[15]

These same elements of purposively tackling climate change are reflected in many other plans. The 2020 plan for Amsterdam begins with an imaginary future: "Can you picture it? By 2030, Amsterdam's streets will be free of exhaust-emitting cars. By 2040, every home will have switched from natural gas to sustainable heating. By 2050, we will have ended our dependence on coal, gas, and oil." The city's alderperson for spatial development and sustainability Marieke van Doorninck is cited as saying: "Climate-neutral is becoming the new normal. And that can't happen without you." Over a crisp forty-five pages, the city lays out twenty pillars of a vision and a strategy for an "energy transition journey" in the domains of the built environment, mobility, electricity, and industry to achieve a "green, healthy, prosperous, and sustainable city for all."[16]

Across the board, global cities put forth a series of remarkable solutions that define what it means for them to act to the benefit of the natural environment. These ambitions range from reducing waste, gases, and cars to amping green jobs, space, and energy to achieving complete carbon neutrality, as shown in textbox 2.1.

Most of these cities' strategic goals have been highly ambitious. Copenhagen's CPH 2025 was not the only city planning for carbon neutrality: from Lyon to Nottingham to Madrid, cities have promised to reduce their carbon emissions drastically. In Brussels' Climate Plan, Mayor Philippe Close writes, "Cities are on the front line when it comes to climate change. Today, the effects of global warming, which are more intense in urban environments, are already being felt through heatwaves,

TEXTBOX 2.1: SOME SUSTAINABILITY PROMISES
IN CITY STRATEGIES

In the mid-2010s, many cities promised a more sustainable future in their city strategies. Most have since followed up with updates, reports, and successor plans that shift the attention to new topics, including climate adaptation:

"If you don't measure it, you can't mitigate it. That's why San Francisco requires every City department to track and *report greenhouse gas emissions* in annual [reports]" (Resilient San Francisco, 2016, 72).

"The City of Athens has implemented and will continue to implement *energy retrofits* and soft energy-saving actions within its buildings to improve efficiency and reduce energy costs" (Athens Resilience Strategy 2030, 2017, 120).

"To meet our goal of *Zero Waste*, we will expand the NYC Organics program by increasing curbside organics collection and convenient local drop-off sites" (OneNYC, 2019, 120).

"Establish a Green Enterprise Zone [that] would make the Downtown Eastside and False Creek Flats the *'greenest place to work* in the world' by focusing green companies and organizations, green infrastructure, as well as innovations in building design and land use planning in one location" (Vancouver–Greenest City Action Plan, 2011, 15).

"By 2025, Copenhagen will be the world's first *carbon neutral capital* and have a leading edge on green technology and innovation in Europe. Also, Copenhagen will be the no. 1 bike city in the world" (Copenhagen Municipal Plan, 2017, 25).

"By 2030, the share of *green spaces* must be kept at over 50 percent. Especially in a growing city, additional recreational areas must be safeguarded to keep up with the rising population figures" (Smart City Vienna, 2016, 37).

floods, droughts, and even a decline in biodiversity."[17] In light of the Intergovernmental Panel on Climate Change's (IPCC's) recommendations, he writes that "local authorities, with their localised, social and economic skills, are key players in combatting climate change."[18] Many contemporary plans include bold and explicit goalposts about carbon emissions. In its 2022 plan, Chicago promised to reduce its carbon footprint by 62 percent by 2040, problematizing disparities in affordable energy access, climate financing, and tree equity. The city reports having engaged 2,145 Chicagoans from seventy communities to create the 2022 Climate Action Plan, including a word cloud of keywords from survey responses that includes bike lands, climate, energy, community, building, recycling, and trees. The plan lists numerous actions aiming to achieve five pillars, outlines the GHG impact and city partners for each plan, and features a detailed implementation table.

Even when carbon neutrality is not yet within reach, making a meaningful contribution to the Paris Agreement's 1.5°C goal appeared to be a priority for many cities. Cited alongside representatives of the city climate change network C40, Kuala Lumpur Mayor Mahadi Che Ngah reports that "an undeniable truth that is all the more pressing in present times is the fact that cities are a major source of carbon emissions. In the same vein, it is cities that are heavily affected by the consequences of climate change."[19] Kuala Lumpur's plan explains the city's vision, offers data about how climate change affects it, makes radical plans for a "low-carbon [and] resilient city," develops priority actions, and plans for implementation. In this plan, the city shows the consequences of global warming for Kuala Lumpur—despite the fact that Malaysia is a major oil producer.

To be sure, sustainability is not the only way to frame a forward-looking comprehensive plan; resilience and smartness were other common frames.[20] Often quoted alongside 100 Resilient Cities CEO Michael Berkowitz, some mayors proclaim that the city will do everything in its power to stay resilient to impending dangers. As Rotterdam reported in 2021: "Resilient cities are prepared for the future. Rotterdam is working to become . . . prepared to engage with the opportunities and challenges of the future."[21] Detailed figures on economic, social, and environmental performance report a growing gap between rich and poor, the declining lifespan of companies in the S&P, and structural changes to the climate despite a welcome energy transition. Striving for a "balanced society," Rotterdam seeks to be "ready for the 21st century." In other words, "resilient" cities are equally concerned with climate change, but they choose a broader frame—for reasons we will explore in the next chapter.

Other frames appeared entirely unrelated to climate change at face value. For example, Hiroshima City's comprehensive plan for the period of 2020 to 2030 emphasizes the city's position as an "international peace culture city."[22] The plan uses colorful formatting and illustrations to show the city's commitment to the Sustainable Development Goals, a series of high-level goals identified by the United Nations (UN) as shared grand challenges. For Hiroshima, these goals include "well-rounded citizens," a "vibrant internationally inclusive city," and "a world-renowned City of Peace."[23] The plan states that its goal is to share the "Hiroshima city basic concepts . . . with our citizens, and to promote Hiroshima city planning together. We're working in cooperation and co-production. May we all build the future of Hiroshima together."[24] Although this plan is clearly committed to the overall idea of sustainable development and explicitly mentions the conservation of the natural environment and construction of the urban environment, sustainability is not the master frame of this document. Instead, Hiroshima was a rare example of a city that frames its future around peace and culture.

INSIDE CITY STRATEGY

Cracking open some of these documents showed that one should not judge a strategic plan by its cover. I found enormous variation in style and content among these plans. Some plans are only ten pages long but outline specific goals, whereas others are several hundred pages long and meander through the details of city planning. Some documents are stone-dry, whereas others use sophisticated images, scientific figures, and even manga drawings to illustrate their points and to make them accessible to both the general public and expert audiences. Not all plans are talking about the same issues, although in general, they do appear to share a number of core pillars, such as environmental, economic, and social considerations. It is also not clear that cities have identified their milestones, goalposts, and benchmarks to the same extent. Which plan fits the city very much depends on the city. As a planning expert at C40 told me: "It's a tricky space because what is a good plan in one city, it really depends on what they need. And there's nothing wrong with having a simple plan if you're just getting started, and what you need are some simple steps to get you moving. The problem would be if that's where you stop and you never go any further. But you don't need to do a 200-page PlaNYC. . . . That's very intense."

All plans had in common that they outline a more or less ambitious vision for what the city might look like in the future; but they do so from a diverse set of starting points and diagnoses. Most plans offer both an appreciation of current strengths and an articulation of opportunities for doing better in the future. As I argued with my colleagues from Vienna and Sydney, these different emphases reflect the historical governance configurations in different cities along such dimensions of time and space, what is seen as matters of public concern, and who is included in the cast of actors of local governance.[25]

City strategy may be what organization theorist Barbara Czarniawska calls an isopraxis: a homogeneous form, even if the content may be all over the place. This heterogeneity in content is meaningful because isomorphism—similarity in form—is a familiar phenomenon among organizations: Following a common (although not always correct) recipe of success, organizations adopt a certain set of practices to please their audiences and to appear like a good, cutting-edge organization.[26] Isomorphic practices do not always make sense, at times, because the needs of a company (or city) may diverge from what leaders in the field are doing. Isopraxis leaves open the possibility that the organizational reality is not quite so rigid: Of course, organizations adapt the content of plans so that they fit the local context.

After combing the sample for all documents that could reasonably be viewed as strategic plans in the sense of city action based on their title and format, we were left with more than 650 strategic documents published by some three hundred cities over the course of two decades. Some of these plans are ambitious and creative proposals to make cities more resilient or sustainable. But not all of these documents established a strong position about the climate. What can they tell us about city climate action? To find out, I worked with a small group of talented research assistants under the leadership of Stanford students Olivia Rambo and John Zhao and Foothill College students Donnesh Farmanfarmain, Francesca Leventhal, and Aitran Doan. We divided one hundred plans—all clear cases of a comprehensive, city-authored, and public strategic plan with names like "Sustainable City 2045" or "Resilient City"—and coded their content.

Remembering the proposition of urban scholars that what is happening to urban governance is part of a larger trend, we examined a similar question that one might ask about, say, a corporate strategy. In fact, we modeled the coding guideline after a comprehensive coding guideline that organization scholars Amanda Sharkey and Patricia Bromley had used to code companies' corporate reports to assess the extent to which they behave like "corporate citizens."[27] Using an online survey to organize the work, we coded the core concepts, tables of contents, tools, and aesthetics of the plans in quite some detail to get a better sense of the core features of a modern city strategy. My interviews with city leaders about the strategic planning processes they had spearheaded also helped put the insight we gained from coding and reading the documents into perspective. What makes for a strategy that knocks your socks off? A look at their time frame, topics, and tools confirms that the majority of plans are indeed visionary by setting goals for the long-term future; holistic and cognizant of sustainability challenges including climate change; and purposive in how they propose to achieve these goals.

Time Frame

Perhaps the most obvious feature of strategic plans is that they move cities' time horizons. Most plants are looking at least fifteen years into the future, with certain decades, such as 2030, being represented in the majority of titles. Occasionally, plans take an even longer view up to 2040, 2050, or, in one particularly prescient case, even 2060. Although these city plans are characterized by their long-term nature, they must strike a delicate balance between articulating a shared vision to address looming challenges and being practical. In Cleveland, for instance, strategic planning gives a general direction, but that direction can change. As the sustainability director told us after the George Floyd protests of 2020, without the plan, "it would be really easy to get off track and, you know, go in a different direction. . . . But it's also important that every four to five years, we look at it again because, especially as we've seen this year, things change, and we can't predict the future, so we may need to redirect our priorities based on context, like now." Making strategic thinking relevant to daily challenges is difficult. Cities, therefore, try to ground their plans in what is underway in the city by involving stakeholders and providing regular reports and updates. As one of the masterminds behind many cities' resilience plans at 100RC explained to me: "That ultimately leads to a strategy that we hope has, you know, the right mix of ambition and a thirty-year time horizon with enough concrete things to be able to demonstrate success, right?"

Topics

Sustainability is one of the key topics, with three in five plans either mentioning the concept on the title page or in the introduction. Other concepts that are well represented include green (33 percent), innovative (32 percent), livable (29 percent), inclusive (29 percent), global (29 percent), resilient (22 percent), and smart (19 percent). The topics most commonly discussed in introductions were the economy and the environment. Of the plans we analyzed, 83 percent talked about the environment or climate change specifically, and 72 percent talked about the job market and economic development. Inclusion and inequality were more minor issues, appearing in 46 percent and 40 percent, respectively, of the plans—certainly with an upward trajectory in recent years. Although these documents are local, one in five explicitly mentions global challenges and transformations. Other central concepts include urban renewal, public-private partnerships, terrorism, safety, trust in government, and education.

These topics are typically backed up with various sections (often called "pillars") that develop topics in greater detail, as we can see from an analysis of the thematic emphases in the document shown in figure 2.4. Many, if not most, of these topics are in some way connected to the environment—either directly (e.g., climate, energy, and nature) or indirectly (e.g., soft transportation, sustainable land use, green jobs). The most common priority is transportation, which is expressed in some 85 percent of plans. Climate change is another thematic emphasis of many plans, with around 65 percent of plans reporting that climate change mitigation and adaptation are primary goals of the city. Other traditional planning topics rank highly, with around 60 percent of plans speaking to economic development and land use. Although topics that are central to contemporary urban studies such as smartness, social and racial equality, education, entrepreneurship and innovation, or financial

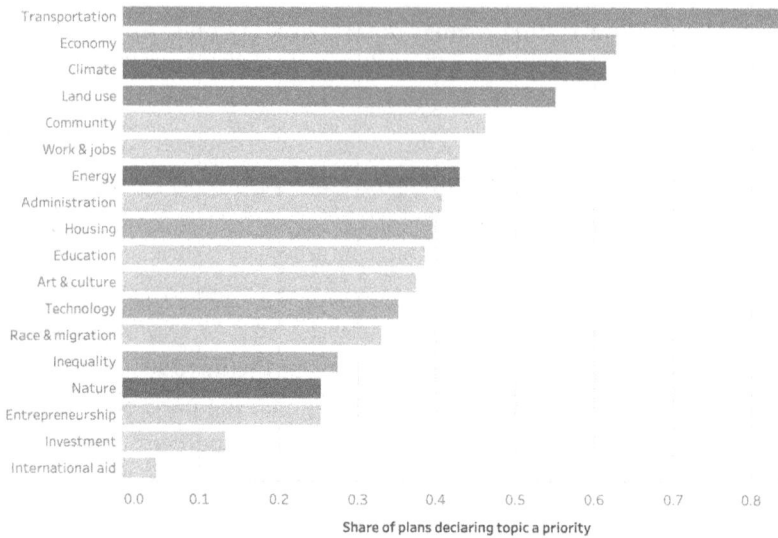

FIGURE 2.4 Thematic emphases of cities' strategic plans.

capital, come up occasionally, it is noteworthy that city leaders do not use these issues to frame their overall goals: The visions of green cities and resilient cities are more common than those of innovative cities, fair cities, or high-growth cities.

Without doubt, the topics and emphasis of strategic plans have continuously evolved. Plans published after 2020 tend to include more extensive discussions of equity, equality, and immigration than they did when the Paris Agreement was fresh in people's minds. In fact, there is no reason to believe that sustainability, resilience, or even economic development have to be perennial frames of strategic planning at all. As the Hiroshima example shows, it is entirely possible to frame a city's future around a different set of issues than climate change while still attending to environmental concerns. Nevertheless, a certain set of issues is likely inherent to planning and managing a city, as conventional policy domains of urban planning, such as land use, mobility, and housing, show. No matter how ambitious global cities may have become, these plans ultimately deal with local and fundamentally urban concerns.

Tools

In terms of their language and style, city strategies are, as the name suggests and figure 2.5 shows, profoundly strategic. Two in three strategies define milestones: cut emissions by 30 percent by 2020, create twenty thousand jobs in the green economy by 2025, develop a seven-hundred-kilometer bicycling network by 2030, become net zero by 2050. More than 60 percent of these plans identify specific performance indicators that allow them to track their goals, including average commute times, unemployment rates, number of small and midsize enterprises, carbon footprints, percentage of population in informal settlements, livability indices, air toxin metrics, and waste diversion rates. These indicators are often directly linked to the UN Sustainable Development Goals. One in two plans include an assessment of the challenges faced in the near and distant future, often accompanied by a list of strengths and opportunities perceived by both city administration and citizens (or what Rotterdam endearingly calls "selfies" of the city). These SWOT analyses are essentially straight out of a business strategy introductory

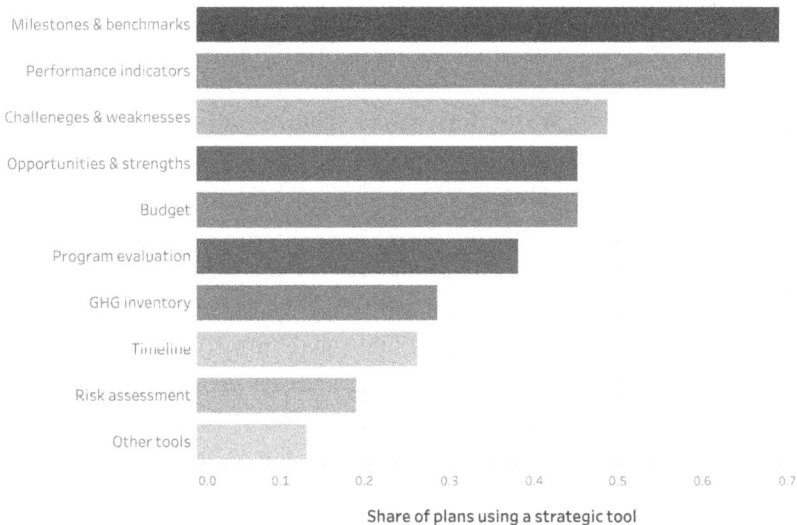

FIGURE 2.5 Tools used in cities' strategic plans.

class. Although these considerations are often based on consultations with the general public, consultants, or experts within the city, one in five cities also includes a formal risk assessment. Around 45 percent of cities explicitly include a budget, either in abstract terms (in two of three cases) or with exact amounts (in one of three cases). One in four cities has an explicit carbon budget or GHG inventory. Timelines and timed goals, as well as benchmarks that indicate a city's performance, are also popular tools.

Not all cities follow a standard set of strategic considerations, as one would, for instance, expect from the annual report of a publicly listed corporation—no binding legislation or single standard dominates the formulation and design of a city's strategic plans. Even if a clear universal standard of a high-quality strategic plan existed, not all cities possess the technical knowledge and expertise to implement details, such as a risk assessment or GHG inventory, as I explain in chapter 4. Nevertheless, these plans differ from traditional technical planning documents or policy white papers. This is obvious even from the aesthetics of the documents, with some 80 percent showing images and many documents being of very high publishing quality. Strategic plans are often endorsed by the city administration's upper echelons. About half have the signature of a major decision-maker, such as the mayor, city council members, a chief administrator or city manager, or sustainability director. In some cases in which cities received outside help to produce the plan, representatives of private organizations—such as 100RC or C40—cosign the documents. The overall message is clear: The city has a plan, this plan is supported by the city's higher-ups, and we want you to know about it.

Tracking Change

Remember that the reason strategic plans and their content are important evidence is their potential to show that cities *increasingly* embrace climate action. Looking at a selection of plans, however, does not allow systematically showing trends over time. Although it appears as if strategies were more sophisticated and increasingly dedicated to issues related to climate change, this could be due to a blip of having oversampled plans particular to environmental sustainability after the

Paris Agreement. The entire corpus of 650 documents thus becomes useful in this scenario. Rather than coding every single one—a useful but exhausting proposition—I found another way to slice the content of these plans, which was to crunch numbers about the words they use.

Analyzing "text as data," I took the full text of all documents and examined the prevalence of certain keywords.[28] Language became a bigger problem than when we were able to split up a handful of non-English plans—in English, French, German, Italian, Portuguese, and Chinese—among a multilingual bunch. When we manually coded, we also had to exclude plans that none of us could understand—including Azerbaijani, Romanian, Japanese, Bosnian, and Hungarian. This time, I made all of the plans machine readable and translated them into English so that I could analyze the prevalence of clusters of words that stood out about each strategy plan.[29] What did the text reveal about time trends?

Figure 2.6 shows the moving average of mentions of keywords related to climate, environment, resilience, and sustainability. What stands out

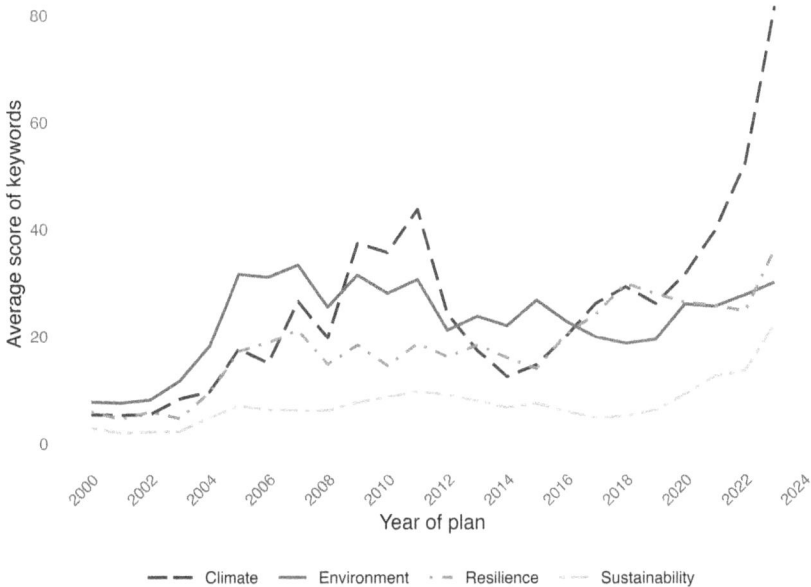

FIGURE 2.6 Frequency of selected themes in cities' strategic plans over time.
Note: Line shows three-year moving average of scores of keywords using term frequency–inverse document frequency (TF-IDF).

is that climate as a concept is up strongly from pre-2006 levels and has particularly increased in recent years since the 2016 global mobilization for city climate action. Environmental topics, in fact, have always been present in these plans, but they are now more likely to be framed in terms of sustainability or climate change. Although resilience has become an increasingly common frame, particularly after the Rockefeller Foundation's effort to boost resiliency plans (to be discussed in chapter 3), the concept is secondary to climate change. Although these plans as a whole are holistic and try to cut across complex issues, we can see that climate change has indeed become a master frame for forward-looking cities and has increased in prominence over time. If city strategies are any indication, city climate action is indeed on the rise.

IN STRATEGY WE TRUST?

These ambitious documents show a broad spectrum of aspirations coupled with a variety of innovative solutions that can function only because of the relative flexibility of cities compared with much more inert administrative bodies like those of nation-states or supranational unions. These strategic plans tell us something important about the nature of city climate action more generally: City climate action breaks out of narrow problems or tasks, cocreates an imaginary of the future together with the public and other stakeholders, and proposes concrete goals and projects that can be tracked and evaluated. In brief, contemporary city strategy is holistic, visionary, and purposive. This is not to say that all conventional urban planning has been fragmented, myopic, and haphazard. What, if anything, can city strategy contribute to city climate action that conventional urban planning could not?

To understand whether we should give credence to these strategic plans, it is helpful to look up from the archival documents and turn to those who wrote them. The chief sustainability planner of Washington, D.C., underscored the difference between old-school planning and the new way of strategizing in the context of D.C.'s strategy Sustainable DC, which he described as offering "a holistic way that's an attempt to wrap our heads around how do we approach sustainability." He juxtaposed the strategy with the planning office's nine-hundred-page comprehensive plan—a common tool of urban planning with a much longer history than strategy. "The comprehensive plan breaks things down into silos of environmental protection. It's very clinical. It's not as much about people." In contrast, Sustainable DC offers a more "holistic look at sustainability. . . . It approached sustainability from really a people perspective, and so there's health and wellness as a big piece of this plan, as I mentioned, climate, employment, etc. These things are in the comp plan too, but the way that this plan takes a more holistic look at those issues and kind of focuses in more on people."[30]

Like city climate action more broadly, many cities' current strategies try to escape entrenched ways and patterns of thinking by combining the work of multiple departments, moving away from specific issues, and providing a citywide vision rather than solutions that apply only to a specific neighborhood. One of the main goals of city strategies is to break out of siloes that characterize much of city administration. Their general nature creates a broad mandate for the city's strategic positioning, as the D.C. planner suggested: "It's not just environmental protection anymore, which means clean water and clean air, which is very important. It's our attempt to understand the connections between those issues. Quantifying our goals a little more scientifically." New York's CRO spoke about OneNYC in similar tunes:

> Our resilience strategy was really framed around the major challenges we face as a city. We have a growing population. We have an aging infrastructure. We have an inequality crisis, and we have climate change, and we pulled together a program that's comprehensive in addressing those together and not siloed. Not just like all climate is just GHG, but here's how to make that work in our neighborhoods or for our people and bring an equity lens to it. The fact that we cut across issues but also deal with it in a coherent and comprehensive way was really helpful.

This attempt to broaden cities' mandate, clearly visible in my content analysis of strategic plans, suggests that many cities claim to be delivering on calls from climate scientist Cynthia Rosenzweig and her colleague to make climate change a priority.[31] The global diffusion of holistic, visionary, and purposive city strategies suggests that cities around the world have turned to action and that this action often involves environmental sustainability broadly or even climate change specifically. My first inclination upon seeing the rise of plans that are out to fix climate change was to be skeptical of their ambitious proclamations. After all, Chicago's chief sustainability officer had reminded me of the quintessential social science adage, "In god we trust; all others must bring data." Although cities bring data on past and current challenges to the table, there are also many reasons not to trust that it can accurately predict future successes.

For one, city strategies have a pragmatic tone that eschews divisive climate politics: As nearly all of them will remind us, considering the current conditions, immediate action is inevitable. The urban politics of "getting stuff done" are all but neutral. As one New York activist engaged in Resilience by Design, the reconstruction initiative after Hurricane Katrina, put it aptly: "The Environmental Justice Alliance is not protesting, you know, 'we want resilience!'" Urban sociologists and activists alike have been deeply critical of any slick, ostensibly apolitical approach to running a city as if it were a corporation. The main concern is that a neoliberal logic of letting the market sort things out is inappropriate for rectifying the entrenched inequalities of contemporary cities and, in fact, could introduce new segregation, gentrification, and a shift of power away from communities. Although it sounds appealing to "just fix climate change," such pragmatism can belie the deeply political nature of deciding who gets what, when, and how—the bread and butter of urban governance. Critical urban theorists are skeptical of any planning intervention because it is often intended to favor property owners and other members of the "growth coalition," who have an outsize influence on urban politics.[32] It will take more than just acknowledging the presence of promises to convince us that the rise of strategic planning is real—with lofty aspirations of carbon neutrality and resilience. We will return to these questions.

Furthermore, even if one trusts that city administrations have the public's best interest at heart, the question remains whether these plans can be taken seriously. Some of them are, frankly, hard to believe. As National Geographic reported in 2017, Dubai, known for its indoor ski slopes in the desert, plans to go from having the highest ecological footprint of any city in the world to zero carbon by 2050.[33] In 2011, the Singaporean and the Chinese governments formed a joint venture to build a "socially harmonious, environmentally friendly, and resource-conserving" city for 350,000 people from scratch—although building a new city is probably not the best way to go about conserving resources.[34] And the idea of being better able to fend off terrorist attacks by filling out a risk assessment is questionable as well. As one urban planner put it: "Strategic planning has to claim to initiate transformation. It has to *hurt*." Are these plans just for show or do they hurt?

To answer this question, it helps to look to organization theory. Organizations are often thought to present certain rhetoric and policies to make them look legitimate. We know from decades of studies of sustainability efforts of corporations, for instance, that corporate social responsibility is often just a ruse to distract from the irresponsible activities that are occurring under the hood. This kind of "decoupling" is one of the oldest insights from institutional theories of organizations, first articulated by John Meyer and Brian Rowan in 1977.[35] Cities might behave in the same way. Perhaps the real reason to lay out big plans is to show that one is headed into the future faster and farther than one's neighbor, as a signal to potential companies or talented individuals. What if strategic plans really were just about gaining a competitive edge over the Melbournes and Lyons of the world by appearing to be as legitimate as Sydney or Paris?

If city plans are just empty public relations (PR) stunts, nobody has told that to the people writing them. "It can't be, right?" said one senior planner in New York City. "I mean, we actually have to show that we are making life better for [our residents]." These plans "can't just be highfalutin statements or commitments. They have to have money behind them. They have to have public accountability." When I asked him whether there would be consequences for failing to hit a benchmark, he said: "It's not like if you don't hit 90 percent, you know, I'm firing the whole team and replacing you. I mean, maybe, but that isn't sort of the stick that the mayor uses with us. . . . [We] want to continue to deliver for New Yorkers, and so we have to hold ourselves [accountable] and figure out ways that we're going to deliver on these goals." This sincerity resonated through many of my interviews. For those dedicating their careers to getting cities ready for the future, the notion that plans are for nothing is a practical question, if not also an existential one. A policy director at 100 Resilient Cities, one of the organizations encouraging cities to publish resilience plans later renamed to the Resilient Cities Network, told me that strategic planning "can't just be aspirational, or otherwise everybody's going to ignore it. But if it's only tactical, it's hard to inspire—to get any traction."

We should not automatically assume that city administrations put their money where their mouth is any more than, say, companies that have discovered that purpose can be monetized. In fact, the lack of conflict reported in plans is similar to how companies claim to be able to heal the world while making a profit. According to such a "win-win discourse," organizations can have their cake and eat it, too.[36] These conflicts, of course, are inherent to any sort of planning. On the one hand, fundamental targets like growing the economy and fostering equitable communities may simply be at odds. On the other, goal conflicts can be even subtler. As the D.C. planner told me, looking through a stack of different planning documents, "sometimes goals are competing with each other. There are so many of them!" He explained:

> There's goals about greenhouse gasses and renewable energy use, and those aren't necessarily aligned completely. . . . [One of our plans] promises to cut greenhouse gas emissions by 50 percent by 2032. But then there's also a goal regarding renewable energy: increase the use of renewable energy to make up 50 percent of the district's energy supply. Those things don't necessarily have to occur together. Some people assume that if you increase the use of renewable energy, you'll cut the citywide energy use and vice versa, but those don't necessarily have anything to do with each other. Those things are competing as far as priority. [We have to get] a little more focus on: are these numbers realistic?

Bad intentions or hypocrisy may not be the only source of inconsistencies; once again, insight from organization theory comes in handy. Organizations often pursue actions that are not, in fact, suitable for reaching the goals they set out to meet. As my colleagues, organizational sociologists Patricia Bromley and Woody Powell, argue, instances of "means-ends decoupling" have become quite a lot more common now that external evaluators and the critical public—empowered by the panopticon of social media and omnipresent rankings—make it hard for organizations to hide their inner bearings. As a result, organizations are less likely to brag about things they do not do, but that also does not mean that what they do actually comes true.

I came away from these interviews thinking that the critique that city strategies are nothing but PR or, even worse, BS, is lazy. Nobody who has spoken to the "bureaucrats with a heart" who are working toward a better future in urban planning departments could seriously think that these people conspire to keep their mayor in office or dupe the public.[37] If their calling were to tell fairy tales for a living, they could make more money doing that in the corporate world.

Although the question of whether we should believe what plans say about climate action looms large and will continue to occupy us in this book, some correlations exist between strategic planning and resources allocated to the issue. An examination of web pages with Stanford student Paloma Hernandez showed that the majority of cities (roughly three in four) that put out a strategic plan also have a department dedicated to sustainable development, environmental protection, or sustainability more broadly. As we know from earlier research on city climate action in the global context by urban climate adaptation expert Linda Shi and colleagues, having resources available to establish GHG emissions and produce programs to lower them is one of the best predictors of climate-related performance among cities.[38] As we will learn, the majority of small cities can only draw on a small contingent of social sustainability advocates to pursue superficial practices aimed at addressing climate change. A megacity like New York, in contrast, has a significant apparatus and resources dedicated to sustainability and resilience, which puts weight behind the city's words.

Professionals are a key driver of city climate action. City administrations are run by civil servants—professionals who are usually deeply committed to the common good. The fact that cities do not only have plans but also the people who write them—and whose job it is to do more than write these plans—is very good news. We know from many other cases that professionals tend to keep organizations in line with their expectations. For instance, university recycling programs are implemented more thoroughly when a team of dedicated professionals is responsible for doing so.[39] Nonprofit organizations that are run by professionals tend to be more consistent with respect to how they say they are managed and how they really are managed.[40] In other words, having departments, budgets, and professionals means that we should not wave off the fact that cities go through months- if not yearslong efforts to create comprehensive master plans that contain serious commitments to the climate.

Whether or not one buys what city administrations are selling, implementation may not be the ultimate goal of city strategies. Instead, it is important to view city strategies in light of where this chapter started: Public administrations are not omnipotent central planners. They are but one node—albeit an important one—in the network of organizations that contribute to governing cities. Strategic plans can establish what public economist Mariana Mazzucato calls moonshots: ambitious goals that would benefit the public and are to be implemented by cross-sector partnerships, including governments, companies, nonprofit organizations, and communities.[41] This approach is common when public actors understand themselves to be part of a larger, complex network of partners and, therefore, take on a role as "orchestrators" of private action.[42] City strategies are exactly about defining big goals and getting everyone on board to accomplish them.

Any strategy scholar will stress that the importance of strategy does not lie in a plan; it lies in the *process* of arriving at the plan: defining a shared vision and identifying pathways to reaching it. As Cleveland's sustainability officer said, without a plan, "it would be really easy to get off track and, you know, go in a different direction. With the plan, we have a common direction with our stakeholders so that we all know which way we're headed." Stakeholders are crucial when it comes to getting anything done. Many plans explicitly state this coproductive nature of the city's efforts. Amsterdam's 2020 plan, for instance, proclaims that "we want to reduce carbon emissions by 95 percent in 2050." And "we're asking every citizen of Amsterdam to play their part."[43]

The participatory production of the plans suggests that the city may use strategic planning to "manufacture consent."[44] Obviously, who is given a voice in the strategic planning process has significant variety. One of the key features of strategic plans is a mix of administrative expertise and public input. Take the plans of Sydney and Vienna, for instance, which were the starting point of this chapter. When I first spoke to Vienna's planners in the early 2010s, the chief planner told me that "these are primarily expert reports that are addressed to actors in the most essential stakeholder groups: administration, research, business." Dozens of the city's departments were included in the production of the plans, often by submitting projects that were already underway for consideration.[45] The Sydney document, in contrast, discusses a lengthy public consultation process in which both the general public and specific stakeholders such as Indigenous representatives, interest groups, and, of course, corporations had a say. Almost every city official I spoke to told me about the lengthy public consultation process for their plans. In the end, these plans are aimed at creating consensus among the many stakeholders involved in urban planning. As New York's CRO put it, the city tries to make plans "as approachable as possible so that anybody can pick it up and understand it, and read it, and know why we're doing what we're doing. But I'm sure if you looked at our plan downloads, 8.5 million New Yorkers did not download our plan. But that's okay." In short, strategic plans still may not be a product for or of the people.

Cities use strategies to position themselves in relation to other stakeholders in the city at what Martin Kornberger and colleagues—who have studied the strategy process in Sydney in detail—call the "institution nexus." They show that the reason Sydney 2030 was considered to be a success by city managers was not because of roads built or departments created. In fact, much of what the city had set out to do in 2007 never became reality; as of 2018, only one out of ten projects the city had planned for was implemented.[46] The strategy process created shared sensibilities and, as the authors put it, helped the city and its stakeholders "learn to dance together." City strategies help the city articulate and even manipulate otherwise latent "institutional configurations": What time and space should cities control, related to which topics or "matters of concern," and with whose involvement? Ultimately, city strategies reveal a proposition about who should get what, when, and how—the essential question of urban governance.[47]

In other words, the lack of implementation of long-laid plans does not mean that plans do not matter at all. Even if we assume that cities proclaiming their ambition to become carbon neutral make no actual dent into climate change with any of their rampant activities and plans, they do create an institutional context for the decisions organizations and citizens make. The incentive structure surrounding the ambition to create a greener and more sustainable global economy surely has effects on organizations in their local context. For instance, decisions about zoning and land use crucially influence the legal context in which organizations, from small bakeries to large factories, operate. When local governments craft and enact strategies, they set a vision that inevitably influences other organizations' actions in the community. We should, therefore, be deeply invested in understanding what shapes the actions of local governments and, therefore, the local institutional context for people and private organizations. If you own a business, for instance, the requirements and incentives related to a city's strategic priorities may determine which cars you purchase for your fleet, whether you locate your headquarters close to public transit, or whether you incentivize employees to bike to work.[48]

To be sure, taking everything we can read in strategy documents at face value would be naïve, given the tendency of organizations to say whatever will bring them laurels and legitimacy: Conduct a GHG inventory while laying new gas pipes? Try to get more sustainable to attract eco-tourism? Help develop electric vehicle infrastructure while underfunding public transit? Talk to citizens about their needs but then distill it all down to a single page? Of course, many plans and initiatives are insufficient in the face of today's challenges, and some are downright hypocritical. But as I will show in chapter 4, such a cynical take is misplaced because planning is only one among several steps in city climate action. Almost any proclamation of action against climate change is better than having nothing to say about the topic. Regardless of how much one believes that cities' proclamations will, in fact, translate into more sustainable and resilient cities, they are important evidence if understood as an expression of priorities and self-image. Cities' strategic plans are cultural artifacts that carry information about the subject and process of strategy, similar to a professional's résumé or a corporation's annual report.

ACTORS PLAN

Strategic plans are an indicator of cities doing more than just the bare minimum that is legally required of them. In fact, having a strategic plan with goals, milestones, and measures for implementation is a great indicator of the ideal type of the city as an actor that I defined in chapter 1. Strategic plans are full of voluntary goals that nobody forced them to have—in fact, some bread-and-butter tools of urban policy, such as zoning ordinances, make little to no appearance in these plans. Strategic plans also contain many innovative ideas and existing projects to reach the goals because, otherwise, the visions sound hollow. As I show in chapter 3, these plans are also deeply relational because they require cities to think about their peers and competitors—in sociological terms, their reference group. Of course, no matter how visionary, not all plans are holistic in that they think of the city as an entity and are authored independently by the city. The data I collected and studied, however, captures which cities had at least one such plan, when that plan was published, and who authored it. In other ways, we're getting closer to some empirical answers about which cities act and how. And this is exactly what we need to develop a potential explanation of cities in action.

As indicators of purposive action, strategic plans point to an institutional sea change in the role of cities in society. Were there earlier instances of cities acting strategically? Undoubtedly. Think of Chicago's famous Burnham Plan, which preserved Chicago's lakefront after the Columbian Exhibition, the plans to build national capitals from the ground up like Pierre Charles L'Enfant's Washington, D.C. (yes, the US capital was designed by a Frenchman, as was Detroit) or Oscar Niemeyer's Brasilia, or the concours to redo the magnificent Imperial Ring Road Ringstraße in Vienna. Urban planners and even business elites had a hand in crafting comprehensive plans with metro lines, boulevards, and skyscrapers with public rooftops long before Michael Bloomberg came along. Today's cities would not look like they do without visions, plans, and tight relationships between the public interest and private capital. But the puzzle is not who invented it; it is how such a way of thinking spread the world and landed almost, but not quite, everywhere. Contemporary strategies follow in a long lineage of planning that has evolved over centuries, and a historical recounting of this evolution is beyond the remit of this book.[49] What is without question is that the occurrence of city strategy has multiplied manifold since the early 2000s and has increasingly incorporated climate action as one of its core pillars.

From this point of view, the burgeoning of city strategies should be seen in the wider context of what it means to "plan" for society. Although national representatives are often among the signatories of international treaties, from Kyoto to the Paris Agreement, remember one city leader's sentiment reported in the introduction: Cities have to step up because the federal government is letting everyone down. In fact, very few countries remain with a solid national planning regime in place. Among them is, of course, China, which also sees a much lower uptake of globally oriented city strategies. The five-year plans that do exist are compulsory rather than voluntary and directed at a Chinese— typically governmental—audience. These plans fulfill, in other words, the purpose of an old-school plan, but without signaling city action.[50] What can the patterns in planning tell us about the rise of "actorhood" among the world's largest cities?

The rise of city action is the real question raised by strategic plans, their purposive style, and their preoccupation with sustainability. Speaking to Vienna's sustainability coordinator in a quaint and charming coffee house (at least until we were interrupted by an aspiring pianist), I learned that the city's pathbreaking Smart City strategy was, in his view, not only a terrific opportunity to weave the UN's Sustainable Development Goals—a global mission—into local planning but also an effort to show people that "the city is *doing* something."

Having published a strategic plan is a sign that the city is taking action in light of its many challenges and opportunities. What the sustainability coordinator was explaining to me over a Viennese Melange is that strategy plans are a *rationalized practice*, which means that they follow a legitimate and highly standardized script and are imbued with essential meanings that can even take on the form of a myth about how the practice will contribute to goals and outcomes in important ways. Such practices are a means for organizations to signal to the wider world: We understand the rules of the game, and we're playing it. As John Meyer and Brian Rowan argued, such public displays of legitimate behavior are often detached from the actual goals of the organization and its people—rationalized practices can be *decoupled* from the technical core.[51] This is why the projects and goals in strategy plans matter even if they are barely implemented.

Why is it essential that city strategy is a rationalized practice—artifacts of an attempt to wrestle complex problems under control through systematic procedures that is perceived as legitimate but not necessarily functional? Because rationalized practices in the management, governance, and constitution of organizations are neither a new topic nor specific to city administration. In fact, institutionalists have studied this development in many contexts: Meyer and his colleagues in the sociological institutionalist tradition, for instance, found that nation-states tend to adopt departments of education or the environment even, in an evocative thought experiment, when they are newly formed.[52] Corporations in a liberal world order create positions for diversity, ethics, or sustainability, as research with Patricia Bromley on the "hyper-organization" of companies and Frank Dobbin's work on the rise of human resources managers in diversity concerns shows. Universities, likewise, tend to document and record everything that happens in them to demonstrate their impact on society.[53] My colleagues and I at the Civic Life of Cities Lab, where we study the transformation of civil society organizations since the 2000s, were among the first to study how nonprofit organizations are trying to be perceived as professional and, in recent years, open to the needs of their constituents.[54] The list goes on and on; organizations seek to attempt to live up to a widely shared standard of what it means to be a good organization.

The strategic turn among cities is therefore most likely not isolated but rather linked to the rise of organizational thinking among nonprofits, companies, national governments, and even social movements. According to John Meyer and Ronald Jepperson, who brought the Weberian idea to the twenty-first century, rationalized practices are indicative of organizational actorhood.[55] Their adoption rarely happens in a vacuum—because actors genuinely invent them de novo in response to a local problem—but rather results from the exposure to what the authors refer to as global scripts of legitimate behavior. All these transformations, so goes the institutionalist narrative, are linked by shared expectations of what it means to be a good—that is, effective, fair, and well-governed—organization that has blurred the boundaries between the public, nonprofit, and corporate sectors.

As a rationalized practice, city strategic planning can take on a mythical nature, trying to talk a future into existence that nobody explicitly agreed to. It is a sign of a wider transformation, however, that means that cities take certain responsibilities (to be resilient to the next shock, adapt to different temperatures or extreme weather events, or mitigate climate change) very seriously, at least to be a good, legitimate city. This, according to institutionalists, is the power of the liberal world order that we will examine next.

WHAT WE SAW

In this chapter, I reveal that strategic planning for urban futures has been on the rise. Only a handful of cities had isolated plans at the beginning of the 2000s—often technical, inward-focused, and frankly boring documents that help the city take stock of what all its different parts have been up to. Twenty years on, almost all cities have some sort of strategic planning device—setting ambitious goals, defining milestones and benchmarks, and purposely defining projects that have the potential of reaching those goals. These plans are not the kind of isolated exercise they once were, but are a coordinated effort to create standardized metrics and collective action to reach lofty goals—such as carbon neutrality, sustainability in the broadest sense of the term, and resilience not only against climate change but also against terrorist attacks, extreme weather events, and demographic risks. Although there is tremendous variation in what future imaginaries, risk assessments, and even GHG inventories actually look like, we would be remiss to think of city strategy as happening in a vacuum. The plans do not mirror each other's methodologies and format by coincidence. The strengthened role of cities in tackling societal problems is the bellwether of a profound shift.

What we have not yet addressed is the question of what drives this transformation. In chapter 3, I will show that the trend to strategic planning for resilience and sustainability is not orchestrated by a single actor but rather is the result of an increasingly dense network of associations that provide technical support and symbolic resources needed for such a sea change. By articulating an expectation that cities take discretionary, purposive, and relational steps to address the social and environmental problems the world faces, this broader institutional superstructure *attributes* action to cities. I will show that the shift from national to subnational actors is not an isolated trend but part of a larger transformation of how organizations function in a liberal economy—which, of course, means that it is not guaranteed that this trend will continue. And I will offer a first answer as to why it is that some cities are "in action" whereas others have been left behind.

CHAPTER THREE

Learning

The alarm clock of a Californian chief sustainability officer (CSO) rang to wake her up at five o'clock in the morning for weeks when the White House announced the United States' withdrawal from the Paris Agreement. She wanted her city to be the first to publicly respond to a major policy change in Washington, D.C., and the West Coast time zone did not work in her favor. Big cities fiercely compete for media attention. As I learned from a chief resilience officer (CRO) on the East Coast, the cause for the West Coast official's insomnia: "There's a lot of big egos in politics, right? Generally, our peers, we all work really well together. But I can always tell if people are holding something back, and sometimes we do too. It's healthy competition. There's no backstabbing going on. But it's a matter of who gets to do the press release first."

They, too, wanted to be the first to get the press release out the door. But no matter the race, city administrators see themselves in a generative competition. The early days of the We're Still In campaign, a joint response to the US withdrawal from the Paris Agreement discussed in chapter 1, illustrates this air of competition well: In a tough contest to position the city as a leader, mayors are keen on issuing their press releases before their colleagues. Dallas and Atlanta have joined the usual suspects, such as Seattle and San Francisco, in wanting to earn the title of the greenest city.

As discussed in chapter 2, myriad cities have published proclamations that they are aiming for a more sustainable future. This trend is an essential indicator of cities that are "in action." What explains which cities are among the front-runners and which avoid the global debate? In this chapter, I turn to the open question of how cities' embeddedness in a global *institutional superstructure*—the broader context in which they plan and proclaim things—may account for this disparity. As I will show, however, the way sustainability officers understand competition is not quite the cutthroat rivalry they tease; instead, these officials respond to cultural pressures and possess a knack for collaboration.

City climate action is deeply *relational,* and cities were looking to each other to decide their future. I was intrigued by the idea that structural features—related to the position of cities in a globalized society—may explain what creates the necessary conditions for cities to turn to the practice of strategic planning. The three features that came up quickly were *culture*—adapting to a standard script of being a good city proselytized by international organizations, *competition*—fighting for power and influence in an economic hierarchy, and *collaboration*—learning from each other in professional networks. But how do these factors play together, if at all, in explaining global cities' strategic turn to climate action?

I will introduce what each of these features looks like, summarize their potential implications for the uptake of strategic planning, and present quantitative evidence about how each one matters for the timing and content of the strategic plans of the 360 largest cities worldwide. As figure 3.1 summarizes, this chapter shows that although all three accounts play a role, collaboration has the greatest—and increasing—explanatory

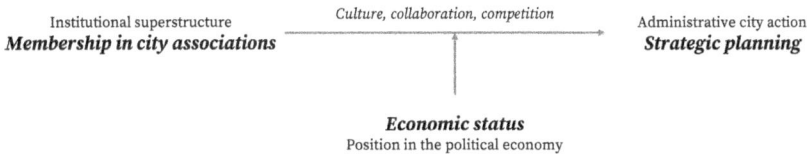

Institutional superstructure   *Culture, collaboration, competition*   Administrative city action
Membership in city associations   *Strategic planning*

Economic status
Position in the political economy

FIGURE 3.1 Visual summary of chapter 3.

power for the uptake of strategic planning. Collaboration is how the institutional superstructure shapes expectations, priorities, and decisions in the offices of mayors and urban sustainability managers, albeit how exactly it operates depends on the city's position in the highly globalized political economy.

WHY SOME CITIES PLAN: CULTURE, COMPETITION, COLLABORATION

If some cities had strategies for climate change, but not all, some cities are apparently more prone to making rhetorical commitments to climate change than others. As I argued in chapter 2, these commitments are among a slew of *rationalized practices* that do not necessarily accomplish what they promise but that indicate that the city perceives itself to be an actor. What may explain such disparities in who speaks up for the climate, and who remains silent? Given the significant variation in what cities have proposed to do exactly, what explains what goes into one of these plans? Considering the strategic plans from more than fifteen years of city strategizing that we discussed in the previous chapter (holistic, visionary, and purposive statements in which cities make big plans) can be a helpful first step to understanding what drives variation in city climate action. We have seen that these plans often go hand in hand with material actions, even though not all projects and ideas are implemented. Some strategic plans may reflect nothing more than good intentions. Others have a prognostic quality. But all create consensus among stakeholders, including the government, the administration, the general public, businesses, and advocacy groups, about shared visions and indicate that the city administration claims to be an actor in response to social and environmental challenges.

Let's bracket for now the important insight that some cities have greater capacity to plan ahead, as well as more robust political will on the part of decision-makers like mayors and city managers. I will explain in the coming chapters how both state capacity and civic capacity of cities determine the ambition and successful implementation of climate action. Whether or not cities take a stance is, of course, not a simple matter of choice, and every city has its own history, political considerations, and specific circumstances that lead to plans. Perhaps the city administration needed a new coat of paint and publishing a polished, corporate-looking document seemed like a good way to make a big impression. Perhaps a key decision-maker was swayed by Fridays for Future or another courageous social movement. Perhaps a newly elected mayor wants to overhaul their successor's outdated vision of the city's future. For now, my focus is on figuring out which structural context amplified these motives and made it more likely that a city adopted a strategic plan and got cracking on climate change sooner rather than later.

I begin my explanation with the city that first piqued my curiosity for trying to become "the world's green and blue capital" and to be completely carbon neutral by 2025: Copenhagen. I had reason to believe that Copenhagen's story was not just a local affair but relevant beyond Denmark. In 2012, Copenhagen published a policy primer titled "Solutions for Sustainable Cities" on the English-speaking part of their website to encourage other cities to follow suit. The document not only lays out feasible steps toward achieving the coveted goal of carbon neutrality, such as bicycle highways, district heating, and wastewater management, but also explains the benefits of sustainability to constituents. City administration has made the CPH 2025 plan available not only in Danish and English but also in Spanish, Chinese, and Russian. A dedicated email address is given to set up meetings for international delegations—like a 2022 visit from the San Francisco Bay Area Planning and Urban Research Association (SPUR).[1]

I traveled to Copenhagen to speak with the people in charge of implementing this plan. Where did *they* get their ideas from? It turned out that even the pioneers had not started with a blank page. Copenhagen's planners looked to cities such as Barcelona, Vancouver, San Francisco, and even Sydney to prepare their climate plan, as they explained to me. This list of cities became a recurring experience in my interviews. Whoever I asked about their city gave me a list of other cities with whom they compared themselves: their peers. And these peers are not always who you might expect. As is the case so often, who you consider to be in your peer group can say more about you than about the actual peer group. Some think of their peers as neighbors: Sapporo looks to Fukushima, Naples to Rome, Hong

Kong to Shenzhen and Macau, Melbourne to Sydney, and Kampala to "other West African cities" like Lagos, Abidjan, and Accra. Some look sideways: Austin specifically mentions best practices in sustainable planning from Asheville, North Carolina, to Santa Fe, New Mexico. Philadelphia looks to New York, Boston, and Chicago. Vienna mentions that it "maintains continuous contacts with other forerunner cities," including Amsterdam, Copenhagen, Hamburg, and Stockholm. And others yet look up: Pretoria talks about New York City's green infrastructure innovations as a template. Most cities, at one point or another, look to the Big Apple—whether it looks back is another question. As the city's CRO told me, New York City sees itself in a league of its own—but it is not entirely immune to a "vanity competition" either. The city has what they tongue-in-cheek call a "foreign minister" on staff, whose job is to keep track of all the many affiliations with intercity associations. This is a considerable task considering New York City's centrality in city climate networks.

The former city manager of Palo Alto takes the prize for making the most ambitious—and revealing—statement about the global nature of city learning: "Before I discuss our problems with anyone on another level of government in the United States, I pick up the phone and call the mayor of Barcelona," he told me. That is not to say that cities like Palo Alto are not highly dependent on figuring out the politics of land use and development with nearby local governments, the county, or the state. In the Danish capital, I asked whether other cities faced the same problems as Copenhagen, which seemed so unique in its commitment to the climate: "If you go down to basics, yes. Retrofitting of buildings, for example. It's not just a problem in Copenhagen. It's a global problem, at least in developed countries. There is a good economic case for energy retrofitting. But investors focus on bringing value in new kitchens instead of new windows or insulation. That's the same problem in Vancouver, New York and Copenhagen. And in Sydney." In thinking about their future, cities look across an ocean rather than just across a bay; local visions are profoundly global. This insight is the next step in understanding all three potential explanations for variation in strategic plans as indicators of administrative city action: culture, competition, and collaboration.

CULTURE: GLOBAL SCRIPTS RALLY TO ACTION

The global context of running a city—what I referred to as the institutional superstructure of cities—has changed quite a bit since the 1990s. At this point, it makes sense to take a step back to examine what this change looks like and why it is likely connected to how cities, and city climate action in particular, are organized.[2] The wider institutional context of the organization of cities—for example, how professionals are educated and how their decisions are informed, influenced, and evaluated—may be a source of the thematic emphases on sustainability that I observed in cities' strategic plans discussed in chapter 2. City climate action is only one form of proactivity, one that requires knowledge, technologies, and data that reshape how cities are understood and, more broadly, run.

In this cultural sense, city action can be seen as the result of institutional change—which, similar to sociologist of the political economy John Campbell, we might understand as a slow, evolutionary process rather than as something that changes from one day to another.[3] For instance, the presence and influence of professional expertise in how to write a strategic plan or how to measure a city's performance makes these tasks integral to city management. In modern organizations, the act of managing requires formal departments, conferences for knowledge exchange, and executives such as CSOs and head strategists, not unlike the many creative job titles by executives in the C-suites of corporations identified by organization scholars José Luis Alvarez and Silviya Svejenova.[4] In other words, how cities are understood has implications for people's careers and structural change in the economy. Transformations in the expectations of cities are evident in the context of higher education, industry, and associations.

One domain that illustrates the changing institutional superstructure of city action is higher education. Many universities have adjusted to the renewed attention to cities. For instance, leading graduate schools and policy institutes offer professional programs such as Regional and Urban Strategy (Sciences Po Paris), Sustainable Urban Development (Oxford University), City Management (Arizona State University), and Sustainable Lands and Cities (University of Edinburgh); many related programs in universities such as the Massachusetts Institute of Technology, Cornell University, and the University of North Carolina; and US policy schools such as NYU Wagner, USC Price, and Syracuse Maxwell. These programs are mirrored around the world, for instance, in the Governing the Large Metropolis program offered by Sciences Po Paris and the London School of Economics, a cities engineering and management master's degree at the University of Toronto, and many others.

These programs establish urban management and governance as a profession whose foundation only partially draws on traditional urban planning. A professional field of the study of cities means that a body of expertise exists about how cities can deal with diverse problems. This change is similar to general management education, whose shift to ethical thinking is well documented in work by organizational sociologists Rakesh Khurana and Michel Anteby.[5] The responsibilities of urban managers range from handling well-established issues, such as transportation planning and land-use zoning, to tackling thornier ones yet, such as inequality, gentrification, political apathy, and climate change. In addition to dedicated professional degrees, mayors and top municipal executives provide a promising source of executive education, which, in the case of the Bloomberg Center for Cities at Harvard University, offers opportunities to share a particular interpretation of what cities need these days.

Changes in what it means to run a city are also evident in cultural artifacts, like books. Based on mentions in the corpus of books archived by Google, figure 3.2 illustrates the growing interest in developing knowledge on "urban management," "urban governance," and "sustainable urban development." The overall share is not as important as the timing: All began to grow significantly in the 1990s around publication of the 1987 Brundtland Report—which offered a first definition of sustainability as meeting the needs of the present without undermining the ability to meet the needs of the future—and the 1992 Rio Conference on Sustainable Development—the Earth Summit that generated a variety of important declarations and a commitment to sustainable development known as the Agenda 21. This shift coincided roughly with the growth of strategic planning for sustainable development by cities worldwide documented in chapter 2. In comparison, traditional terms like "city planning" appear at a higher rate than the noted phrases but stagnate around 1980.

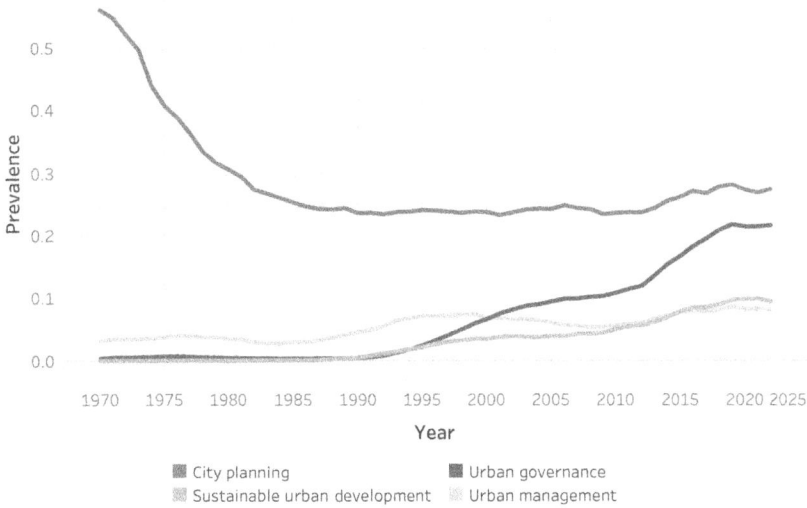

FIGURE 3.2 Prevalence of phrases associated with urban management in books, 1970–2022.
Source: Data from Google Ngram Viewer, March 2025, https://books.google.com/ngrams/.

The transition in cities' goals—and the expertise this transition requires—creates opportunities for private industry. Consultants in urban management and planning seek to facilitate strategic planning and organizational learning between cities. For instance, several plans I encountered were codeveloped and designed by Gehl Architects, a "networked urban design and research consultancy" located in Copenhagen.[6] Corporations, government agencies, and scholars interpret the globalized free flow of labor, capital, and ideas as intensified competition between locations.[7] Technology companies advise cities to adopt sustainable infrastructures and "smart grids" as product packages. In 2011, for instance, IBM launched the Smart Cities Challenge—a $50 million grant program for entrepreneurial municipalities. Google's Sidewalk Labs (unsuccessfully) sought to transform the Quayside waterfront district in Toronto "from the internet up." National and supranational governments support local sustainability initiatives as well. Part of the European Commission's information and communications technology (ICT) strategy, for instance, is to support cities in becoming "smart" through a European Innovation Partnership for Smart Cities and Communities, just as the US Department of Transportation declared a US Smart City Challenge in 2016. These initiatives also have an international dimension. For example, the World Bank targets cities, not nations, in some of its resilience-related programs. A division of the China Sustainable Energy Program seeks to "reduce carbon emissions and air pollution in new and existing Chinese cities by promoting sustainable urbanization and transportation system development."

The most significant change in cities institutional superstructure, however, is the rise of a substantial network of international nongovernmental associations seeking to facilitate learning and collaboration between its city members. Much of this network is held up by mission-driven nonprofit organizations. Three among many such associations, the International City/County Management Association (ICMA), the City Climate Leadership Group (C40), and Local Governments for Sustainability (ICLEI) are incorporated as 501(c)(3) public charities in the United States and have an annual budget of $32 million, $10 million, and $2 million, respectively, compared with the American Planning Association, the traditional professional association of urban planners in the United States, with its budget of $18 million. These associations are significant in size and reach and shaped city climate action in important ways. For example, the Rockefeller Foundation's 100 Resilient Cities (100RC) program offered $100 million in technical assistance to cities worldwide that sought to produce a resilience strategy and also introduced private contractors who helped implement such a plan. Textbox 3.1 discusses the initiative in detail.

International organizations dealing explicitly with local rather than national issues have proliferated more generally. The number of international nongovernmental organizations (INGOs) concerned with the municipal level illustrates that the international aspect is relevant. Figure 3.3 shows the cumulative number as well as the annual growth rate of city-related INGOs, defined as any INGO explicitly dedicated to urban, municipal, or local affairs, since 1900.[8] As of 2022, the Yearbook of International Organizations has registered 578 organizations with subsidiaries in at least three countries. The largest increase in international organizations with an urban agenda occurred in the 1990s, after the field-defining Rio Earth Summit in 1992. The presence of cities

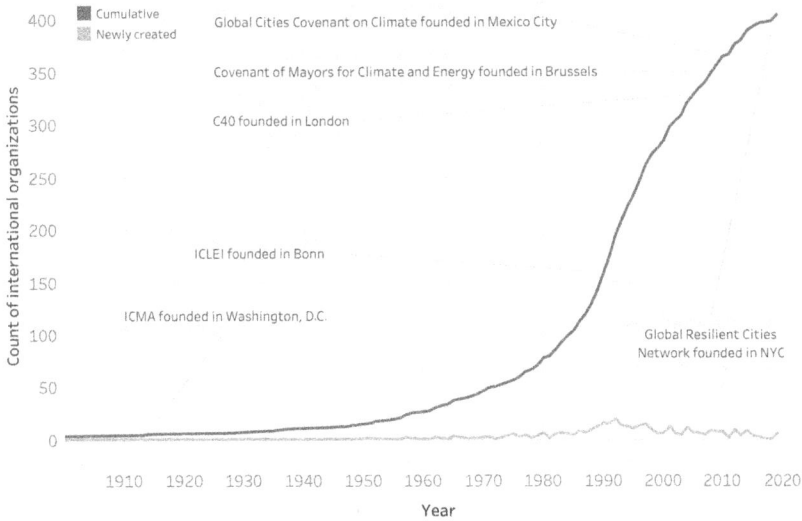

FIGURE 3.3 Number of city-related international organizations worldwide, 1900–2010.

Note: N = 417, 161 additional organizations did not register founding year; keywords were "city," "cities," "urban," "municipal," "mayor," and "local" in English, French, Spanish, and German. Organizations highlighted play a significant role in the city networks mentioned in chapters 3 and 4.

Source: Data comes from the Yearbook of International Organizations, https://uia.org /en/yearbook.

TEXTBOX 3.1: THE EXAMPLE OF 100RC

The connections among cities have matured much since the days city climate action was still in the "urban laboratories" of New York and Copenhagen, as the example of the 100 Resilient Cities (100RC) initiative illustrates.[9] 100RC was an impressive attempt to marshal some of the most promising cities in different parts of the United States and the world at large to push the notion of urban resilience. The premise of the organization was that cities supported with a $1 million grant would hire a CRO who worked directly with the mayor. In addition, these cities were expected to leverage the $1 million by working with private companies from Rockefeller's lineup.

Successful applicants had to have both "commitment and promise," as I learned from one of the architects of 100RC, a policy director at the Rockefeller

(continued on next page)

(*continued from previous page*)

Foundation.[10] This effort, of course, also has relevance for climate change, even though the Rockefeller Foundation explicitly conceived of the effort to go beyond such a polarized issue as the natural environment. As the director explained, resilient cities are "certainly in a better place to deal with climate change, but they're also in a better place to deal with an earthquake or a health pandemic. . . . It's not just about levees or sea walls, but it's also about what is the human and community resilience in a city and how do those connect to the physical."

I got an insight into this initiative just six weeks after the announcement that the US would withdraw from the Paris Agreement when I was able to visit the offices of the Rockefeller Foundation and 100RC on the upper floors of New York's famous 30 Rockefeller Plaza. There, I had a chance to learn about the work of what ended up becoming the Resilient Cities Network after the Rockefeller Foundation reconsidered what it meant to give a floor of one of the most iconic buildings in the world to an organization trying to marshal mayors into doing better. Behind heavy mahogany doors, a buzzing sense of urgency was palpable. You usually only get that feeling from being in the kitchen of a revolution. People with advanced degrees like MBAs and masters of public policy were on the phone with city officials and mayors from all over the world. The goal was to help write and implement strategic plans specifically aimed at improving cities along a roster of resilience understood as "the capacity of individuals, communities, institutions, businesses and systems within a city to survive, adapt and grow no matter what kinds of chronic stresses and acute shocks they experience."[11]

How did this initiative work? 100RC was a massive media effort. In return for the financial support, cities were expected to provide a plan and an officer who was available to tweet. The foundation's interest in media attention was perhaps not entirely unrelated to the fact that Rockefeller was seeking to create not only enthusiasm for resilience but also a brand that clearly distinguished its program from climate action, a topic already coined by the fellow New York foundation Bloomberg Philanthropies. Competition among cities is ostensibly at the heart of this initiative. 100RC is one of many efforts that operate this way. Others are even more explicit in rewarding cities that are already at the top of the status order. More visibility signals impact—as desired as much as it is elusive—both for the grantees and the grant makers. At first glance, 100RC's entire premise of boosting urban resilience relied on a national competition for money. But there is more than competition under the hood.

and their representatives in such events is a strong indicator of a city's involvement in the global discourse on sustainability.

Why not stop with global cultural change in our search for explanations? Even though research in the tradition of sociological institutionalism gives us very good reasons to consider the influence of national institutions on local organizational practices, the argument may only partially play out in the context of cities.[12] As I laid out earlier, cities are nested in other levels of social organization, such as regions, states, and countries—what urbanists call multilevel governance. Of course, cities are exposed to international influences relayed by international organizations and INGOs, and their level of exposure may depend on whether or not, say, a country is a paying member of the World Health Organization. As a result, it is reasonable to assume that the question of timing and the extent to which local actors take responsibility and *act* will also depend on the national context, the capacity of the nation-state in which cities are located, and the local organizations that are aware of global blueprints—something sociologists David Frank, Ann Hironaka, and Evan Schofer refer to as "receptor sites."[13] These features can explain country-level differences in the presence of national parks, environmental ministries, and memberships in environmental organizations. According to Frank and colleagues, concern for the natural environment became a constitutive interest of nation-states over the course of the twentieth century.

There is, however, also some reason not just to blindly copy these powerful but high-level explanations to cities. Institutional theories of globalization would have little to say about variation in the propensity of adopting rationalized practices *within* a country. But exposure to globally circulating ideas about what is expected of rational organizations (or scripts) differs greatly within nation-states—say, between New York City, which is a global center of the knowledge economy and houses multiple influential international organizations, and Detroit, which has been largely abandoned by international and highly educated workers after its industrial decline. These considerations point to an opportunity to add to established institutional theories of globalization that the global integration of cities likely outweighs the influence of more indirect national pressures. More simply put, a cultural sea change is underway about what it means to run a city well—from technical planning to participatory strategizing—but how this high-level transformation translates to the local level is far from obvious.

COMPETITION: UP AND DOWN THE ECONOMIC ORDER

The culture of city management is not the only thing that has changed since the 1990s—so has the global economy. From a political economy point of view, city strategies are a response to vibrant competition among themselves and a reflection of each one's response to economic globalization. Competition among coalitions of interested parties like politicians and developers is a central mechanism in critical perspectives on urban politics.[14] It is, of course, an interesting question what role political coalitions play in shaping the content and direction of local strategy. For our purposes, however, the political-economic view suggests that cities—ultimately motivated by growth coalition interests—will pursue capital investments, brand themselves, and devise strategies favoring business interests to outcompete their peers.

The same critical logic is central to the notion of entrepreneurial cities, which try to create favorable conditions for businesses within states and regions competing for resources. As critical geographer Bob Jessop writes, entrepreneurial cities have a "self-image as being proactive in promoting the competitiveness of their respective economic spaces in the face of intensified international (and also, for regions and cities, inter- and intraregional) competition."[15] In other words, the perception of cities as rivals trying to outcompete each other to attract capital and resources is a potential contender for explaining some of the purposive nature of city strategies.

Social scientists examine the position of cities and the global economy through the lens of economic status.[16] Cities with many headquarters of corporations that maintain branches throughout the world are at the top of this order because all flows of capital alternately go back to the headquarters' cities. As urban sociologist Saskia Sassen has argued, the structure of the global service economy is highly stratified and gives rise to global inequalities amplified by unequal access to capital. We can measure economic inequality by examining the city's position in the global network of service firms, including management consulting companies, law firms, and the like. These data allow for a frequent assessment of which cities are the most central in the global economy and dominate others, from alpha cities such as London, Hong Kong, and Paris to beta cities like Nairobi, Kenya; Beirut, Lebanon; or Chengdu, China; and even gamma cities like Detroit, United States; Phnom Penh, Cambodia; or Managua, Nicaragua, that are on the receiving end of the influence of these powerhouses. A research group led by the British urban geographer Peter Taylor, at Loughborough University, has created an impressive collection of research and data about this global political economy of cities under the label of Globalization and World Cities (GaWC). Their thoughtful and evocative cumulative work opened my eyes to the incredible interconnectedness of cities in the world economy.[17]

These accounts show that, in the late 1990s, a narrative took hold according to which cities are structured by a global hierarchy. According to this order, cities differ in terms of their influence on other cities, their economic reach, and their status. Vying for economic standing, cities engage in entrepreneurial actions and try to stand out through distinctiveness, such as through a particular industry, composition, or strategic emphasis. According to this narrative, the global economy of cities is highly stratified, giving only the most central cities access to resources, talent, and representation in an increasingly multinational corporate world. As a result, the ability and propensity to plan for social and environmental problems may also differ because not all cities are competing to the same extent. This logic undergirds the never-ending flow of rankings and league tables that put cities in order and considers a variety of data points about how well a city is doing relative to others. These league tables have turned into a manifestation of the global competition among cities featuring a liberal logic according to which people and companies can move anywhere in the world if they so choose.

The tendency to quantify performance and publish the result in the form of rankings has shaped organizational life in all realms, not just cities. Platforms like Yelp, Rotten Tomatoes, Glassdoor, and the Dow Jones Sustainability Index influence consumer and investor decisions. Law schools, which have reacted anxiously to being ranked by *US News & World Report*, are a typical example.[18] Rankings—alongside sibling "evaluative devices," such as ratings, awards, and certifications—have redone what it means for organizations to be successful, and they have not stopped short of shaping the organization of cities. The idea of comparing cities to each other is not an entirely recent invention. The principle for the scientific comparison of cities is linked to the economic transformation of the twentieth century. Livability rankings, for instance, date back as far as the post–World War II period, when inter-city mobility surged.[19] Airplanes and fast-rail tracks allowed white-collar workers and tourists alike to move quickly between cities, raising the question of where to live.

The idea that cities can take responsibility for social and economic progress generates a new demand for proof that the city's actions are paying off.[20] An explosion of rankings, ratings, awards, metrics, league tables, benchmarks, and best practice case studies provide this evidence.[21] In 2014, the International Organization for Standardization (ISO) published an international standard for sustainable cities. Central to most of these evaluations is the comparability of different cities across a wide range of contexts, for instance, how livable, sustainable, or economically competitive a community is. These evaluative devices pursue diverse goals, from providing benchmarks for informing cities of where they stand (e.g., Economic Intelligence Unit) to selling proprietary data that helps companies calculate expat wages (e.g., Mercer). Other entities use rankings for product marketing (e.g., Siemens or IBM) or as a social movement tactic, creating artificial competitions related to some concept (e.g., UN-Habitat or ICLEI). Rankings underscore the belief in global competition for capital and the headquarters of multinational corporations.[22] Periodic, systematic comparisons of cities also solidify the impression that cities can control their performances in ways similar to rankings and ratings of firms and universities.[23]

The Economist and its Economist Intelligence Unit are at the forefront of publishing these rankings. In partnership with multinational insurance groups or banks, the consultancy puts out a variety of different rankings, often following complex methodologies. Just like the *US News and World Report* or *Times Higher Education* rankings for universities, these lists hold a lot of sway in the public discourse about cities. Depending on the focus and methodology, the outcome differs, but the same cities tend to win. For instance, in 2024, *Economist Impact* published the Resilient Cities Index in cooperation with the insurance company Tokio Marine Group as a "global benchmark of urban risk, response, and recovery." On four dimensions—critical infrastructure, environment, socioinstitutional, and economic—the report generated a benchmark of twenty-five cities "to help policymakers and stakeholders understand risk and design effective policies for urban resilience." The report also lists influential informants like the C40's climate resilience director, the global director of programs and delivery of the Resilient Cities Network; representatives of Lagos, Mexico City, and Cape Town; risk managers of the World Bank; and academics. As the authors put it, cities are "dynamic and progressive powerhouses, brimming with innovation and vitality," but "repeated heat waves hit urban tourist hotspots, harming people's health and disrupting their daily lives."[24] That's certainly one version of the story. Rankings are not only a neutral way of measuring an objective truth, but they also transmit a certain understanding of excellence that the ranker happens to believe in. In other words, rankings can be an ideological business.

The view that rankings are imbued with values is what motivates organization theorists Martin Kornberger and Chris Carter to claim that city rankings are "performative," meaning that they shape the very thing they are attempting to capture. According to them, league tables not only capture a city's performance like a camera but are engines of social action that generate strategic responses among city managers.[25] In my view, however, the many city rankings do not have the powerful impact of law schools or college rankings because they largely complement each other, creating many niches for cities to thrive.[26] City managers indeed monitor their city's performance in various ranking publications and may be inspired by performance metrics. One department head in the City of Vienna was highly skeptical of the methodology of most city rankings, and downplayed their influence by joking about the fact that, one year, Hamburg won a prize for being the greenest city in Europe and the most car-friendly city in Germany. But even ranking-skeptical Vienna celebrates that the global human resources (HR) consultancy Mercer keeps ranking Austria's capital as the world's most livable city—for expats, but that fact is typically omitted. Viennese press releases, as well as Austrian media, are full of references to how well the city has been doing in terms of its livability. The department head nevertheless instructed someone on his staff to keep track of all the credible rankings (at least those that are affordable—one ranking report costing €2,500 was excluded) to see whether the city could improve on any of the dimensions. At the same time, politicians and administrators use select rankings and ratings extensively to justify and argue for policy choices. Vienna is not alone in its appreciation of a good ranking position. As Copenhagen's chief planner told me: "We are quite keen on being on top. We want to be. And we know that there will be more competition and we cannot always be number one. But we want to be in the top on indexes that are related to environmental issues, livability, sustainability."

This shows that the language of competition is not alien to city managers and planners. They compete not only for who is the most powerful but also for who is greenest. Many cities around the world feel compelled to respond to economic pressures by making sure that elites find them appealing, with tax incentives, subsidies, and a rosy outlook. Despite the competitiveness anxiety, evidence suggests that people and companies have limited mobility when it comes to economic incentives because they are embedded in these places. Thus, cities may establish their competitive advantage through their strategic positioning. If the global economy is a game, in other words, cities need a *strategy to play it*. Strategic plans are the blueprint for differentiating one city from the others and for moving up the ladder of global city rankings.

To better understand how city action happens as a response to other cities and their wider environment, I asked an expert in urban planning at the Royal Danish Academy about whether Copenhagen compares itself to other cities in their massive push to sustainability: "They do. . . . The municipality believes it's necessary to compare itself with other cities because they believe that investors could be that way." This is certainly the impression I had gotten from reading early city plans— which often mention not only other cities, as we have seen, but also their "competitiveness" and their attractiveness as a company location and a place to live. Many places have competitiveness agencies whose job is to attract big business. Could intercity competition be the key? I took the question to the city managers I interviewed.

"Sure," the Chicago CSO readily admitted, "we are competing for funds, we are competing for notoriety, you know, events, things of that nature. There's always a tinge of tension there." And this sense of competitiveness can even inspire some cities to make bold claims about their ability to tackle climate change. Several interviewees backed the idea that cities' turn to climate change mitigation, adaptation, and resilience is not unrelated to their desire to compete. The chief strategist of the City of Vienna put this into clear terms: "We're in a vanity competition, if you want, with Amsterdam, Stockholm, London, Copenhagen, Hamburg, Singapore, New York, Rio, Barcelona. Who's smartest, who's the high-flyer?"

So, yes, "there's always a level of healthy competition between cities," as the Chicago CSO put it. Cities are competing for federal and foundation grants, vying for rankings and recognition, and at least in the case of this one CSO, even engaging in a mild form of industry espionage. He peruses other city's strategies regularly. "I spend a lot of my time reading plans from other cities, stealing ideas . . . what are they focused on, what are they not? What can we learn from them?" Some of the city's ideas indeed came from elsewhere: "Some of it is competition and some of it is just learning. You know, the mayor talks a lot about the LED streetlights, smart lighting project we have, how you know, he stole that idea from Vancouver and seeing the work they've done on LED lights. And so I think there is a lot of idea sharing or idea stealing that occurs, and that's a good thing, right? So we see what other cities have done."

COLLABORATION: LEARNING FROM PEERS

"Some of it is just learning." The word *learning* got me curious. I wanted to know if what my informants from both coasts of the United States called "competition" indeed had anything to do with the kind of cut-throat rivalry among companies that master of business administration (MBA) students learn about in strategy courses. A "third coast" CSO in Chicago pointed out the middle ground:

> I think that we are dealing with elected officials, folks who run cities, and so they often want to be first. You'll see, you know, often in press releases we're the first city to do this or the first city to do that, but at the same time, you know, we're also like looking at what other cities are doing as well, and so actually I think

they feed off of each other, and I think that's actually a positive. More work gets done. It gets done faster and more efficient[ly] because of it.

He cited a tax on disposable bags as an example: "We spent a lot of time looking at models in San Francisco, looking at Seattle, looking at New York, D.C., other places around the country that have those." These existing laws served as a role model for a similar policy in Chicago. "I would hasten to say there are not a ton of brand-spanking new ideas that often, right? They're ideas that we're taking pieces and good parts from other cities, and that's good. That's good for us."[27]

In other words, what the Chicagoans called "stealing" was not just a transfer of knowledge from A to B; it's what organizational theorists since the Carnegie School have called organizational *learning*. Copenhagen's chief planner explained this phenomenon along similar lines: "I like that way of working. Learning from each other. A real focus on learning, common knowledge . . . By cities!" For Chicago's CSO, learning is what connects competition and collaboration. In fact, idea sharing was not just done on a bilateral basis. Learning involves direct knowledge of what others are doing: "Sometimes it's just picking up the phone and saying, 'I saw you're working on X. We did something similar. Can I put you on the phone with our team that worked on that with your team and talk about what we've learned through that process?'" So part of this network is informal, but the quality of collaboration is more formal than that, in a way that explains the "tinge of tension":

We understand the necessity to rely on each other in terms of generating good ideas, and so then when you have groups like Urban Sustainability Directors Network, they can often serve as a clearinghouse for those good ideas, and groups like C40 do the same. They do a lot of that. [One major initiative] has been funding a lot of that work with mayors around the country, mayors around the world, and help[ing] them think about innovation and share in ideal sharing and whatnot. There's always going to be a little bit of tension there, and that ultimately drives more work at the end of the day, so I think that's a positive, and particularly because every city is different, right?

Like the Chicago CSO, many interviewees stressed the importance of the Urban Sustainability Directors Network (USDN). Quite unlike a massive bureaucracy stifled by national politics and diplomatic decorum like the United Nations, USDN was founded as early as 2008 "by and for local government sustainability professionals as a safe place to learn from each other."[28] USDN is a prototypical example of an association of sustainability officers that operates through horizontal ties of learning rather than vertical ties of coercion. With around a dozen staff, USDN is a nimble organization primarily driven by its members and relationships. Although the association was founded to primarily support sustainability officers in North America, one of its projects is the Carbon Neutral Cities Alliance, a bottom-up, worldwide network of some of the ambitious cities with particularly aggressive greenhouse gas (GHG) goals. On its website, the organization boasts best practice cases as well as free webinars for urban sustainability officers.

The association also helped publish a book called *The Guide to Greening Cities*, "the first book written from the perspective of municipal leaders with successful, on-the-ground experience working to advance green city goals."[29] The book, published in 2013, features articles on a variety of topics from "Greening City Fleets in Raleigh" and "Funding Sustainability Through Savings in Asheville" to "Growing Green Businesses and Jobs in San Antonio" and "Sustainability Performance Management in Minneapolis." As the authors write, a "new urban imperative" requires cities to become greener and do everything within their power to adapt to and mitigate climate change. The *Guide to Greening Cities* stands on my bookshelf next to another book published by urban economist Ed Glaeser in 2015: *The Urban Imperative Towards Competitive Cities*, which also features several essays about "how to achieve more successful cities by making cities competitive." Funded by the World Bank, the book gives clear instructions to city leaders that, yes, there is a new imperative—and it is to outcompete other cities to be able to attract top talent. Although these two books illustrate the changes underway in how experts imagine cities to be run, they perhaps disagree somewhat with respect to how to go about these changes.

The observation of collaboration among cities shows that competitiveness and culture are two sides of the same coin: The culture of the institutional superstructure involves collaboration among competitors. City networks, I found, are doing much to activate cities through three mechanisms: bridging gaps in expertise by transferring knowledge, creating social bonds by routinizing social interactions, and raising the bar by setting new standards.

The first mechanism is to *bridge gaps in expertise*. City associations feature collaboration by convening city officials around the topic of sustainable development. Many host annual or biannual multilateral conferences, such as ICLEI's World Congress, the World City Summit in Singapore, or UN-Habitat's World Urban Forum. The World Urban Forum has attracted more and more people since the first meeting in the organization's hometown of Nairobi, Kenya, attracting more than twenty thousand participants to Medellín, Colombia, in 2014 and twenty-three thousand to Kuala Lumpur, Malaysia, in 2018. More recently, seventeen thousand people trekked to the massive fair in Katowice, Poland, in 2022. But the more impactful meetings happen, most likely, in a more intimate setting. A policy director at a grantmaking association told me that these networks "bring together these sustainability staff together in meetings that are incredibly fun and productive. And it's just so clear that some of these staff experience similar challenges. That's a huge problem we see in cities taking on climate change!" As a result, most associations follow the same logic that the director of the Southern Sustainability Director's Network (SSDN), a more regional version of USDN in the US South, laid out: "We provide as much peer-to-peer interactions between directors as possible. . . . Some of that capacity building for members is through peer learning. . . . We like to share as many best practices as we can. And some of that capacity building is through technical assistance, or coaching, or expert-driven teaching."

These exchanges seek to fill a gap in technical expertise that many city officials have yet to develop. According to 100RC, "there's a couple of different buckets of knowledge and experience" that CROs must be able to access. The websites are packed with videos, white papers, toolkits, and news, and not just for their members. At C40, I was told that "we share all of our case studies and information online. . . . That then allows them to spread more widely over the world." C40 has several topic-based subnetworks that teach and encourage an exchange of technical expertise required to solve issues like adaptation and water, air quality, waste management, food systems, and energy and buildings. A major funder of USDN told me: "Certain solutions and programs and ideas should be spread. We like to think of it . . . not as stealing each other's ideas. Because these are ideas . . . they should be stolen! They should be replicated, and we have found that cities find these networks to be incredibly valuable and useful for them." This is why, in these networks, knowledge spreads easily: "We're going to say, here are three communities that are doing that really well. Let's have a workshop where they're sharing their experience. . . . That way we're influencing what the thinking looks like, but we're doing that sort of in the service of what members are telling us they want."

The second mechanism is *routinizing social interactions.* As a policy director at C40 told me, the conferences are just the start: "Once they've met in person, their e-mail interactions are sort of more natural and stronger." Indeed, an official evaluation of the program by the Urban Institute—a D.C. think tank—highlighted the importance of these intercity networks. "Most cities noted that access to a global network of CROs was the most valuable program offering, as they could learn from others and share knowledge."[30] For 100RC, this networking was not limited to in-person meetings and conferences: "Many CROs noted that they remain in communication with each other through formal and informal mechanisms." One CRO reflected: "The best pieces of those networks are the city-to-city interactions. It's one thing for cities to think that they have peers, but having developed really strong relationships with CROs and sustainability officers in other cities, they are people that I text and call, and we bounce ideas off each other. . . . When it was telegraphed that the Paris withdrawal was coming, we were all on the phone weeks in advance saying how are we going to respond, what are we going to say." These relationships are an intended by-product of formal networks that, according to a USDN officer, is baked into what she called the organization's theory of change: "Let's get folks talking to each other in the network, help them understand what these best practices look like, and we think they're going to carry them forward." Meeting at a webinar was only the first step.

Many relationships among CROs and other city officials persisted far beyond the original purpose. As the sustainability officer of a

Midwestern city told me, this reach is a result of the fact that "the networks are really useful for learning and sharing, and there's an inherent trust that comes if you're part of that network."[31] Such trust-based interactions are the true core of the intercity network, and they are painstakingly created. For C40, "there are several mechanisms that we use to bring together our cities. . . . Increasingly we use forms like WhatsApp. For sort of quicker questions and also just sort of informal interactions and between the members of the networks so that they build strong and causal relationships." As we learned in Chicago, sometimes collaboration is just knowing who else is working on the same issue and connecting the teams. So the networks are not just set up for formal exchanges but also for creating lasting social relationships for quick "consultations among colleagues," as sociologist Peter Blau called it in his important treatise on what greases the wheels of public bureaucracies.[32]

The third mechanism is *setting new standards*. Encouraging collaboration through learning by facilitating exchange and convening participants to share experiences and defining new agendas is the stated goal of this emergent network. In fact, the mission of most intercity associations involves facilitating exchange and convening participants to share experiences and define new agendas. Not only do these networks provide an opportunity to learn from others about effective policies but also, according to the CRO, "give us a platform to show off what we're doing." In contrast to many other social networks, the city network was constructed with the purposeful input of some of its members: Early adopters of plans such as New York, Toronto, and Copenhagen realized the global coordination problem and founded networks such as C40, the USDN, and the Carbon Neutral Cities Alliance. These networks facilitate networking and learning among administrators to benchmark performance across dissimilar municipalities.

By creating best practices, like a GHG inventory or risk assessment, and legitimating certain goals, like the 1.5°C goal that seeks to reduce global warming to that threshold, city networks also engage in standard setting. The Global Covenant of Mayors is another example that has had a positive effect on activity in Chicago: "They asked for an emissions inventory sometime in the next five years. We knew that we were doing—we were going to do one. They asked for emissions targets, which we talk about on an action plan and a plan to get there, so you know, we felt . . .that the bar was pretty good for us." Like the "We're Still In" campaign discussed in chapter 1, these intercity agreements require a certain set of minimum requirements, which often are not ambitious, but they do create a shared floor that national—or even global— regulations do not provide. As the Chicago CSO reflected, these goals are intended not to push the boundaries of what is conceivable but to float all boats:

> I think that one of the interesting things going forward is, particularly for smaller cities, how do they get there, right? So they know they want to commit to Paris. They know what they want to do, so what are the tools that are going to be necessary to help them get there? How do they build a climate action plan? How do they create an emissions inventory, which is run-of-the-mill stuff for cities like us? New York has a staff that does it every year, but that isn't necessarily true in Peoria, Illinois, or other parts of the country.

Among these emergent standards, we find a familiar practice: strategic planning. Even in the absence of any legal requirement or political obligation to adopt strategic planning, membership in city associations provides both the necessary expertise and a strong expectation to do so. We saw that 100RC wants cities to develop and publish a resilience plan. C40, the Bloomberg-supported organization connecting ninety major cities, has a similar requirement, as one of their directors told me: "We require all of our members to have a mitigation and an adaptation plan . . . the format is discretionary." The goal is not just to encourage strategic planning as a best practice but also to help define and implement specific benchmarks: "[We] are asking that our cities do a certain assessment to have comparable figures on a mission across the board that we can aggregate and compare."

MODELING HOW COMPETITION, COLLABORATION, AND CULTURE MATTER

There is good reason to believe that the institutional superstructure has influenced cities profoundly by proselytizing sustainable development, encouraging competition, and providing networks of learning. All three of these factors could be at least partially responsible for the rise of city strategies, which we have seen often deal with sustainability and climate change. The broader takeaway is that the wider environment of cities has radically reshaped what it means to run a city in the twenty-first century—through culture, competition, and collaboration. Today's cities have come a long way from Friedrich Engels's Manchester, Émile Durkheim's Paris, Max Weber's Vienna, and even Robert Park's Chicago. They are dramatically more integrated in both the global economy and society. Some aspects have just intensified, such as multinational corporations operating out of more or less any major capital in the world inhabited by both high- and low-skilled immigrants who have increased cities' international outlook. In addition, the young people who are now joining municipal administrations seeking to help govern cities are more likely to have been to, worked in, or even trained in cities in other parts of the world.

What do these insights about the impact of globalization on cities tell us about the timing and nature of city climate action? The three Cs of a city's institutional superstructure—culture, competition, and collaboration—all look like they matter for the likelihood of starting to take action about climate change, with a plan. These dimensions provide us with three reasonable hypotheses about how the global institutional superstructure of cities influences their uptake of strategic plans that, in a final step, can be tested using the quantitative data on the global diffusion of strategic plans related to city climate action examined in chapter 2, with the goal of not only observing but also *explaining* the rise of city climate action.

Through the political economy lens, entrepreneurial cities compete about who gets to be at the helm of an economic world order. Through the institutional lens, a global cultural norm has made cities aware of world society's rules to adopt strategic planning as a rationalized practice. My expectation, based on how both perspectives intersect in organizations, suggests that collaborative cities picked up and shared a rationalized practice to engage with others, which they often learn from other cities. These hypotheses are not mutually exclusive, but whether or not they play out should tell us what motivates "cities in action" and what holds them back. To examine whether these potential explanations hold water, I examine how the timing of adopting strategic plans among the 360 largest cities in the world aligns with the complex networks of culture, competition, and collaboration.[33] To reiterate the three ways in which we might expect the institutional superstructure to affect strategic planning:

Culture: Cultural connectedness accelerates the adoption of a strategy.
Competition: Economic status accelerates the adoption of a strategy.
Collaboration: Professional centrality accelerates the adoption of a strategy.

Before I discuss the findings of this comparative analysis, it is worth spending a moment on how cities' positions and interactions are measured. Table 3.1 provides an overview of how I operationalized these dimensions using a mix of archival data and original network analysis. Further detail is available in the methodological appendix.

TABLE 3.1

Three forms of cities' embeddedness in the institutional superstructure

	Culture	Competition	Collaboration
City's position	Cultural connectedness	Economic status	Professional centrality
Mechanisms	Proximity to world society, receptors of global norms	Sites of global economy, economic interests, stratification	Situationally shared norms of conduct, collaboration
Measure	INGO headquarters in city (Yearbook of International Organizations, International Congress Calendar)	Shared service firms (Globalization and World Cities)	Membership in city associations (two-mode network of membership)
Carriers	Empowered individuals, professionals, organizations with international ties	International corporations, financial institutions	City administrators and public officials

We already have well-established measures for cultural connectedness from research on the world polity, which tracked country memberships in international associations and even in which cities these organizations are headquartered. We can also draw on existing measures of economic status from our geographer colleagues at the GaWC network, who modeled the headquarters and branches of international service firms. No such network model of intercity associations existed when I started my work. Even though we now have some indication of which cities are affiliated with international organizations, we still need to learn more about the emergence of the collaborative city network. Much has changed since the National League of Cities started lobbying Congress in 1926 in the interest of local governments—including its nineteen thousand US city members—and most of the network is barely two decades old. ICLEI, the global organization promoting Local Governments for Sustainability, grew from two hundred city members in 1997 to more than thirteen hundred in 2015. Seeing how the network formed is crucial to understanding its relevance to city action.

The starting point to showing how collaboration shapes city action is to bring the entire collaborative city network into the computer. Cities' interactions through city associations are yet another way in which cities are interconnected, like through trade, travel, and immigration. Researchers like the GaWC group, with respect to the global economy, have used such linkages to assume fundamental, relatively static networks among cities. For instance, a firm in city A may have a partner in city B. Taking one hundred such firms adds up to a fundamental network of interrelations in the global service economy that explains the status of some cities as "global cities."[34] Based on these networks, one can infer a structural position (e.g., in "tiers" or some form of regular equivalence) of each place. The network includes both cities and intercity associations, and both can be more or less central than one another.[35] The result is a massive collaborative network of cities that stratifies cities in a way not unlike the world economy—some cities are more central than others, some are on the periphery of the network, and some are entirely isolated from the network.

We can model the collaborative city network, too, using social network analysis. Joint membership in these intercity associations creates professional affiliation networks between cities. This focus on learning through wirings between administrations complements existing studies that conceive of cities as being linked through economic trade and commercial culture, as we have seen earlier. The city network is the result of intercity associations whose constituting members are cities rather than nations. For instance, Copenhagen and New York's City's joint membership in C40 suggests that the two cities are connected: their administrators are more likely to meet, interact, and exchange ideas than those of unaffiliated cities.

Are these different networks multiplex (carrying different information) or simply overlapping (carrying redundant information)? City networks illustrate that fundamental ties are complemented by many particular wirings between organizations that can be meaningful in their own right and operate separately from the thicket of global status orders. But are a few cities, like Paris, New York City, and London, high on all three dimensions, making them analytically worthless? In some cities, these dimensions overlap, but in others, they do not. Porto Alegre, a Brazilian city that is a minor economic player on the global scale, illustrates varying hierarchies. Because the city was the first to adopt an innovative democratic practice—participatory budgeting—its representatives are frequently invited to speak at global summits. In the professional network of city administrators, the importance of Porto Alegre outweighs its position among its peers in the fundamental economic structure of cities. Rather than simply viewing culture, competition, and collaboration as mutually exclusive, taking into account both fundamental and particular networks shows that they can intersect, and how.

The evolution of interactions among the 360 largest cities reflects an ever-denser network with intensifying opportunities to interact along multiple channels that connects global cities, one snapshot of which is shown in figure 3.4. One reason to look at this network is to examine which cities interact with each other, a core feature of cities' institutional superstructure. The network analysis reveals a number of different clusters—meaning that not just one conversation, but multiple conversations, are happening. Whereas some cities are members in more than half of these networks—Barcelona, Tokyo, Rio, Porto Alegre, Vancouver, and Mexico City—others only participate in a select number of prestige networks like 100RC, C40, and USDN—Copenhagen,

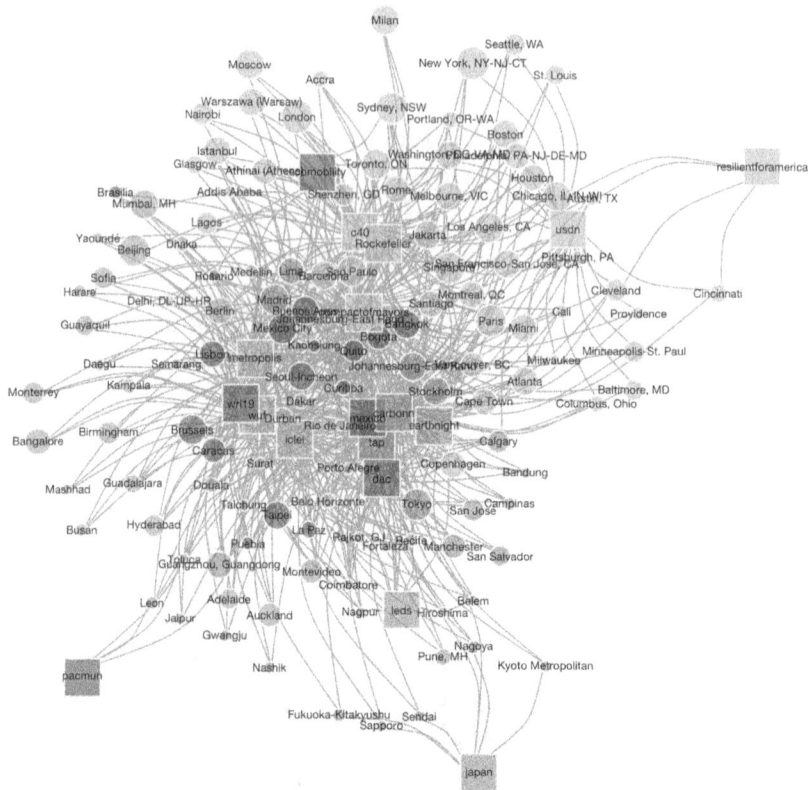

FIGURE 3.4 Global network of cities connected by intercity associations.
Note: Two-mode affiliation network based on 360 cities' membership (shown in circles) in seventeen professional associations (shown in squares) in 2015, shaded by leading eigenvector clusters. Snapshot from 2015 showing only the core component (excluding isolates).

New York City, and Seattle. Most cities in the Global South often have a sole tie to the United Nation's UN-Habitat but are locked out of these elite conversations. Many smaller cities engage only in programs that focus on particular regions, some of whose networks I did not include, like the Great Lakes Climate Adaptation Network or the SSDN in the United States, Eurocities and the European Green Capital in Europe, or the Asian Cities Climate Change Resilience Network and ICLEI Japan in Asia. As so-called isolates, about one hundred cities are completely out of the loop. Treating them all as the same risks glossing over organizational variation among these associations. Some associations have an office and versatile goals, whereas others are designed to address a particular problem. We will return to these clusters in a moment.[36]

In some sense, intercity associations mirror the world polity of international governmental and nongovernmental organizations studied by institutionalists of globalization, but when concerning their primary mechanisms, sanctioning power, and history, they do not mirror this polity.[37] When the League of Nations was founded in 1920, people took note. In fact, its constitutive document, the Covenant of the League of Nations, was signed as part of the Treaty of Versailles, arguably one of the most important multilateral treaties of the twentieth century as it ended World War I. The same was true when the charter of the United Nations was signed in 1945, promising peace after World War II, or when countries got together to sign the Bretton Woods treaties establishing an international monetary system that led to the establishment of the International Monetary Fund (IMF) and the World Bank. But the creation of the municipal world polity happened slowly over time, with many different networks layered on top of each other to create a thick mesh of interaction, learning, and collaboration. In most countries, cities and states are not legally enabled to enter into multilateral treaties with each other—diplomacy, like national defense, is reserved for nation-states and supranational bodies whose members are nations.[38]

Nevertheless, collaborative city networks have become central to what defines cities in the current age. Together with many others analyzed in this chapter, they have come to create a network that spans the world. In the inaugural editorial to the newly created journal *Nature Cities*—intended as an interdisciplinary outlet for urban science— published in January 2024, the editors write:

Indeed, cities are increasingly recognized as being crucial to addressing major global and societal challenges. All of the United Nations 17 Sustainable Development Goals are relevant to cities, and SDG 11 focuses explicitly on "Sustainable Cities and Communities." Just as cities have networked historically for commerce (among other reasons), cities are networking with each other now to address such challenges of the Anthropocene, as with the C40 Cities network for the climate crisis and the Resilient Cities Network.[39]

Cities interacting with each other is not a new phenomenon. Greek city-states like Athens, Corinth, and Thebes had formal alliances and wars, Italian communes like Venice, Florence, or Milan engaged in trade negotiations and power struggles, and city teams competed in sports leagues from the Oxbridge Boat Race to the Champion's League. The contemporary collaborative city network also has historical roots in the Sister City Network, dating back to a 1956 initiative by US President Eisenhower to create friendship and cultural exchange between municipalities. Strictly bilateral sister cities have become a friendly but antiquated nod to other parts of the world. These affable exchanges among twin towns, however, are nothing like the sophisticated collaborative network of 100RC whose beating heart I visited at 30 Rockefeller Plaza.

Contemporary networks are dynamic, multilateral, and characterized not only by personal relationships but also by serious political coalition building. As I learned from a USDN director, "It's starting to evolve towards thinking about opportunities for collective influence and collective action. . . . How do we—as a network—work together to collectively influence some of these systems that we know need to change to achieve the goals that we've set." The city manager of a Californian city took this thought to another level by noting that "increasingly you're seeing cities taking responsibility in their own hands, not waiting. And building relationships with each other." The philosophically inclined official, who had clearly spent much time in Silicon Valley and was writing poems about sustainability in his free time, likened the intercity associations to "a neural network being built around the world." He explained: "In the same way that neural networks connect and build these neural pathways that include learning and intelligence and all of those sorts of things." And as a Midwestern sustainability official predicted, this network "is going to accelerate and deepen." Why? "There's no time to let up."

We have seen that cities turn to each other and engage in learning, collaboration, and exchange—all indications of the *relational* nature of city action. But how does a city's position in the collaborative city network relate to the proclamations and plans discussed in chapter 2? I found that the centrality and position in the institutional superstructure are related to the rise of a homogeneous practice—strategic planning—with heterogeneous content reflected in varying thematic emphases. Modeling which city features are associated with the timing of adoption helps identify which of the noted explanations makes the most sense—and shows that collaboration plays at least as important a role as culture and competition. Textbox 3.2 and the methodological appendix offer a more detailed look at these findings.

TEXTBOX 3.2: HOW NETWORK POSITION AFFECTS CITY STRATEGIES

City strategies are affected by network position in the following ways:

- **Cities linked to the national institutions of the world polity adopted strategic planning sooner:** Cities closely tied to the world polity—most notably through ties to active INGOs and international organizations—and located in more affluent nations are more likely to adopt a plan.
- **But a city's network position matters more than the country in which it is located:** The effect of cultural connectedness is fully mediated by the effect of economic status: Core cities such as New York and London are more likely to adopt a plan than are cities at the periphery of the global service industry. Status matters.
- **Economic status has a curvilinear effect:** I also examined the idea that the competitive pressures to adopt a plan are most substantial in the middle tier of this world order. There is support for this proposition. This so-called middle-status conformity suggests that a city's structural position relates strongly to whether or not cities present elaborate strategic plans,

(continued on next page)

(*continued from previous page*)

    controlling for such factors as city size, internet access, economic capacity, and a global time trend.

- **Professional centrality matters independently of economic status, and the effect is similar in size:** Considering cultural embeddedness and economic status, a city administration's centrality in the professional affiliation network has an independent effect on adopting strategic plans. The model shows that professionally central cities are indeed more likely to adopt a plan than those at the fringe of the city network.

- **The effect of professional centrality is strongest for lower-status cities:** The normative pressures appear to be most pronounced for cities in the middle of the status order. Even controlling for the effect of economic connectedness, the professional centrality of the city administration is an equally strong predictor of the adoption of strategic. This finding holds upon including the middle-status effect in the full model.

Centrality in the Collaborative City Network

We can use statistical tools to examine whether there is indeed a relationship between membership in intercity associations and the prevalence of strategic planning. To do this, I draw on so-called event-history models, which predict the probability of adopting a plan in any given year. These models were initially developed to examine why some people are at higher risk of the hazard of experiencing a particular event quicker, such as death or, less terminally, a promotion. Thanks to sociologists, including Nancy Tuma, Michael Hannan, and David Strang, they have also been commonly used to study how contextual factors affect the diffusion of practices and policies among organizations and countries.[40] That is precisely what I want to examine. What do these models tell us? The big picture is that cities with deeper ties to intercity associations are more likely to adopt a strategic plan, and they will do so sooner. This pattern holds true even when considering our two likely explanations: the country's embeddedness in world culture as well as the city's status in the global economy.

    It turns out that cities with greater *centrality* in the collaborative city network have adopted strategies faster. The centrality in city networks has a bigger effect than good old cultural connectedness because the global scripts emanating from world society do not affect cities indiscriminately.[41] Similar to management consultants and professional creeds, administrative associations prescribe normative conceptions of proper management or—in this case—planning.[42] Is this ironclad proof that city associations are responsible for the rise in planning? Not quite, because city associations require cities to actively apply to join them, which means that purposive planners are more likely to join in the first place. But even though cities can strategically select into alliances that serve their goals, not all playing fields are accessible to all players. Some network clusters— sets of nodes that constitute a community characterized by mutual ties— are selective and specific to structurally equivalent entities such as developing cities, financial centers, or affluent midsize cities.

This also means that networks were not equally important for all cities. The extent to which centrality matters declines as status increases and eventually becomes statistically insignificant for highly competitive cities (somewhere around the economic status of Tel Aviv, Chongqing, Nairobi, or Geneva), as the estimated marginal effect sizes of centrality by economic status in figure 3.5 shows. Professional centrality turned

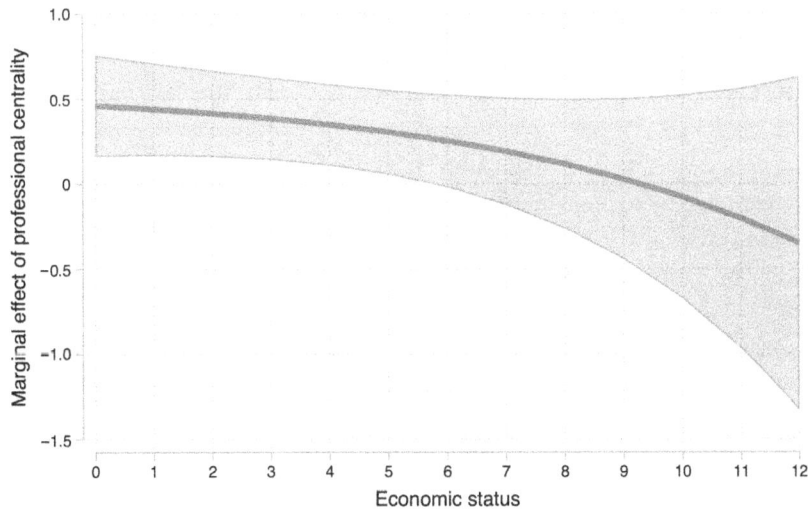

FIGURE 3.5 Estimated effect of administrative centrality on the hazard of adopting a strategy plan.

out to be less important among high-status cities. In other words, normative pressures were most pronounced when they were not redundant: High-status cities do not need professional stimulation to make their role in the collaborative city network known. For leading cities, professional pressures can be redundant in part because those leaders *created* these pressures in the first place. Alpha cities like New York, London, or Chicago are often among city associations' founding members. For medium-status cities, membership in global associations is an important distinguishing factor.

Positions in the Global Network

Another piece of evidence supports the insight that collaboration differs depending on economic status—meaning that competition and collaboration interact rather than exclude each other as explanations. As we have discussed, the collaborative city network does not look like a single mesh, like the suburban sprawl of Los Angeles, but instead it is a larger structure made up of communities, more like the Sydney's "city of villages." These network clusters—and therefore the specific cultural content that cities are exposed to through membership in the affiliation network—also differ by status. It is thus not surprising that cluster membership is directly related to the probability of adopting a plan. The cluster around C40 has the highest average degree of economic status, except for the cluster of regional US cities and their networks. The messages cities receive differ depending on whether they are at the periphery and the core (e.g., UN-Habitat versus C40); mid-status cities are exposed to various institutional contexts from regional networks.

Most cities are not involved in all conversations simultaneously, but rather in a subset of exchanges. The six network clusters that had solidified by 2015, shown in figure 3.4, are distinguishable by both purpose and approach. The first cluster centers on the UN agency for cities—UN-Habitat—which pioneered the idea of urban sustainability in the Global South in direct response to the dire conditions in developing cities in the early 1990s. UN-Habitat helps local municipalities in developing their capacity and hosts the biannual World Urban Forum. The second cluster includes both leading cities and associations that rally leading cities. C40 stands out as one of the most influential groups in advancing sustainability among large, global cities. Among other strategies, such as regular webinars, C40 strongly encourages the adoption of a climate strategy. For comparison, the average gross domestic product (GDP) of the thirty-six cities in this cluster is twice that of the ninety-nine cities primarily associated with UN-Habitat ($38,612 to $16,825). A total of thirty-seven cities are associated with Metropolis, which was founded in 1985 and formed one of the core components of the collaborative city network before the mid-2000s. It is one of the oldest associations and became active before theories of cities sharing responsibility for climate change–related efforts emerged. The three remaining clusters contain a total of eighty-four cities and include international efforts led by ICLEI and two regional clusters for mid-tier cities in the United States and Canada and in Latin America. In all of these clusters, membership in an association is focused on the implementation of a specific practice or activity, such as Earth Day or the symbolic commitment to a climate goal. ICLEI is an exception, because it encourages both GHG reporting and local capacity building among smaller cities—we will revisit the organization in chapter 4. Notably, ICLEI was a central cornerstone of the affiliation network in earlier years and many cities outside the "international" cluster are also nominal members in ICLEI.[43]

Homogeneous Practice, Heterogeneous Content

Different positions in the professional affiliation network emphasize different aspects of strategic planning, not only in form but also in content. Some associations, like 100RC, actively proselytize planning and elaborate risk assessments. Others encourage reporting, learning, or expression of identity. Both UN-Habitat and C40 encourage holistic planning to build local capacity but for vastly different cities. Other widely diffused frameworks offer tools for measuring GHG emissions and action plans for periodic or specific activities. The clusters also tell us something about the emphases discussed in the previous chapter. Membership in the C40 cluster increases the likelihood of having a strategic plan that acknowledges climate change by some 80 percent. In sum, a city's location in the professional network of intercity associations determines not only the form of planning, but also the content of the subsequent actions. Although all networks have some activating effect, cities prioritize different topics, frameworks, and buzzwords in their plans—resilience networks focus on adaptation, whereas sustainability networks are more likely to stress mitigation.[44]

Although the collaborative city network has become a denser scrub with time, it is not one single professional network; cities are exposed to multiple professional networks at different status levels. These networks pursue different agendas: All encourage some form of planning to address urban problems, but the suggested content of action differs depending on whether networks are global-high status, global-low status, or regional-middle status. Cities are stratified within each geography and are connected through a field of intermediary organizations with other cities that are structurally equivalent to them. More central entities, such as first-tier cities like New York City and London, face common institutional influences. Less central entities, in our case, second- and third-tier cities from Adelaide and Bologna to Yekaterinburg and Zhengzhou, are subject to regional influences. In addition, some ideas trickle down from high-status cities to middle-status cities; or rather, middle-status cities pick higher-status role models and emulate what they do.

This model tells us something about the patterns of "peers" that we noted at the beginning of this chapter. The most central entities share a reality with other highly central cities (e.g., London and New York City), but less central cities share a reality with other middle-status entities in their reach (e.g., Portland and Oakland; Vienna and Copenhagen), and isolated or peripheral cities live in a world of their own (e.g., Almaty, Kazakhstan; Baku, Azerbaijan; Detroit, United States; Marseille, France; and Shiraz, Iran). This idea is not necessarily specific to cities: It is similar to many business schools copying Harvard Business School, whereas Harvard Business School has a more selective perception of who its peers are. This is important because it tells us something new about how culture—through professionalization, rankings, and INGOs—creates not only homogeneous practices but also heterogeneity. The content of rationalizing pressures can differ greatly over time and depending on what actors pay attention to.[45]

These findings explain and compound some of the inequalities among cities explored in chapter 1. Elite cities from around the world convene around abstract and expansive problems, whereas developing cities share many concrete problems, such as slums and sanitation, in common. High-status actors offer broadly conceptualized plans, but mid- and low-tier cities often specialize in issues that allow them to differentiate themselves from others.

WHAT WE SAW

In this chapter, I explain the institutional superstructure to be a primary source of cities in action. City networks have an independent effect on whether or not cities plan strategically. Of course, many features specific to cities may explain both why a city joins an intercity association and why it plans strategically. It would be too soon to drop the proverbial microphone and conclude that city associations and their program managers have caused city action or that they are the sole cause; that is unlikely. The insight about the diffusion of city strategy, however, backs the idea that the wider institutional context of cities and city administration significantly affects how cities think about challenges, such as climate change and resilience.

Starting to think about others and our relationship to them is a central feature of social action—and is yet another box we can tick in our list of questions of what constitutes city action. I found something distinct and heartening in how my respondents reported on their relationship with other cities. No company, research lab, or charity is routinely happy when their idea gets "stolen," as I have heard repeatedly. As the Chicago CSO stressed, a finite amount of good ideas circulates in the world of sustainability and CROs, and once you have one, you want to ensure that others have access, learn from your mistakes, and do an even better job than you did. This is not cutthroat competition, in which one city is trying to run another out of business for their own gain. Petty instincts of big egos like the "vanity competition" obscure the genuine willingness to work together. Although it is clear that culturally reinforced expectations are circulating, I did not encounter any mindless buffoons who structure their organizations in silly ways just because culture dictates it. Sustainability chiefs are an impressive and reflexive bunch and have actively cocreated the network that influences them.

We have also seen that a city's position in the institutional superstructure affects climate action in the largest cities, which have taken on powerful functions rivaling the climate policies of states and even countries. At the same time, the lack of activity on the fringes of the collaborative city network can lead to geographic disparities, tension, and complex relationships with other levels of government that we have yet to examine. Doing so will require peeking at how cities differ concerning the actions they take in response to the plans and pledges reviewed in chapters 1 and 2. When environmental evaluators and sustainability advocates encourage local communities to create a strategic plan, that plan's content still depends on the city's status and local governance constellations. How the duality of local and global pressures interacts in shaping city action is an open question. In chapter 4, I hand the mic back to these city officials to learn *how* city networks matter to them. As I will show, sparks fly when the institutional superstructure reacts with the local organizational infrastructure.

CHAPTER FOUR

Leading

> We are called upon to lead. Many would say the United States has lagged
> in response to climate challenge. . . . Many would say that California has
> led in response to climate change. . . . Many would say that Palo Alto has
> been a leader in this process, with our early climate action plan, our car-
> bon neutral electricity, and our actions to support green buildings and
> electric vehicles.
>
> —SUSTAINABILITY AND CLIMATE ACTION PLAN OF
> THE CITY OF PALO ALTO, CA

Despite having a population shy of seventy thousand people, the city in
the heart of Silicon Valley felt the world's eyes on it as it devised a new
strategy for sustainable development in 2016. Palo Alto's plan was meant
to be more than a technical document, like those master plans required
by California state law stacked in the city archives; the plan was both
a statement and a commitment when it perceived a lack of leadership
at the country's top: "It's time for us to lead again, with a new sustainability and climate action plan that sets a new bar for leadership, that
builds quality-of-life, prosperity and resilience for this community, and
that sets an example once again for other communities to emulate."[1] In
accordance with the plan, Palo Alto adopted an ambitious 80×30 goal:
an 80 percent greenhouse gas (GHG) reduction by 2030 compared with
the 1990 baseline, twenty years ahead of California's 80×50 climate goal.
"The time to act is now," the city justified its commitment to becoming a
carbon-neutral city. In this chapter, I examine how plans like Palo Alto's
relate to action, and how discretionary this action really is.

A year earlier, I had biked over to Palo Alto's first town hall, which
was meant to solicit public participation in the comprehensive planning

process. The city's leaders were eager to consult with the public to learn what should appear on the city's agenda. Experts from community organizations and companies were represented as well. While waiting in line to register, a man beside me gave me his business card and pitched his business to me: preparing companies for EnergyStar accreditations. Recently, his company had started considering cities as potential clients. "We love Palo Alto," he says. This was not the man's first town hall. "We *love* Copenhagen," he added as I mentioned the city's promise of carbon neutrality.

The entire city council and several hundred citizens came out to Mitchell Park Community Center to listen to City Manager Jim Keene give a keynote about the city's bright future. Jim—a seasoned civil servant and hobby poet who had cut his teeth as a city manager in Tucson, Arizona—simultaneously emphasized the challenges related to housing prices and homelessness that are "happening around the peninsula" and reassured the many owners of single-family houses in attendance that nobody was going to challenge their privilege in a city with 85 percent low-density housing. Who were those people in attendance? Most people who showed up were the usual suspects—white, affluent retirees or homemakers that tend to be overrepresented in public deliberation compared with workers, minorities, and youth.[2] People fumbled on their phones to participate in a text message poll about how they got to the event: 45 percent of the attendants drove, some walked or biked, and only 2 percent took public transit. Regardless of their mode of transportation, the citizen roundtables seriously engaged with the city's perceived problems all afternoon. The attendees, however, were not primarily concerned about the planet.

That spring, California was at the height of what turned out to be a 376-week-long drought. Even though the heat was a constant topic of conversation, I did not overhear much debate about climate change per se. The only place where sustainability loomed large was at one of the many information stalls the city put up in the community center's courtyard. The fanciest one was about the city's sustainability work. Glossy flyers announced the city's Sustainability and Climate Action Plan and a flat-screen television showed a video of the city's charismatic chief sustainability officer (CSO) and self-declared "sustainability OG" Gil Friend explaining why sustainability matters and what the city was committed to doing about it. I started wondering why the City of Palo Alto was so gung ho about climate action even though neither the country's federal government nor the participants of its town hall cared much about it.[3]

This town hall raised a question central to this chapter: What explains why some city administrations embrace sustainability, while others do not, and what role does its public play in this process? I had already seen by then that global cities engage in purposive planning (see chapter 2) and that the relational ties to other cities and intercity associations shape both the form content and timing of these plans (see chapter 3). Palo Alto is not exactly a backwater either—this is where social media giant Meta grew up, Palantir develops secretive security technologies, and Stanford University produces cancer cures. But it is still a much smaller city than Chicago or Cleveland.[4] Are Palo Alto's ambitious plans just a response to a global (or otherwise higher-level) imperative to explain oneself, or do they translate into discretionary city climate action—that is, voluntary sustainability practices that do more than meet existing legal requirements?

In this chapter, I offer a road map through the urban sustainability initiatives of cities in all parts of one country—the United States—to understand the relationship between administrative initiatives to boost sustainability and the local context in which they take place. Focusing on a particular context holds some regulatory and political contexts constant, and it also allows us to examine cities that face steeper barriers to sustainability. This analysis closer to the ground is a crucial step toward understanding variation in city climate action and inaction, as previous chapters considered major global cities that are exceedingly visible and exposed to global pressures; these pressures may not trickle down to suburbs and the "hinterland" of large metropoles.[5]

To understand what explains why some municipalities accept the cost and hassle of trying to become more sustainable while others hold back, we need to examine the antecedents of city action—the fellow travelers, guides, and tools that help city administrations follow the uncertain path to sustainability. As I argued in the introduction, local conditions like the participants in urban governance, the mechanics of urban politics, and the interests of stakeholders shape city action, but they are not always the wind in the sails of sustainability proponents.

Institutional superstructure
Membership in city associations

Diffusion of expectations

Administrative city action
Sustainability practices

Organizational infrastructure
Nonprofits, managers, firms

Receptivity to expectations

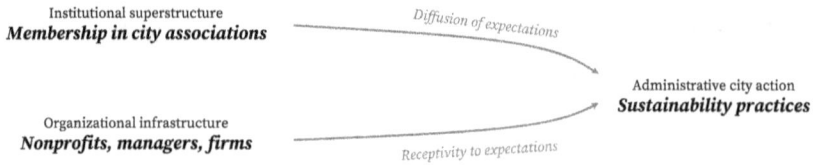

FIGURE 4.1 Visual summary of chapter 4.

    The chapter flips the table from how cities respond to pervasive top-down pressures from state, national, and global institutions that encourage certain local actions—the *institutional superstructure* that they share with many other cities—to the *organizational infrastructure* that is specific to them, as figure 4.1 summarizes. I will begin by reviewing survey data of US cities, small and large, to map out the leaders and laggards in city climate action and draw on interview data to understand these disparities. I find that even within the same political context, meso-level influences stemming from the organization of cities rather than individual politicians or the general public matter a great deal. These factors are additive, covering different aspects of the cities' organizational infrastructure that all contribute to whether cities lead or lag within the structural constraints discussed in chapters 2 and 3. As we will see, detachment from international associations, starving for financial and political support, and civic degradation stifle sustainability efforts, explaining the wide gap in cities' actions as well as their profundity.

SURVEYING SUSTAINABILITY ACROSS ONE COUNTRY

Instead of solely focusing on the presence of a sustainability plan, in this chapter, I draw on survey data about dozens of practices that a team of experts at Arizona State University (ASU) and the International City and County Management Association (ICMA) identified as the state-of-the-art repertoire of sustainable cities as of 2010, ranging from gray water use to tax incentives for energy efficiency and preservation to encouraging reusable water bottles. The data are by now best thought of as historical rather than as describing the current state—but that's the point. These data not only are a tested means of providing a sense of geographic patterns related to sustainability practices but also help us understand the process by which sustainability practices (beyond mere planning) first took hold in city administrations around the time we saw cities' concern for sustainability burgeon in chapter 3.[6]

To analyze these data, I teamed up with public administration scholar David Suárez to examine what explains which features of cities are associated with more or fewer sustainability practices. David had long been concerned with the functional explanations that public administration scholars draw on to explain how administrations optimize some utility function when determining what to do. He knew as much as I did by then that many administrative commitments are rationalized practices that are seen as legitimate without necessarily bringing functional improvements and whose adoption is shaped by the institutions in which administrations are embedded—an assumption we planned to test.

The data David and I drew from came from a sample of 1,796 cities with more than twenty-five thousand inhabitants in the United States. The 2010 Sustainability Survey, collected by the ICMA, provides information about local government environmental policies and programs, including a wide range of sustainability-related practices for the largest set of cities presently available.[7] The range of sustainability practices continued beyond planning. It also included various policies and programs or activities like collaborating with other municipalities to reduce carbon emissions and encouraging green building certifications. In total, the survey included 123 sustainability-related practices across ten contexts, ranging from planning, policy targets, and recognition programs to substantive areas like water, recycling, energy, consumption and conservation, transportation, buildings, and land use. Textbox 4.1 expands on what exactly these practices are and how to interpret them.

City climate action is a global phenomenon rather than one limited to the United States. However, countries differ significantly in their legal and political context. Focusing on a single, large country allows us to hold federal legislation constant, unlike when studying other world regions, such as Asia, Europe, or Latin America. Do the lessons derived from this case apply to all other countries? Not universally, because growth coalitions and market orientation are

TEXTBOX 4.1: SUSTAINABILITY PRACTICES FROM THE ICMA SUSTAINABILITY SURVEY

This list of practices included in the sustainability survey are important indicators of different types of mitigation and adaptation efforts. The fact that they cohere strongly in a single factor—even with additional indicators, such as initiatives for social justice—indicates that they reveal an overall propensity to take action rather than specific and mutually exclusive sustainability strategies. Does the city act, and how much?

1. *Planning*: Does the city have a resolution with policy goals related to sustainability, energy conservation, resilience, emission reductions, or similar concerns? Is a specific budget provided? Any staff?
2. *Policy targets*: Are there GHG emissions targets for the community? For businesses? For residences? Is there a plan for tree preservation?
3. *Recognition programs*: Are recognition programs in place, like Tree City USA? An EPA Smart Growth Achievement? Or others, like a LEED certification?
4. *Water*: Are there gray-water systems reusing reclaimed water? A water price structure to encourage conservation? Any actions to conserve water from aquifers?
5. *Recycling*: Are there internal programs to recycle glass and plastic? Recycling of hazardous household waste? Community-wide collection of compost?
6. *Energy*: Is there a fuel efficiency target for the government's vehicle fleet? Are there EV charging stations? Solar panels or a geothermal system on government facilities? Energy audits of government buildings?
7. *Consumption and conservation*: Are there energy audits or weatherization of individual residences? Subsidizing solar equipment or HVAC upgrades for businesses?
8. *Transportation*: Are local government employees encouraged to bike, carpool, or take public transit?
9. *Buildings*: Does new government construction require LEED or EnergyStar certifications? LEED standards? Incentives to build more densely?
10. *Land use*: Are there Brownfields Programs to revitalize abandoned sites? A land conservation program? A program to purchase or transfer open space or preserve historic properties?

powerful, and governments are hollowed-out in liberal economies such as the United States, the United Kingdom, and Australia, suggesting that green laws and regulations likely carry greater weight in other places. Notwithstanding the US-American values and governing principles that have spread worldwide since World War II, focusing on the United States provides an in-depth case study of one nation.[8]

The United States is an insightful case for a couple of reasons. First, it has plenty of variation among many cities because it is both highly urbanized and dispersed. Consider in comparison my home country of Austria, where a quarter of the country's population lives in Vienna; 30 percent of the Japanese population lives in the Greater Tokyo Area; and 15 percent of the British population lives in London. In contrast, only 6 percent of the US population lives in New York City. Second, the role of cities in the governance system of the United States is comparatively weak because of its history. Cities were conceived as "creatures of the state," which means that US states both create and, as a result, may control the actions of each municipality. The 1868 Dillon's rule states that "municipal corporations owe their origin to, and derive their powers and rights wholly from, the legislature. It breathes into them the breath of life, without which they cannot exist. As it creates, so may it destroy. If it may destroy, it may abridge and control."[9] Assuming that states would legally curb—preempt—sustainability practices is a foregone conclusion, but without doubt, state policy is a constant concern for urban sustainability officers.[10] I will explore cities' relationships to the fifty states in which they are located later in this chapter, showing that states support rather than shackle their cities. Finally, the United States offers a trove of data about the extent to which city administrations prioritize and respond to climate change, a question that, in light of organizations' tendency to decouple their actual practices from their official policies, has been the elephant in the room since it was raised in chapter 2.

The political context of the United States amplifies the question of whether cities lead on sustainability, and if so, why? On the one hand, the federal government's flailing commitment and the global mobilization for local climate action puts cities on the spot. On the other, their populations do not necessarily want them to spend scarce resources to make up for national inaction. What we found not only underscores the fact that city climate action is indeed discretionary and frequent but also offers a surprising response to the seemingly intuitive assumption that only progressive mayors drive the push toward sustainability.

Of course, self-reported practices have certain limitations because we do not know how far-reaching the consequences of each practice are. The same environmental practice, such as recycling, can be implemented more or less profoundly—I will leave questions about the consequences of cities' efforts to chapter 5. For now, these practices have a crucial advantage over the strategic plans examined in chapters 2 and 3. The data include indicators of concrete actions that are typically under the control of the city administration—rather than abstract measures heavily confounded by contextual factors, such as the overall carbon dioxide emissions of a city. Sustainability practices thus represent more substantive measures than the hypothetical commitments to climate change that we have seen in strategy and climate action plans. The survey data also cover several practices throughout several domains and administrative departments. This breadth means that rather than looking at the binary adoption of a plan or practice, we can examine the extent to which a city has embraced different facets of sustainability.

Surveys like this can be instructive because they offer a snapshot of the landscape. In ICMA's somewhat more recent 2015 urban sustainability survey, about 32 percent of US cities reported adopting a sustainability plan, and 40 percent of cities reported that public participation has played some role in shaping this sustainability strategy. This average uptake among hundreds of more domestic cities is somewhat lower than that among the 360 largest cities in the world (see chapter 2). Around 53 percent of these cities have adopted a strategy of some sort—typically with sustainable development in mind. Among the motivations for pursuing sustainability rank the potential for fiscal savings, the leadership of local elected officials, funding opportunities, and the potential to attract development projects—all of which are listed ahead of a concern over the environment. Cities cited a lack of funding, higher-level government legislative restrictions, and staff capacity as the major hindrances. Only half of the cities reported that environmental protection is a priority in their jurisdiction, compared with nine in ten that care about economic development. (One can only guess what the tenth does for a living.) Three in four cities reported having had to respond to at least one major disaster, such as a hurricane, earthquake, wildfire, or flood, in the past fifteen years.[11]

The first thing David and I checked was whether strategic planning was related to other sustainability practices. We noticed a very strong relationship between all sorts of activities within municipalities. This finding suggested that cities that report to set goals for climate change also tend to present related planning documents and are more likely to adopt sustainability-related programs and policies. We drew on a technique called principal component analysis to extract a latent factor (something like the underlying propensity to do sustainability-related things) from all possible practices adopted by the city administration.[12] The factor based on municipal sustainability practices showed that municipalities' self-reported activities varied across a series of departments and policy issues, from planning to water and energy.

This analysis suggested that cities with a sustainability or climate action plan tend to have more sustainability practices throughout all domains. This consistency does not necessarily mean, however, that the activities correspond to what cities laid out in their plans or that a causal relationship exists between planning and urban sustainability. This is partly due to confounders—unobserved variables that can distort the estimated relationship between two others—between city activity and such city-level performance indicators as GHG emission scopes or individual attitudes of the city's population.[13] The association between political preferences and sustainability practices suggests that administrations that serve an environmentally conscious electorate may also be more likely to pursue a progressive environmental agenda.

Selection is another concern. We know that cities already performing at a higher level are more likely to adopt a climate action plan in the first place. Urban planner Adam Millard-Ball has argued that most climate action plans are a consequence of above-average environmental performance rather than a cause.[14] The tendency to put such accomplishments on paper is part of the search for legitimacy and evidence of the city's competitiveness relative to other cities, as discussed in chapter 3. When cities have positive performance to share, they are more likely to put them in writing. Even when we consider various features of the population, as we did here, sustainability practices are endogenous (meaning that their adoption may be driven by the same factors that explain environmental performance). This endogeneity leads to imperfect estimates of the extent to which voluntary, and even required, city strategies or any city-level activity contributes to such generic and overarching outcomes. Although the data will not prove any causes of city climate action, they are useful for showing which cities embrace urban sustainability and for examining why.

LEADERS AND LAGGARDS

Not all cities reported action for sustainability to the same extent—this much is old news. The problem of inaction looms large, considering US cities' drastic variation in sustainability practices. The late public administration scholar Kent Portney, one of the pioneering surveyors of urban sustainability among US cities, started aggregating various sustainability measures into a comparative index of "seriousness about sustainability," in which higher-scoring cities move faster and in more areas, as early as 2003.[15] His research showed that some cities have emerged as leaders who take sustainability seriously, and others are laggards who do not want to or, more commonly, cannot afford to do so.

Although studies like Portney's blazed a trail to understanding sustainability, they barely drew a complete picture, partly because their sampling frame left out many cities—and ended up with an in-depth understanding of few. This chronicling helps examine how big-picture forces, such as political and administrative culture, shape urban sustainability, but it does not allow for developing or testing a theory of city action. So, despite some shortcomings, survey data of hundreds of cities of all sizes and regions opens new doors to comparative research.[16] We linked the responses to various sources of city-level data, which allowed us to assess different potential explanations of how local and global contexts shape sustainability practices. In other words, we could test whether city administrations' dual embeddedness really shapes climate action.[17]

Explaining who leads and lags in city climate action requires first articulating *why* we expect cities to adopt sustainability practices. We learned that the public was still catching up with cities' sustainability efforts, even though administrators tried to convince them that responding to a changing climate was the right way. What about the cynical take that experts are running the show, with little concern for the public interest? This possibility offers another, simpler explanation: Cities experience real problems and city leaders solve them, often maximizing their self-interest. These functional explanations are not without merit. For instance, larger cities with a thriving economy may be more likely to invest in innovative labor market policies and tax regimes. Cities with greater environmental threats may worry more about the effect that flooding or extreme weather events will have on their livelihood. Cities' actions are a simple response to a demand, and they exist because they ostensibly help cities perform their duties better—by fixing climate change.[18]

The reality is often more complicated. Politicians and administrators do not optimize their utility—budgets, hires, votes, fame—by choosing specific policies; instead, they devise and implement such policies in response to institutional pressures that are sometimes orthogonal to a city's natural and economic realities. For instance, Frank Dobbin's analysis of railway systems in the United States, Britain, and France shows that industrial policies depended on the country's institutional context and political framework rather than just how far train stations were apart or what technologies were available. The United Kingdom took a much more dispersed approach than did highly centralized France, the consequences of which one still sees when crossing from one side of France to another—usually through Paris. It is neither functional nor optimal but an empirical reality.[19] Similar insight is central to the institutional approaches to understanding the global influences on national environmental policy that we encountered in chapter 2.

Thinking about sustainability as a rational choice risks giving in to necessary but ultimately incorrect assumptions about what motivates people—none of the civil servants I spoke to saw climate change as an opportunity to grow their budget. Institutionalists have long argued that efficiency and effectiveness are not the only, and sometimes not even the primary, organizational pursuits.[20] Not only do the motivations for embracing innovations differ across cities, then, but also many practices spread for reasons that have little to do with their effectiveness.[21] Nevertheless, we need to consider commonsense explanations before turning to new ones. It is indeed quite likely that large cities have more per capita capacity to deal with an obvious challenge than others, and, as a result, they have a larger array of sustainability-related practices. It is also possible that sustainability is a luxury problem that only wealthy cities can afford to pursue. And then there is the crucial issue of politics. We paid attention to all these factors, ensuring the statistical and substantive relevance of our insight beyond common explanations, and still found evidence that cities' dual embeddedness in their institutional superstructure and organizational infrastructure mattered a great deal.

What do the survey data tell us? As the map in figure 4.2 reveals, most of the laggards regarding a standard suite of sustainability practices (the light gray dots) tend to be small, suggesting that less populous cities are struggling significantly more to overcome institutional barriers to urban

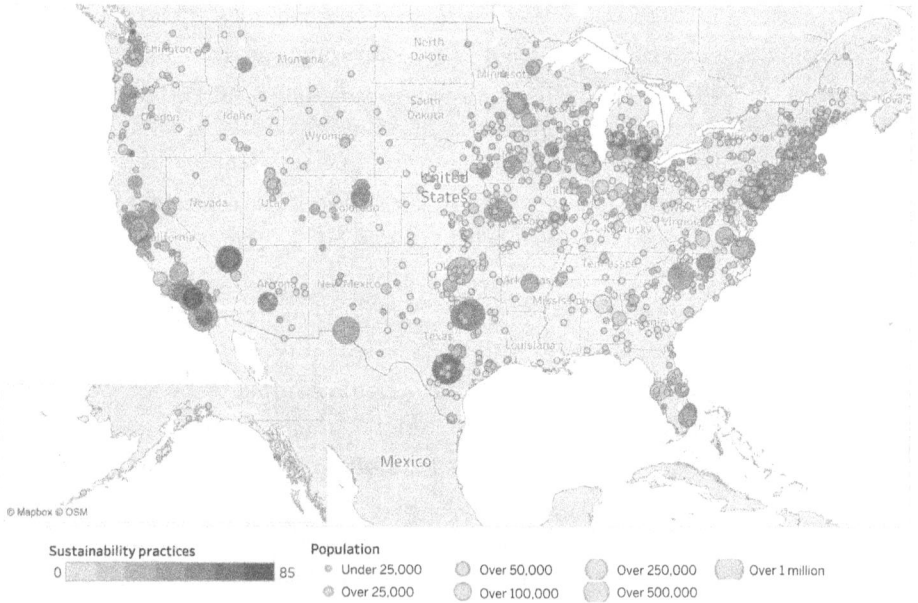

FIGURE 4.2 Map of cities' sustainability practices in US municipalities, 2010.
Note: N = 1,590.
Sources: Based on the 2010 ICMA Sustainability Survey; map uses map data from Mapbox, https://www.mapbox.com/about/maps; OpenStreetMap made available under the Open Database License, https://www.openstreetmap.org/copyright.

sustainability. It is also immediately apparent that states, regions, and loca-tions matter—whether for political or geographic reasons. Coastal cities seem more preoccupied with sustainability concerns than cities in the country's heartland; in many other respects, the West Coast of the United States looks noticeably different from the country's Southern states. Some of the leaders in sustainability practices (the darkest dots) denote more prominent cities, such as Las Vegas, San Antonio, and New York.

Is variation in city action just about population numbers? Even major city networks like 100 Resilient Cities (100RC, later Resilient Cities Net-work) pride themselves on being home to a wide range of cities, as I learned in an interview with one of 100RC's vice presidents: "Our cities range in size from 50,000 all the way to Bangkok and Mexico City. . . . And so the challenges are different but . . . the possible solutions and the desire to exchange no matter what the city's size is there." A policy director of Local Governments for Sustainability (ICLEI) sees partic-ular beauty in the small: "The larger cities you hear about quite a bit. Boston, New York, San Francisco, and so on. But I'm struck by cities like Fort Collins, Colorado. They are doing incredible work on sustain-ability. Eugene, Oregon, has been a leader in terms of looking at their climate action work on a scientific basis. They worked with local climate scientists to develop their climate action plan and targets." These small communities sometimes go about their work differently, too: "There are communities that have tremendous environmental justice communities involved in their work . . . to ensure that their work is inclusive."

We have already seen how size and status shape who cities consider their peers: major networks tend to have cities with much higher gross domestic product (GDP) and economic status than more regional ones. Size matters a great deal for who can be considered a peer. Chicago looks to Toronto, as their CSO told me: "Toronto's almost our perfect peer city to us in terms of climate. We're both just under three million residents. We are both on a Great Lake. . . . There are a lot of cities in Europe that we have good relationships with, so Paris, London, Copenhagen. Obvi-ously, their politics and their environment are different, right? But we still have a lot in common. We spent a *lot* of time thinking about them." Lateral relations have their limit, however: "Shanghai has been a sister city to Chicago for a long time. I was on a panel with a counterpart from

Shanghai, and simply the scope of the problems they're dealing with. . . . A city that's eight times the size of Chicago is a very different animal." In other words, Chicago worries about Toronto and New York and may even consider what's happening in Shanghai or Tokyo. But cities like Evanston and Springfield, under its nose, are beyond the eye.[22]

If smaller cities scored so much lower on the sustainability programs, does this mean that the bulk of cities smaller than the global cities previously discussed are entirely disinterested in sustainability issues? Is climate change ultimately an issue reserved for the largest and highest-status cities, as chapter 3 suggested?

To better understand what distinguishes ostensible leaders and laggers in the survey data, and how they seek to overcome their respective barriers to urban sustainability, my colleague Ana Gonzalez and I interviewed sustainability officers from vastly different cities on the West Coast, East Coast, the Midwest, and the South of the United States. Our emphasis was to expand the data to cities that earlier research found less active regarding the sustainability agenda, because their experiences may illuminate the causes and nature of city *in*action. We drew on various sources to identify such cities that ran the gamut on the political spectrum and concerning how seriously the city government took sustainability.[23] Ana and I conducted twenty-one additional interviews with the sustainability directors of thirteen cities throughout the United States. The cities we virtually traveled to were not just large ones like the ones we mentioned so far—Chicago, New York, and Los Angeles—but also significantly smaller ones in predominantly rural parts of the state—second- and third-tier cities that you may have, for example, rented a car from to drive to a national park. See the methodological appendix for details.

We found that cities on both ends of the sustainability spectrum grapple with issues related to sustainability on a daily basis and with significant concern for the city's well-being, but the path that each city walks in doing so differs significantly. The proactivity of global leaders— wanting to be the first in carbon neutrality, smart technologies, or even environmental justice—was less common among our local interviewees.[24] The CSO of one Midwestern city, for instance, admitted that their approach is merely reactive: "We try to plan strategically but often do the things right in front of us. It still is harder to get to things that maybe have a bigger return on investment, for instance, in terms of reducing inequality or improving air quality or electrification opportunities. These things are just hard for us to prioritize when we're trying to negotiate an agreement for something." What cities are active about also runs the gamut. Whereas some cities pursue "green" strategies that feature minimal interventions, including trees and recycling, others go for "gray" strategies focused on impactful investments in infrastructure and housing.[25] Claims that all cities are "first responders" to climate change mask this variation in discretionary action among cities. So what explains it?

FROM CREATURES OF THE STATE TO CITIES IN ACTION

The first potential explanation for this variation is that some cities are too weak to wield discretionary action because they are bound by strict institutional forces—like state politics. This explanation is compelling but curious.[26] In fact, cities have earned a reputation as hotbeds of innovation regardless of the state in which they are located. This not only is true in the context of typical outcomes of research on the geography of innovation, such as patenting rates or business foundings, but also is reflected in *social* innovation in response to tricky social and environmental problems.[27] Sociologists and economists have examined local responses to challenges such as climate change, poverty, sustainable economic development, and the maintenance of civil liberties. Mounting evidence from scholarly work on city administrations shows that cities are at the forefront of innovations related to open government data, new public policies for economic growth, and public service provision.[28]

The notion of cities as mere "creatures of the state" clashes with many defiant actions that fly in the face of ever-more-reactionary national and state politics. Consider the example of initiatives that investigate how cities can respond to ever-stricter national immigration laws, such as the sanctuary city movement, municipal identifications (IDs) for undocumented immigrants, or the United Cities and Local Governments (UCLG) Lampedusa Charter, which advocates for a migration policy based on dignity and human rights.[29] Some observers have gone so far as to declare an "urban revolution" in which cities are a new locus of social innovation and progress.[30]

At the same time, state or national regulation can be more effective and democratically legitimized than leaving matters to the voluntary choices of individuals. For instance, European countries have more green buildings than those relying on certification because of higher standards. In California, state policy came *after* the certification had spread, as I discuss in chapter 5. The Civil Rights Act was infinitely more powerful than a couple of municipal libraries that defied Jim Crow laws, although the latter may not have been possible without the former.[31] A well-funded federal administrative apparatus, including agencies like the US Environmental Protection Agency (EPA), the National Oceanic and Atmospheric Administration (NOAA), the Fish and Wildlife Service (FWS), and so on, is more effective than hundreds of understaffed local environmental protection offices.[32] To curb cities' carbon footprint, bold national legislation could clinch a green deal, as critical environmental sociologist Daniel Aldana Cohen and his colleagues have demanded.[33] In most cases, however, global, supra-level, or national legislation is not in the cards: High-level compromises are often criticized as late, uninspired, or far removed from local needs. Understanding why cities differ in their ability to solve collective problems is particularly urgent when national governments struggle with polarization or austerity.

Take the example of the COVID-19 pandemic, which the US federal government failed to address at the onset. Cities responded quickly to the public health threat by issuing shelter-in-place policies to close schools, cancel public events, and order their population to stay home to confine the virus's reproduction. By reacting faster than states and creating a template for legislation in the first place, cities may have saved innumerable lives by counteracting the increased spread of the virus in urban centers. As I found in work with colleagues from the Mansueto Institute of Urban Innovation at the University of Chicago, civil servants in 136 counties saved critical time in enacting measures against COVID-19 by passing shelter-in-place policies before their state leadership took action. Larger, more metropolitan counties such as Denver, Colorado; San Francisco, California; and Philadelphia, Pennsylvania, were most likely to react swiftly.[34]

The second finding added a surprising wrinkle to this expected dynamic. Counties that were soon to adopt a discrete policy tended to be especially liberal counties in conservative states. The best predictor besides county size for an early shelter-in-place order is the difference between the county-wide result in the 2016 presidential election compared with the state, even while controlling for liberalness altogether. Counties whose local election results deviated from the state result by a standard deviation were more than twice as likely to have a shelter-in-place policy. Cities like Saint Louis, Missouri; Salt Lake City, Utah; Cleveland, Ohio; Dallas, Texas; and Tulsa, Oklahoma, opted to urge their population to stay at home despite inaction or tardiness on the part of their respective Republican states. In states like Oklahoma and Missouri, county orders remained the only official shelter-in-place policy throughout the pandemic. Overall, these local shelter-in-place policies protected 66.9 million people.[35] This context illustrates the important role cities can play in responding to crises and the significant variation in their responses depending on their location. Across these examples, cities appear to fill what urbanist Neil Brenner refers to as "state spaces" that were left void by higher-level governments.[36]

Tensions and competing interests between different levels of government are not rare. In fact, they are the norm; the politics of urbanites typically differ from those of people in the countryside. In the US federal system, cities are particularly weakened relative to states and the federal level, but how frail their position is drastically varies from state to state. This tension is not unique to North America. Copenhagen, for instance, wants to be a citadel of climate change liberalism for *all* people—but the country of Denmark, with its exceedingly conservative immigration policy, has not always made this easy. The historically socialist stronghold of Vienna is surrounded by deeply conservative Lower Austria. You may have noted this gap during a mapping of recent election results in your country, or simply when you drive across the country and notice dramatically different political and cultural preferences on two sides of a city boundary, be it about football versus rodeo, vegan ramen versus jerky, techno versus country music, or Green Party versus far-right populists. Although the urban-rural divide is simplistic and not deterministic, it has become a symbol of our polarized society—and it reinforces the question of the political autonomy of cities.

The relationship between states and their cities is a crucial question for understanding city climate action. That is because discretion is a defining aspect of city action. Discretion does not mean that cities make decisions in isolation—far from it, as discussed in the previous chapters. It suggests, however, that how cities' plans are laid is original and autonomous and exceeds simply being given a legal mandate by higher-level governments. The specific context of the United States raises the question of whether cities are genuinely turning to local action on their own accord or are just reacting to civic and political pressures. Although we cannot generalize from the US context to the rest of the world, the setting also gives us a sense of whether cities' actions result from the receding higher-level influences from nation-states, states, or provinces. Are cities stepping up more when higher-level governments fail?

To be clear, states vary significantly in terms of the prevalent attitudes about environmental policies. On one hand, one in four people in Tennessee, Alabama, and Mississippi believe that climate change is a hoax, compared with but one in ten to twenty in California, Illinois, and Massachusetts.[37] On the other hand, Californians need no convincing that climate change is real, regardless of how combatted the topic is in the chambers and antechambers of federal politics. During an interview at the US Department of Housing and Urban Development, a senior official highlighted that not all states and federal government officials think so: "We don't call it sea-level rise; we refer to it as recurrent flooding." After January 2017, and again after January 2025, most information about climate change disappeared from White House web pages, and the EPA has scrubbed climate data and excised words like "fossil fuels" and "greenhouse gas" from its web page.[38] The Florida state government went so far as to ban the term "climate change" from official business. Different federal legislators and executives, from Republican presidents to Democratic ones and back, vary in their level and direction of leadership. This oscillation between embrace and denial, however, is exactly what requires the steady hand of local administrators who implement not a volatile political agenda but rather a long-term plan.

What has become a political taboo in some places has become an indisputable reality in others. Cities' exposure to extreme environmental conditions in climate-skeptical states illustrates that urban sustainability both transcends and incites petty politics. As the *Miami Herald* reports, Yankeetown, Florida, a small town threatened by rising sea levels and ghost forests, has applied to become an "adaptation action area" to take matters into its hand.[39] Regarding Georgetown, Texas, whose Republican mayor plans to power his city with 100 percent renewable energy, a Fox News commentary decries that a "Texas town's environmental narcissism makes Al Gore happy while sticking its citizens with the bill."[40] As ethnographer Liz Koslov uncovers in a study of a coastal community in Staten Island, some people resort to a method of last resort to deal with persistent flooding: They retreat to higher ground. In these cases, cities act despite their states' reluctance.[41]

State preemption—limiting or reversing city legislation to curb local discretion—is a salient concern from an urban political perspective, which suggests that the decisions of local governments are constrained by their relationship with higher-level governments.[42] As local government law expert Gerald Frug and former US Assistant Attorney General David Barron argue, states often stifle urban innovation by limiting local-level autonomy.[43] This line of work emphasizes the classic insight of political scientist Paul Peterson in *City Limits*, in which he states that cities cannot freely choose policies, particularly those redistributing resources to underprivileged groups.[44] These analyses have established that municipal policies are best understood not as autonomous but as intertwined with those of higher-level governments.[45] Still, some states are more innovative than others, in that they develop and adopt new policies earlier. What does this mean for local discretion? Are cities in less innovative states less likely to engage in city climate action because they are constrained by the local political context? And could the reverse be true as well: Innovative states are home to proactive cities?

A growing body of literature diagnosing an "urban revolution" in the United States, for instance, suggests that innovation-resistant states might see more local-level innovation.[46] From this perspective, city administrations might be expected to intervene to mitigate the lack of action from a higher level of government. Or are cities in less innovative states more likely to engage to compensate for their state's inaction, as the COVID-19 example suggested? By contrast, cultural theories of organizational behavior would ostensibly predict that institutional pressures to tackle social problems proactively pervade government. From this institutional perspective, the political culture and social norms of organizing at the state level contribute to isomorphism—similarity in form—at lower levels of government, at which local practices look similar to higher-level ones. As a result, the states with higher levels of innovativeness also will have more innovative city governments.[47] This would suggest that cities in states with higher state policy innovativeness will adopt a larger number of sustainability practices. Whatever the relationship between state innovativeness and city action, it is not obvious.[48]

Our results analyzing survey data on the sustainability practices in a sample of US municipalities support the latter scenario: Cities in more innovative states are also more proactive when it comes to the uptake of sustainability practices. Cities with more expansive sustainability programs tend to be located in states that adopt novel policies early. In other words, contrary to Frug and Barron's argument that states may stifle urban innovation, a relationship exists between state policies and city action—but state policy is enabling rather than stifling.[49] Of course, we also considered many other factors that may correlate with state policy innovativeness, such as the state's budget, size, and political orientation. State policy innovativeness continued to be strongly associated with proactive cities—without, however, being deterministic. Cities in each state ran the gamut, as figure 4.3 shows. California and Texas, for instance, are home to some of the most *and* least sustainable cities.

The drastic variation within states reflects the insight that cities are in fact poorly portrayed as "creatures of the state." There are some significant outliers and massive variation within states, indicating that state influence is not deterministic. Several cities—from Las Vegas, Nevada; to San Antonio, Texas; and even to New York City—leave the laggard states in which they are situated behind. Some of the smaller places that register as outliers self-identify as environmental leaders. The conclusion that the proactivity of cities is a direct consequence of state-level factors, therefore, would not do justice to the dramatic heterogeneity within states; the more active cities in less innovative states outperform less active cities in more innovative states. Because of this significant cross-sectional variation among cities, I now turn to examining city-level determinants of city action using the survey data: whether or not cities are tied into the global superstructure we saw in chapter 3, whether a mayor or a manager runs cities, and whether cities respond to an organizational infrastructure concerned with sustainability.

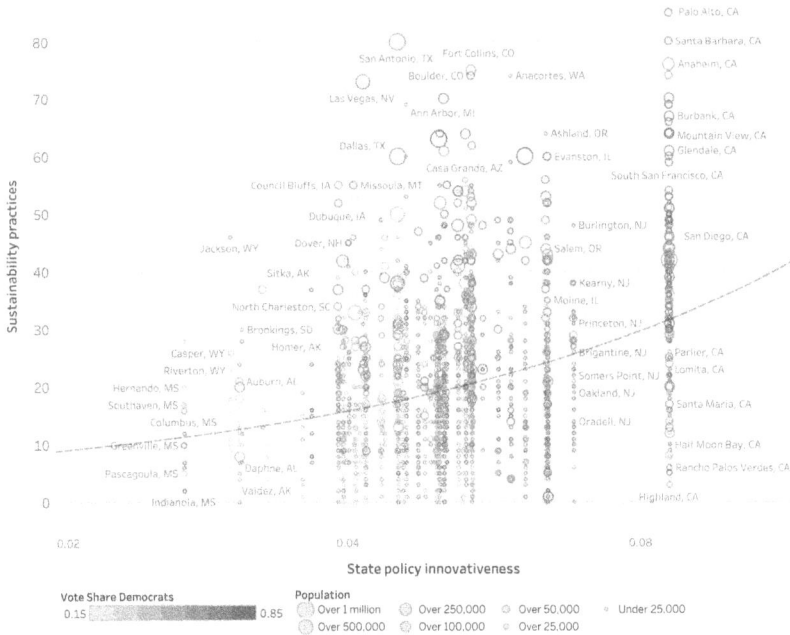

FIGURE 4.3 Sustainability practices of US cities by state policy innovativeness and county politics.

Sources: Sustainability practices measured from the 2010 ICMA Sustainability Survey and state policy innovativeness measures from Boehmke and Skinner (2012).

FROM GLOBAL SUPPORT TO LOCAL BACKLASH

In a federalist system of governance, states represent an immediate institutional context for cities. Of course, cities' institutional environment is not limited to the state, and their linkages with the broader institutional context could be just as consequential for city action as the states in which they are located. To examine the influence of the institutional superstructure explored in chapter 3, I looked at the specific influence of a network that all these cities are eligible to join and has a low entry barrier: ICLEI. ICLEI is a particularly salient organization because it teaches its huge network of city professionals how to track GHGs through an inventory. But what the organization does is quite a bit broader and very much in the spirit of the associations we encountered in chapter 3: They publish case studies, have a "solutions gateway" with local policies that help with all sorts of technical problems from energy to recycling, and offer platforms for peer sharing of solutions among city officials interested in the same issues. Having celebrated its thirtieth birthday in 2020, ICLEI is open to an incredible range of cities, which explains the thousands of members worldwide.[50] As a policy director at the organization told me, ICLEI is distinctly positioned because they work with "those nontraditional cities that are a bit more relatable to the average." Why not look at the City Climate Leadership Group (C40) or another major association with members like New York, Paris, and Copenhagen? "There's oftentimes a resistance from cities when they hear those examples of, you know, Copenhagen. Copenhagen do this or that. Well, don't compare us to Copenhagen, because we can't do what Copenhagen did. You know, so give us an example that's more relatable." In other words, ICLEI specifically targets smaller and midsize cities and prides itself with being equally accessible to cities of all sizes: "100RC is 100 cities. C40 is more than 40 cities, but it doesn't have the broad reach that ICLEI has. It tends to be more of the mega cities or the leading cities who are, you know, innovative and who have been doing this work for quite some time. . . . If you're a city who's never had a climate action plan or never even thought about that, we have tools and resources available for those communities."

Now, do cities that are members of ICLEI, in fact, report higher sustainability activities? David and I found membership in ICLEI to be strongly associated with higher rates of city action, at a rate of about three times the effect of politics. As the whisker plot in figure 4.4 shows, although several larger cities are not members in ICLEI despite having a significant sustainability program, barely any cities that are registered members of ICLEI have a slim program. The ICLEI effect is what social scientists call a conservative estimate of institutional influences; if even paying a small membership fee to ICLEI is associated with more sustainability practices, then being a member of a prestige network like C40 or Urban Sustainability Directors Network (USDN) will undoubtedly have an even more substantial effect. As we discussed earlier, however, being proactive about the environment may also be why someone selects into the group of ICLEI members—so we cannot be confident that ICLEI is a game changer instead of simply a sign that the game has changed.

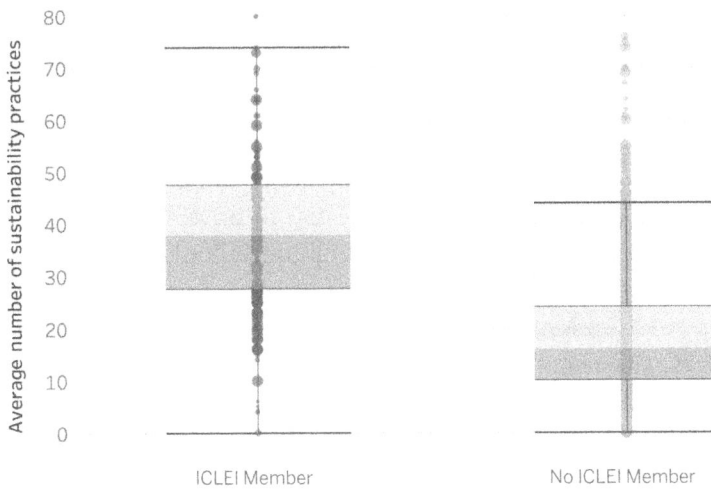

FIGURE 4.4 Whisker plot of the average number of sustainability practices by ICLEI membership.

In either case, the influence of intercity associations certainly appears to hold up upon closer inspection.

Although a membership slip for ICLEI may be quickly sent in, nothing is easy about becoming green in the United States—a country in which one in six people question the existence of anthropogenic climate change.[51] ICLEI also illustrates how contentious local climate action can be in the United States. As one CSO of a major city tells me, ICLEI is a "non-entity" because it has been politically instrumentalized as an example of how the influence of global institutions has shaped local decisions. For instance, in a pamphlet titled *Behind the Green Mask*, a Californian activist spins a popular conspiracy theory that the United Nations uses the topic of the natural environment to support their globalist agenda, disempowering people in their local communities. Many urban planning ideas, such as Carlos Morenos's fifteen-minute city, appear to be pragmatic or, at best, a matter of subjective preferences.[52]

In a skeptical public's eye, however, many decisions essential to combating climate change are profoundly political and even contentious.[53] Morenos received myriad death threats for his modest proposal to ensure that everyone has a bakery and grocery store within a fifteen-minute walk from home. The notion of being able to reach all the food that is required for a healthy and livable life appears to some people to be a globally mandated prescription to remain confined to close quarters. Does the suggestion to rely on one's bike or feet make it harder to go where one chooses? Most urban planners have no intention, or at least not the courage, to limit the liberties of drivers, question the property rights of homeowners, or stand up to landlords. But it would be foolish to think that any decision related to people's homes, mobility, education, or business opportunities was not fraught with the potential for disenfranchisement.[54]

The backlash against some associations and the capacity required to engage with the juggernauts among sustainability networks raises the question of how city networks are experienced on the ground. The interviews Ana and I conducted gave us insight about understanding what cities expect from organizations like ICLEI or their more prestigious counterparts in New York. This look to the ostensible laggards that are trying to "grow green in hard places" is an important piece of the puzzle because it shows that urban sustainability is not a matter of having an encouraging institutional environment—many sustainability officers are playing an uphill battle, and it involves convincing working alongside supportive organizations.

We found that the resources city administrations require to overcome their respective barriers to sustainability are different as well. Leading cities might depend more on symbolic resources, such as opportunities to boost one's reputation, than smaller cities, which we expected to have been held back by a lack of material resources, such as money and staff members. Contrary to this intuition, we found that all cities required some mix of both types of resources. The progressive needs of cities with full-fledged sustainability programs, however, seriously differed from the more pragmatic needs of cities whose leaders needed help to prioritize sustainability. Nearly all cities we considered for an interview had a dedicated officer for sustainability, resilience, or environmental policy.[55] Compare this with the critical contemporary issue of artificial intelligence (AI). Most cities, even those that have embraced the label of a smart city, have no AI director or anyone on staff who is specifically responsible for the digital agenda of the city.[56]

In brief, cities that already had a sizable sustainability program sought to share their successes with other cities to capitalize on their reputation. These cities brought up competition, for instance, for external funding and for prestigious awards and desired spots in city rankings. They also relied on detailed information about implementing certain strategies to fill knowledge gaps related to technical issues. For instance, as one sustainability officer told us: "The work around climate sustainability is becoming more and more technical. . . . You have to be able to find that information or to develop partnerships to leverage that, whether that's through consultants or higher ed."

Cities with more modest sustainability programs, in contrast, sought a safe space that allowed them to share their problems in discussing technical as well as political roadblocks. They also required mechanisms that allowed them to affirm the department's goals as a citywide priority, for instance, by discussing how to get on to the mayor's agenda. Finally, these cities actively sought to learn how to neutralize the politically contentious nature of climate change by reframing the issue and emphasizing the cobenefits of creating a sustainability strategy.

These cities also required more material forms of support, such as a stage for sharing local successes and small wins on a regional or national scale, which gave them leverage in internal debates and power struggles. Furthermore, city networks encouraged cities and their sustainability officers to push the boundaries of what is feasible and to try to do better. One sustainability director told us of a local city network that specifically aims to provide direct support to cities in the US South. This region is challenging to navigate as a sustainability professional: "We do a call every other month where we talk about different issues, and it's also meant to be a safe space to talk about issues that we're dealing with in our communities, so it's really, really valuable." The primary material resources these officers needed were not monetary but political because every bit of progress required a political favor: "It makes it easier for us when we have another city within the state. We can check in on them and learn how they're doing it and how they're navigating some of the challenges that are coming from the state level. We can see how they creatively deal with those challenges in an environment that is not conducive to sustainability."

By speaking not only to the global leaders but also to representatives of lower-status cities (see chapter 3), we noticed some things that we had not previously known from reading the relevant academic literature. For one, it gave us a sense of why lower-status cities are less likely to join large, international city networks. It wasn't a lack of sustainability efforts, but rather the fact that C40, 100RC, and USDN did not have as much to contribute to their mission as the more regional networks did. Furthermore, some networks are not even openly accessible to all cities. Consider the American Cities Climate Challenge, a competitive initiative in which the most advanced cities receive additional material support to implement their plans. Smaller networks, such as the Southern Sustainability Directors Network (SSDN), use the exact mechanisms as their more established siblings to connect cities with what they think are shared problems but on a more regional scale. As one SSDN leader told us, creating safe spaces is part of the organization's mission: "We will use the space to have candid conversations about challenges or opportunities that folks have. We also use our time and space to share work that might not be ready for public consumption. But often, we provide a behind-the-scenes safe space for staff to come together and really sift through their problems."

Cities where sustainability was lower on the agenda also appreciated these networks for the ability to speak and learn among peers: "Our membership in [regional] organizations has been incredibly helpful because everybody has the same challenges, and we can learn from others' successes. I think it's nice because [our city in the US South] is not always on the vanguard of environmental sustainability, so we can learn. Anything that you want to do, someone else has already done it."

Sustainability officers managing cities with weaker sustainability programs face different and more pronounced institutional barriers than those in their counterparts that are more serious about sustainability, and in turn, they look to different networks for support. This perspective on barriers helps explain the insight of urban planner David Wachsmuth and urban sociologist Hillary Angelo that many cities still avoid addressing systemic issues through sustainability, instead opting for easier "small wins" such as trees and green space.[57]

Another reasonable expectation for variation among cities, even those in the same state, is that who leads the city makes a difference. For example, a sustainability officer from a large West Coast city, a political appointee who now works for a clean tech incubator, praised the role his city played in creating the Climate Mayors initiative. If mayors came together on this issue, a lot of progress could be made. Indeed, groups like the Global Covenant of Mayors, Climate Mayors, and the idea of the Global Parliament of Mayors give a strong sense that elected local leaders are behind the move to greater sustainability.[58] Remember that 100RC, as we learned in chapter 4, looked for "commitment and potential" in the one hundred cities whose hiring of chief resilience officers they subsidized with $1 million each. The main criteria for becoming a member of 100RC was to have a strong letter of commitment from the mayor prioritizing resilience. The Urban Institute later identified support from city leadership as one of the critical features of success: "Program uptake is highly dependent on local political support and governance structures." This was partly because chief resilience officers were "important champions for pushing a resilience agenda within city government, but vulnerable to turnover and change."[59]

To select promising cities, the 100RC team did one-on-one interviews with all mayors before selecting their city to probe their commitment. What does commitment mean to 100RC? As my interlocutor explains: "Additional resources to support the office beyond what we were going to be able to provide through our funding for that single line in the city budget. The commitment of additional staff and resources was something we were looking for, the commitment of, you know, imbuing that person with the necessary authority to work transversally in city government." Interestingly, he also understood well that this was not just a matter of finding a passionate proponent but also putting them into the right position: "We don't have to go to Max Weber and talk about bureaucracies. . . . But forcing people to work in different ways and get out of their comfort zone requires a lot of leadership, which is a personal characteristic, but it also requires authority, right."

In fact, support from the public and democratically elected officials is indispensable. As I learned from one policy director: "There are a lot of sustainability staff who are passionate about what they want to achieve, but if they don't have the support of the mayor, and if the mayor isn't putting climate change as one of the—you know, the top of their agenda, whether they're talking about it or not, that's a huge problem. And it's probably an investment that is doomed to fail in some ways." A major grant-maker to city networks offered a similar take of reinforcing existing initiatives: "We try to help mayors who are already committed to fighting climate change or improving the sustainability of their city and connecting them with resources or support or a network of other cities that can help then accelerate achievement against their goals and make progress."

One might think that talking about mayors is just a shorthand for party politics, and that progressive mayors are the driving force behind city climate action. That is not to say that city networks have not sought to recruit new cities and their mayors into their fold. Cities with conservative mayors are not unwelcome in city networks; precisely because they are unexpected bedfellows, Republicans are a hot commodity. A program officer of a national initiative to support cities willing to take the lead on climate change explained: "We have selected cities like Indianapolis, Columbus, and Cincinnati. . . . We see cities in Republican-led states as potential emerging regional leaders. If Indianapolis can prove that it's possible to transition from coal and natural gas to renewable energy, and it's possible to save a lot of money on building energy efficiency, hopefully that will start the conversation and the commitments in other cities in those regions."

Sociologists called this dynamic the Nixon-in-China effect, referencing the famous 1972 visit of the staunchly anti-communist US President Nixon to China that eased global diplomatic relations with Mao Zedong's communist state. In other settings, conservative organizations making bold statements have a greater impact than if the statement came from a more progressive one. Organizational sociologists Forest Briscoe and Sean Safford found a similar dynamic in how same-sex partner benefits spread among corporations—once more conservative firms extended spousal benefits to same-sex couples, the practice became common.[60] The notion of counterintuitive pioneers as a cultural resource applies to cities, too. Another major funder of city climate efforts invoked this logic of including cities whose participation in climate efforts held particular symbolic value: "In some of the states where we're working, you have Republican-controlled legislatures like Ohio or Indiana, but cities who are leading from the front. We're talking with the mayors to make sure they understand that we see this as a priority, but also so that they see it as a priority and will try to elevate the visibility of what we're working on."

As I learned from one major city network: "We have Republican mayors, we have Democrat mayors, you know. We're trying to improve all residents' lives in your cities." But despite the bipartisan—even politically agnostic—nature of this plural network, it is essential to recognize that most existing initiatives tend to support cities that already have sizeable sustainability programs. This search for the cream of the crop leads to what sociologists call the Matthew effect—those that already have receive even more. In other words, part of the elite city network has contributed to a self-reinforcing feedback loop of leaders and laggards in urban sustainability.

Let's not pretend that climate change is anything but a profoundly contentious and political issue. Whether the people in a state voted for Obama or McCain in the 2008 elections was still a strong predictor of the city administration's commitment to climate change. But just like in the other cases we considered, political preferences need not play out through ideology—liberalism and education also shape the organizational landscape, like how many people join the Sierra Club or whether there are local urban planning enthusiasts.

There is a simple challenge to the idea that mayoral politics make a difference—most cities in the United States are not even run by mayors. A large share of cities in the United States follow a so-called council-manager form, which means that administrators are officially appointed to implement policies in the public interest as determined by the city council. City managers founded the ICMA—one member of the Big Seven associations lobbying for local government interests in the United States in 1914—to advocate for greater adoption of the council-manager system. As the association explains on its website: "ICMA's origins lie in the council-manager form of local government, which combines the strong political leadership of elected officials . . . with the strong professional experience of an appointed local government manager or administrator. Under this form, power is concentrated in the elected council, which hires a professional administrator to implement its policies. These highly trained, experienced individuals serve at the pleasure of the elected governing body and have responsibility for preparing the budget, directing day-to-day operations, hiring and firing personnel, and serving as the council's chief policy advisor."

If mayors are the driving force behind urban sustainability, we would expect mayor-council cities to have a greater share of sustainability practices; and if administrators are a crucial driving force behind attempts at urban sustainability, we would expect manager-council forms to prevail. The ICMA's municipal yearbook provides information about a city's form of government going back decades. As of 2018, 38 percent of cities that responded to a regular survey of the city's form of government follow a mayor-council form, and 48 percent have a council-manager form.[61] There are other forms as well, such as government by commission and town meetings. Three in four cities have a chief appointed official, usually appointed by the city council, which means that even cities without the council-manager form can have an appointed official such as a city manager or chief executive officer. City managers are an interesting occupational group in the United States because they have an internal labor market that allows them to move from one city to another. As a result, some city managers in major cities already have experience managing smaller cities.[62]

I already expected that cities with managers would be drawn to urban sustainability, considering the allegiances of administrators versus politicians to environmental values. In a survey among city leaders in fifty of the fifty-four largest US cities, 65 percent of councilors and commissioners and 71 percent of city administrators indicated a very high or high commitment to sustainability.[63] This conviction that sustainability matters means city administrations are even more concerned about the environment than politicians. Although the reasons for support are varied, as some officials support these practices to improve regional competitiveness and others support them to promote environmental and social justice, sustainability has become a legitimate mission for cities to pursue. An ICMA representative put this difference into starker terms: City managers are "not used to boasting." As far as sustainability practices are concerned, they may try to get things done without addressing their efforts publicly. "They're so used to letting the mayor be the front person and stand in the limelight, and they always, just by the nature of our profession and our code of ethics, [are] apolitical and remove themselves from such situations."

Other work by public administration scholars has shown that the municipal government structure significantly predicts the presence of GHG reduction policies.[64] Local sustainability priorities, regional governance, and city network participation all account for variation among cities' levels of sustainability commitment.[65] According to rational choice scholar Richard Feiock and his colleagues, this is because council-manager cities "are oriented more toward efficiency concerns than their elected counterparts in a mayor-council," and the empirical findings support the claim.[66] It may well be that council-manager systems are generally oriented more toward efficiency. Rational choice, however, is not the only potential explanation for why we would expect cities run by managers to be more attuned to the deeply rationalized practices— practices that are legitimate in the eyes of observers without necessarily being functional—included in ICMA's sustainability survey.

Another explanation is related to the intuition about cultural influences on organizations discussed in chapter 3. Because being run by managers rather than elected officials is a feature of rationalized organizations (i.e., organizations that adopt purposive practices to appear legitimate), that display what institutional theorists think of as "actor-hood," cities run by managers are likelier to adopt rationalized practices across the board.[67] The sociologist Gili Drori applied this line of thought to the social construction of nation-states, showing that countries with complex economic and political systems that are more embedded in a global network of international organizations use significantly more rationalized practices in their governance, including hierarchical internal structures and concerted action against corruption.[68] Rationalization orients organizations toward explicit purposes and the formalization of structures, procedures, and processes—all of which contribute to the cultural construction of organizations as actors.[69]

In many respects, council-manager cities embody the spirit of new public management, a global reform movement in the public sector in which public agencies learn from business and embrace managerialism— the reliance on management logic and expertise to make decisions.[70] This

efficiency orientation means doing things like relying on external consultants and performance-related pay to motivate civil servants. In another
sense, the tendency of cities to behave like all other organizations—
demonstrating evidence-based actions, a positive impact on society,
and empowering communities and individuals—has become universal
among organizations tied into a global discourse of what it means to do
a good job.[71] As a result, city managers may also act as symbolic network
ties or "receptor sites" for cultural scripts about rationalization, meaning
that they are more relevant to some of the global pressures to become
oriented more toward efficiency.[72] As a result, we expected cities with a
council-manager form of government to adopt more sustainability practices than their mayor-council counterparts.

This is what we found, as illustrated in figure 4.5. Governments
with city managers tended to have more sustainability-related practices
than those with a mayor. The level of sustainability practices was higher
than those of mayor-council cities throughout the political spectrum,
indicated by the higher dark gray line, and driven by such role models
as Palo Alto, California (with the maximum of eighty-seven reported
practices); Boulder, Colorado (seventy-eight); and Grand Rapids, Michigan (sixty-five). Among the larger cities are San Antonio, Texas (eighty);
Las Vegas, Nevada (seventy-three); and Dallas, Texas (sixty-two); all of
which are manager-council systems. Even major mayor-council cities
such as Philadelphia, Pennsylvania (sixty-five); Orlando, Florida (fifty-
four); and New York City (sixty) reported relatively lower levels of sustainability practices, and a majority of places reporting twenty practices
or fewer were mayor-council governments.

This is *not* because managerial cities are technocratic and less
responsive to the political preferences of their population. As I have
noted, a link exists between political liberalism and sustainability practices, similar to progressive policies such as civil liberties legislation.[73]
Democratic turnout in presidential elections is consistently associated
with more city climate action. In terms of magnitude, the effect of politics and managerial forms of government are about the same size as the
effect of state policy. The association between liberalism and sustainability practices, however, is bigger among cities with a manager than
those with a mayor. In other words, there is no simple story about how
city managers are pushing a sustainability agenda against the interests

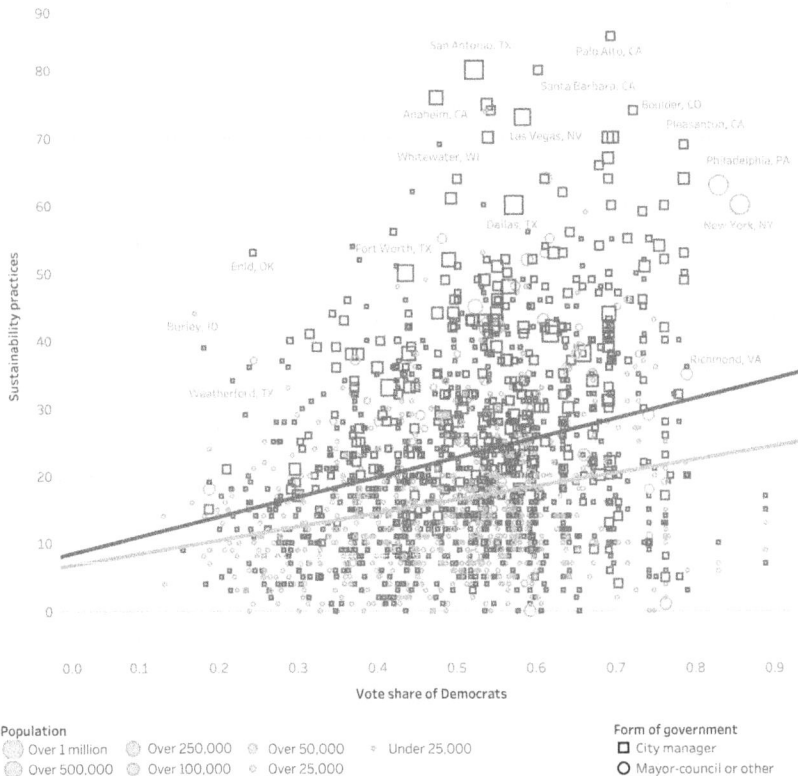

90

80

70

60

50

Sustainability practices

40

30

20

10

0

0.0   0.1   0.2   0.3   0.4   0.5   0.6   0.7   0.8   0.9

Vote share of Democrats

Population
◌ Over 1 million ◌ Over 250,000 ◌ Over 50,000 · Under 25,000
◌ Over 500,000 ◌ Over 100,000 ◌ Over 25,000

Form of government
☐ City manager
○ Mayor-council or other

FIGURE 4.5 Cities' sustainability efforts by their political orientation and form of government.

of their population—they are actually *more* responsive to their population. For instance, Oklahoma City, Oklahoma (manager) reported forty sustainability practices, whereas Birmingham, Alabama (mayor) reported eight. Richmond, Virginia (mayor), with a whopping vote share of 79 percent for Obama, reported thirty-nine sustainability practices, whereas the more conservative Virginia Beach, Virginia (manager) reported fifty-six.[74]

FROM CITIZENS TO EXPERTS

This complex interplay between the professionals who run cities and the constituents to whom they are beholden deserves further discussion.

Recall my experience at the townhall with Palo Alto's city manager, where people of a liberal city were concerned with everything *but* climate change. How does public demand factor into city administrators' commitment to urban sustainability? My informants reported a tension between the ideal of widespread public participation, and the reality of who and what is involved in the production and consumption of city strategies for resilience and sustainability. Considering that for PlaNYC and OneNYC, the city had had to pull together more than seventy agencies that contributed details to the plan, it is a tall order to produce coherent documents that appeal to the general public. As the city manager explained: "We try to write it so it's approachable by just about anybody. But in reality, it is the engaged advocates, the stakeholders, you know . . . the one-issue group that will look for their own thing and make sure you talk about it the right way."

Public participation is a standard ingredient to city action. A larger Midwestern city broke the sustainability program into smaller work groups with key stakeholders whose input they solicited using focus groups and public meetings. "It's not a linear process," the city's CSO admitted, sighing. "If you really want to listen to the community, you have to incorporate their input at different points in the process and then make sure you've got it right." Many other cities in the region also held community forums and sought out community-based organizations and business groups to get feedback. Some of these approaches are creative, as in the example of a city that "developed a climate game where we . . . asked people to prioritize those strategies in small groups at these community forums. We took all that feedback and had a couple of targeted additional meetings with some local environmental justice advocates." Some cities make these events a regular occurrence. A state capital in the rust belt used the process "to help direct our office's goals and activities. Just two weeks ago we held our annual summit. . . . In addition to the keynote speaker and our chief giving an update on what we're doing, we had twenty-four breakout sessions that brought together our partners and addressed many of the actions that are in the action plan, whether that was just to give an update and help, answer questions or in some cases it was, we want feedback from residents." The Chicago CSO was hopeful that there is a broader audience: "Yes, environmental advocates read the plan, but also community development organizations, environmental justice groups, everyday residents and businesses."

Readership aside, there is no question that city strategies were not just a printed version of citizen demands. Even some of the most engaging plans—with a strong commitment to citizen participation—bore the mark of experts. Several of the plans my colleagues and I found were entirely produced by consultants, for instance from the World Bank. Even the Copenhagen and Sydney plans were written with the help of the Copenhagen-based urban consultancy Gehl Architects. Expert involvement—whether internal through agencies or external through consultants—does not mean that the public's voice is not taken into consideration, but they give the process a more technocratic than participatory flair. In New York, a representative of a resilience initiative founded after Hurricane Sandy commented on the involvement of McKinsey, a multinational management consulting company: "They're not exactly a socialist think tank, you know." This also shapes how sustainability is justified, drawing on reasoning that is more market-oriented than it may have been otherwise: "There's an economic argument to social equity. There's an economic argument to migration and refugee integration." In the 2015 ICMA sustainability survey cited earlier, municipalities reported fiscal savings, attracting local development projects, and state or federal funding opportunities as three of the top four reasons to engage in urban sustainability; environmental concerns came in fifth.

Why bother framing sustainability as fiscally sound rather than just necessary for the survival of the human species? Public support, especially in cities that receive less support from political leaders, is indispensable. A representative in the US South mentioned that though some of the sustainability plans are very technical, it's "the public that's pushing. . . . You know, the public votes for our city council members. While climate change is important and something that thankfully our city government finds important—and it's things we're working on anyways—we want the public to be on board with that, and we want the public to see the value that it brings to them in our community." Another CSO from Western New York also cited the public as a major motivation to do better. Their colleague in Toledo County mentioned that, ultimately, the city's programs are "decided by citizens. . . . We have to go back and ask for money from city council, to see if they would be open to it." Saint Paul, which actively tried to involve the public in the production of its sustainability program, mentioned [a group of] high school students who were just *so* hungry for the conversation." We also learned that the city administration expected objections to their sustainability plan, but none came:

> I kept waiting for the other shoe to drop in terms of some organized opposition to some aspects of the plan from somewhere, some parts of the business community or elsewhere, and it just never materialized. We had a public hearing at city council on the plan where, I don't know, twenty people or so came and spoke in support of it and literally no one spoke in opposition, which doesn't really happen in local government. . . . People were already like way ahead of us in being ready for such a thing to exist.

How does the intense need for public support go together with the limited involvement in the actual production of strategic plans and administrative practices? With a whole lot of communication work. But this work can be incredibly hard—even internally. As an interlocutor in Nashville mentioned, it's not easy to communicate with the ten thousand people who work for the city, all the more given that these people are not necessarily on the cutting edge of sustainability. As I saw at Palo Alto's town hall, the public does not always muster the same concern for climate change as some of the cities' officers. The city manager of a Californian city told me that the population did not have "not much a concern for sustainability" even though, on paper, the population is deeply concerned with climate change if public opinion polls are to be believed. As a result, city administrators who take sustainability seriously must engage in a reframing exercise, because whether a practice or policy is good for the environment is not a winning question. Instead, the question has to be "'How is this going to benefit us in the long run economically?,' because a lot of times it's just about our budget and our bottom line."[75]

The problem of making sustainability relevant is of course exacerbated in parts of the country where the environment is not on the political agenda. Some of the less proactive cities were noticeably more guarded in conversation, pitching us that "sustainability is not just about the environment. It's about the economy and social equity. When all of those are working together, it just makes sense. What we do is framed as a benefit for the environment, but it is benefiting our economy and our community members as well. Ultimately, it is for the community." A sustainability director in another peripheral city mentioned that "not everyone knows what climate action is, or why it's important, or how it affects them." In the Midwest, a sustainability officer lamented the difficulty of convincing the city's population of the long-term positive effects of addressing climate change: "If you plant a tree today, you will get all the ecological benefits of it, but for it to count for carbon it takes about ten years. So just to say, this tree will help with carbon management in ten years may not sound very good to some residents. But you say this tree in ten years will have an impact on carbon, and in the meantime we are providing shelter for our butterflies and offering shade to people, that's the key. . . . Adding all these other not-so-obvious benefits."

In other words, opportunities for public participation do not necessarily mean direct democracy. This insight is in line with sociologists Carolyn Lee, Michael McQuarrie, and Edward Walker's observation that democracy is declining *despite* rising participation in the United State, including its cities. This possible, in short, because many public deliberations are managed by consulting companies who seek to improve the legitimacy of their clients without necessarily changing the trajectory of their decision-making. Public participation has become a politically uncontroversial industry whose reach has extended into government agencies, businesses, and nonprofits—all of whom try "to become more open, transparent, accountable, and welcoming of public input than their rigid, bureaucratic, and more hierarchical predecessors of generations past."[76]

FROM SUPERSTRUCTURE TO INFRASTRUCTURE

Despite the rise of public participation in municipal decision-making, the (lack of) interest on the part of town hall participants in the climate has not kept city administrations from acting. Our analyses show that this is because nonprofit organizations call cities to action, attend public consultations, and advocate on behalf of the public interest. This often-overlooked part of the city's organizational infrastructure plays a fundamental role in explaining variation in city climate action, including through the administrative adoption of sustainability practices.

Cities are embedded in multiple institutional fields—regions, states, nation-states, and global cultural norms—that are nested in each other and, therefore, come together to influence local action. As organizational and political sociologists Marc Schneiberg and Sarah Soule have argued, with an eye to social movements, organizations are exposed to more than one singular institutional environment—like the global associations previously discussed. Organizations that are embedded in the same institutional fields are likely to be homogenous because they are subject to the same peers, professions, policies, and financial pressures.[77] Besides the global influence from peers and intercity associations, the second level of embeddedness pertains to cities' local organizational infrastructure. City administrations are only one among many relevant actors in cities that make decisions about sustainability. Cities' organizational infrastructure features a diverse "cast of actors" that determines what topics become "matters of concern" and which remain sidelined.[78]

The "cast of actors" that make up the local organizational infrastructure strongly affects how communities are governed. Although we can attribute organizational action to cities as discrete entities, cities are never a single, coherent unit of organization when it comes to decision-making. We may think of this as an attribution-aggregation paradox: by attributing action to an entity whose behavior is the result of aggregate behaviors, cities act.

Various organizations constitute each city and exert influence on both government and public administration—as part of the city's governance network.[79] Like most policymaking agencies, city planning and strategy departments are primarily designed to interact with other organizations: They invite citizens, interest groups, and companies to deliberations, consult with them to pass laws, and employ people who are members of voluntary associations and interest groups. These organizations, among others, exert bottom-up influences in which local movements and grassroots organizations convince decision-makers to adopt innovative policies.

This line of thinking is inspired by urban sociologists Nicole Marwell and Michael McQuarrie, who lamented in a 2009 essay that urban sociologists "treat organizations as derivative rather than productive of urban social relations."[80] Organizations are a "missing dimension" in urban sociology because they influence "the distribution of resources, the arrangement of social networks, and the constitution of dispositions and identities."[81] In other words, organizations shape many processes that constitute urban life. Among others, as Marwell elaborated in later work, these include the collective decision-making that occurs when city administrations, nonprofits, and the markets determine together who gets what, when, and how (the process of urban governance that we first encountered in chapter 3). This work suggests that taking stock of which organizations dominate the organizational landscape of a city is a first step toward identifying what city administrations respond to when they arrange their "governance configurations."[82]

At the same time, economic geographers have noted the influence of local networks in helping elites agree on a common vision of the region and that organizations often bring together such networks—in such mundane organization as childcare centers and garden clubs.[83] Seeing cities as networks of organizations is not distinct to this trend of urban sociology. Sociologists of space and place have earlier argued that economic growth and the ability to rebound from crises are linked to the relationships among organizations within a community. In the words of economic sociologist Sean Safford, we can see regional economies as "intricate, overlapping systems of inter-organizational relationships."[84] In contrast to the broader institutional environment, organizational infrastructures are specific to each city. They may explain intrastate variation beyond the sociodemographic and political factors highlighted by urban sociologists.[85]

The idea that the institutional configuration of governance regimes matters for administrative practice is a central insight of multiple literatures. For instance, the eminent Chicago sociologist Ed Laumann, who later became famous for his work on sexuality, established as early as 1978 that community structure is determined by "inter-organizational linkages." His student and colleague Joe Galaskiewicz later showed that the "urban grant economy" made up of corporate philanthropies and nonprofit organizations shapes the development and revitalization of cities, thus limiting who gets access to resources.[86]

In contrast to theories about urban regimes and growth coalitions, however, I do not argue that government policies are an outcome of an alliance of interested actors who influence decision-making processes through coercive power or non-decision-making. Cities are collectives whose governance depends on the local organizational infrastructure, irrespective of the influence of specific coalitions and their coercive influence on government. To put this into more institutionalist terms, cities' pursuit of legitimacy depends on the constituents that inhabit their organizational field.[87] As especially critical players in the cast of actors in a city, from the local organizational infrastructure perspective, the size of the nonprofit and business sectors could have consequential effects on the likelihood of action.

After many years of studying the nonprofit sector as part of Stanford's Civic Life of Cities Lab, the first aspect that I expected to matter particularly was the presence of nonprofits in the city's organizational infrastructure.[88] The nonprofit sector has expanded significantly throughout the United States in recent decades, and nonprofits have become more involved in public management and public policy. Not only have decentralization and devolution in the public sector made cities more reliant on nonprofit organizations as service providers, but these shifts have politicized nonprofits and made them more central to "collaborative governance."[89] Nonprofits foster individual civic engagement and community building and mobilize citizens to engage in activities to further systemic change.[90] For instance, Nicole Marwell showed that nonprofits played a significant role in community development in Brooklyn and, in the process, took on a political role—although nobody democratically legitimated their position as neighborhood representatives.[91] Other sociologists have shown that nonprofits also matter for encouraging and enabling social movement activities in which citizens demand social change. Robert Sampson, Doug McAdam, Heather MacIndoe, and Simon Weffer-Elizondo examined data from a longitudinal study of Chicago, finding that the number of nonprofits in a neighborhood had a significant, positive effect on "collective civic action."[92]

How is the presence of nonprofit organizations associated with the extent of sustainability practices? Cities with a denser ecosystem of civil society organizations are more likely to adopt sustainability practices. David and I found that for a single increase in nonprofit density by one standard deviation, sustainability practices went up by about 7 percent of a standard deviation, even accounting for all other factors discussed thus far. Do nonprofits boost sustainability just because mission-driven nonprofits are more likely to send environmental activists to town hall meetings? It turned out that this effect was not just driven by organizations with an environmental agenda. If we consider the number of registered environmental nonprofit organizations (rather than all types of nonprofits together), the nonprofit effect is similar in magnitude. This insight reveals the influence of civic capacity—a profound feature of cities related to the part of the city's organizational infrastructure dedicated to social and environmental problems that I will explore in chapter 5.[93]

The nonprofit effect on sustainability practices could be entirely due to population features that are more likely to give rise to a greater number of nonprofit foundings. For example, the population of cities with more nonprofits could simply be wealthier, be more liberal, or have a more environmentally conscious population.[94] How do we know whether this is the case, or whether nonprofit organizations' actions are meaningful? To make this determination, I examined the effect of another type of organization that previous research had shown to be associated with social capital and liberal values in the population: corporations. Organization theorists interested in local communities had previously established that some metropolitan areas cohere around a certain, shared level of corporate social responsibility (CSR). Organization theorist Christopher Marquis and his colleagues argued that "corporate social actions are commonly oriented towards the locales in which a corporation's executives reside."[95] The authors also found a high level of homogeneity within communities regarding corporate social action—corporations in some places, such as Minneapolis, Minnesota, or San Francisco, California, are more likely to adopt CSR practices than others, such as San Jose, California, and Columbus, Ohio.

This concentration of prosociality may mean that regions that share the same institutional pressures should generally converge around an interest in environmental sustainability and progress. For example, Marquis and his colleague András Tilcsik found that communities with a larger population of corporate headquarters had more thriving community organizations, including both elite clubs and social welfare organizations.[96] They were also more likely to see corporations willing to donate to the community after mega events and certain disasters. We, however, did not find that city administrations in metropolitan areas with a higher level of CSR (measured based on the average CSR scores of companies in the area evaluated by investment research firm MSCI) featured more sustainability practices. Neither more firms nor larger firms had a consistent effect on the uptake of city practices. There were weak indications that cities with more large corporations tended to suppress sustainability practices, and cities with firms whose CSR scored higher saw more sustainability practices, but neither effect was statistically significant by conventional standards. In addition, the greener orientation of certain cities may encourage firms to be more responsible toward their stakeholders. Surprisingly, this means that although most urban scholars agree that companies play an important role either in shaping the community through philanthropy and political influence or by forming developmental-oriented growth correlations, the presence of businesses was not a major indicator of corporate sustainability practices—unlike the presence of nonprofit organizations.

When I first saw these results, I did not readily believe them. My skepticism did not result from their contradicting my intuitions but precisely because they aligned with how I saw the world. A confirmation bias alarm went off in my office. Although the initial stars of my research on city strategies were in the public sector, I was deeply concerned with the effect that private organizations—particularly nonprofits and companies with a social mission—have on society. When I first started studying in Vienna, my entire attention was on understanding what happens when nonprofit organizations become more professionally oriented and, therefore, more concerned with tangible and ultimately narrow goals. The fact that the presence of nonprofit organizations had a positive, significant, and robust effect on sustainability practices meant that they play an essential role in the shape cities take.

At this point, this insight was not entirely new. Scholars have shown that cities with greater levels of nonprofit organizations are also more likely to develop so-called collective efficacy, by which they collectively solve problems that arise in the community.[97] Other work has found nonprofit organizations to be crucial nodes and social networks regarding outcomes such as innovation and resilience. My colleagues Nicole Marwell and Jeremy Levine have both repeatedly argued that nonprofits play an essential role in the fabric of urban politics that earlier urban regime theorists had overlooked because they prioritized greedy developers in their quest to understand the interplay of public and private governance.[98] Nevertheless, many urban theory frameworks—urban regimes, growth coalitions, and even contemporary ones about urban governance—were not particularly concerned with nonprofit organizations but much more focused on elites in town halls, boardrooms, and party offices. This finding, however, was the second revelation that indicated that nonprofit associations play an important role in structuring urban sustainability. The first had been that intercity associations, which are typically nonprofit organizations, played a central role in facilitating city learning.

Nonprofits may have a limited amount of direct power to force certain outcomes. But they can still be a source of institutional change despite the limited direct effect on government practices through coercive power.[99] Many public administration scholars think of nonprofits as substitutes for government activity and providers of public goods when high diversity does not allow states to satisfy the needs of all citizens.[100] In contrast to the expectation that nonprofits will emerge under conditions of government failure, their abundance is associated with more municipal attention to sustainability, controlling for many other explanations.

WHAT WE SAW

Cities have expressed bold commitments to sustainability, and this chapter has shown that despite all implementation challenges, cities with a plan tend to follow through on these plans in one way or another. Survey data allow for systematically examining variation in the practices cities report in various sustainability domains, from water to recycling, while drawing on a well-equipped tool belt, from regulation to recognition. Many cities have attempted to involve their civil society in their efforts. But the romantic image of enlightened citizens pounding on the doors of city halls to demand climate justice turned out to be an illusion. The "usual suspects" that appear at these deliberative efforts do not primarily care about climate change.

So, where do the ambitions of even small and poorly resourced city administrations come from? Bringing some of the macrosociological explanations of chapter 3 onto the ground, this chapter showed that integration in global intercity associations is but one factor. Membership in ICLEI—an international association supporting cities through knowledge and networking and just one conservative example of the myriad associations supporting cities—is, in fact, associated with more city action. This evidence that the institutional superstructure matters should not distract from the fact that urban sustainability is also a partisan issue, which has been made more difficult by the occasional backlash against global efforts to advance sustainable development and reactionary policies by preemptive state governments.

One reason that not all cities are highly sustainable is that urban sustainability is an uphill battle, costly, and fraught with political opposition from inside and outside. One cannot overstate the enormous institutional—financial, technological, and political—barriers to sustainability. Interviews with sustainability officers in cities that lagged behind the field because of serious institutional barriers to sustainability revealed that some cities cannot afford to play the high-stakes game of leaders. Instead, they are confined to sticking with more superficial policies and finding safe spaces in more regional networks. Although political barriers may conveniently explain why some cities are more proactive about climate change than others, the survey data also revealed tremendous variation in cities' efforts within both states as environmentally protective as California and as reluctant as Alabama to take climate action.

The econometric analyses illuminate the vital role of how urban governance and administration are organized locally. We saw that politically oriented mayors were, in fact, less likely to support an environmental agenda than city managers. Managers with a strict professional ethos to effectively solve city problems were more responsive to their climate action imperative. Seeing how international professional associations have played an important role in motivating world cities, the crucial role of city professionals in championing sustainability almost a decade before politicians and diplomats formulated the Paris Agreement's 1.5°C goal in the mid-2010s should not come as a surprise. Nonetheless, this finding corrects the myth that heroic mayors envisioned "cool cities."

Many mayors merely shine a light on plans already in the making or that have even been resting in their city administrators' drawers for a decade.

Finally, these results remind us not to reduce urban governance to city administrations. Cities with a greater presence of nonprofit organizations were much more likely to offer a broad range of sustainability practices. Notably, these practices did not only involve the expected groups of environmental advocacy organizations such as the National Resource Defense Council, the Sierra Club, or grassroots environmental activists. Our analysis showed that the presence of *any* nonprofit organization is associated with greater adoption of sustainability practices. This curious insight deserves greater attention, which I give it in the next chapter, revealing civic capacity as an essential feature of cities' capacity to act. How the organizational infrastructure enables cities to lead is a crucial question because the value of cities' efforts ultimately depends on whether they make a difference in cities' emissions and lived environment.

CHAPTER FIVE

Scaling

The librarians at the San Mateo Public Library were literally in the dark when they decided to go green. Property owners in San Mateo County had just passed a general bond obligation contributing $35 million for a new library building, topped off by the State of California with a $20 million grant. A senior library management analyst recalled the deliberation process:

> Because a large component was coming from the community, we had several public meetings asking our citizens: What do you want in your new public library? It happened to be that many of those meetings took place back in the late 1990s when we were having some energy challenges and [California utility] PG&E at that time had something called rolling black outs. One of the meetings *[laughs]* happened to be in our own library, it was a public meeting, and our meeting room was just abysmal. It was small. It had very small windows. It was hot and stuffy in there. And during the meeting a rolling black out happened.

A teacher from nearby Cañada College who was starting a sustainability studies program at the junior college was in attendance. "She raised her hand and said: 'Why can't we build a sustainable building?' It was like a light bulb went off." The library ended up 20 percent above California's building efficiency standards, Energy Code Title 24, and it became one of the first public buildings in California to receive a Leadership in Energy and Environmental Design (LEED) Gold certification. As local environmental nonprofits took note and featured the library in its campaign to build sustainably in subsequent years, the library "set the tone" for the city. In 2001, San Mateo became one of the first places nationwide to adopt a sustainable building policy, requiring public buildings to apply for LEED. The library was a major cheerleader for green construction; according to the librarian, green building just became "the right way."

Green buildings are the central illustrative case of this chapter, which examines how organizational infrastructure shapes city climate action. According to the US Department of Energy, commercial and residential buildings (39 percent) account for a greater part of energy use than transportation (29 percent) and industry (32 percent), respectively, in US cities.[1] Considering this substantial ecological footprint, the US Green Building Council (USGBC) developed a voluntary certification scheme for energy-efficient green construction at the end of the 1990s. LEED encourages project owners to choose sites with access to public transit and bicycle parking, build using recycled materials and certified wood, and incorporate biophilic elements in the interior design. The framework meant to improve the lived experience of living in energy-efficient, fully insulated concrete boxes that were built in response to the oil crisis of the 1970s. This certification is in constant development and has expanded, over the period of my observation, to include the certification of homes and the energy-efficient operation, rather than merely construction, of buildings as well as entire cities and neighborhoods.

Although some engineering experts may contest LEED as an industry standard as timid, it has, for almost two decades, been the tool of choice for municipalities whose goal is to encourage green construction. According to the Department of Energy, building certifications are twice as common a "city action" intended to affect energy use than public transit expansion and bicycle infrastructure.[2] Early on, it was uncertain whether the promised cost savings and energy efficiency were going to justify the extra costs of construction. As a former foundation president told me of the decision-making for certifying their foundation's new building in the early 2000s, there was a "fair amount of skepticism because of the cost." The foundation went through with the certification to do justice to its significant environmental grantmaking program and achieved a LEED Gold certification, spurring a neighboring peer foundation to pursue LEED Platinum several years later.

In this final empirical chapter, I illuminate the intertwined roles that municipal governments and civil society played in the origin and spread of green construction among US cities. Deepening a finding from chapter 4, the spotlight is on civic capacity—the presence of prosocial organizations in the nonprofit sector that seek to recognize and solve social and environmental problems where they arise. Understanding how civic capacity enables city action is important, because its influence extends not only to climate change but also to a series of other contemporaneous and future problems that cities face. Civic capacity, it turns out, has played as big a role in explaining why LEED took off as the actions of municipal governments. The values of nonprofit organizations prompted them to adopt buildings early and establish proofs of concept that catalyzed action among other citizens and even city administrations. Figure 5.1 summarizes the chapter's argument visually.

In what follows, I first revisit some insight from the introduction about why we need to consider both state and civic capacity, and their association with what I call administrative and distributed city action. I then explain why green construction is a theoretically and substantively

FIGURE 5.1 Visual summary of chapter 5.

meaningful context to study these differences and show how features of cities' organizational infrastructure shaped the diffusion of green building certifications among US cities—through catalysis, legitimation, and scaling.

What can the diffusion of green buildings tell us about city climate action? Cities make bold promises and claims about how they will neutralize the impact on the climate and improve the experience of urban dwellers and the sustainability of urban ecosystems. We have so far seen that the right institutional conditions encourage and amplify administrative actions. The institutional superstructure, including a growing intercity network and knowledge exchanges among their city members, sets lofty expectations from the top; a strong local civil society lights a fire from the bottom. But these newly adopted climate commitments are best understood as rationalized practices, intended to address external expectations about what it means to be a good city without necessarily improving a city's ecological footprint.[3] The practices of city administrations—avoiding bottled water and sending electric vehicles (EVs) to pick up your compost—are just a drop in the bucket of carbon emissions. Who has ever consulted their city's strategic plan before buying a car, deciding what energy mix to use, or switching on the air-conditioning? At the end of the day, administrative city action matters only if it meaningfully goes together with *distributed* city action—the individual behaviors of people and organizations in their city. But does it?

A satisfying answer to explain disparities in city climate action must acknowledge that sustainability is not a matter of adoption but rather a transformation of an ecosystem of organizations and individuals that constitute each city. As we have seen, the average Jane and Joe is not among the readers of plans like PlaNYC or Copenhagen's ambition to become the green and blue capital—most people pay little attention to such highfalutin statements. But Jane and Joe must be involved in any lasting ecological transition. Sustainable practices make a meaningful difference only if they make it into the heads and hands of *most* people and organizations living in a city.[4] How, then, do urban ecosystems tilt or even tip toward more sustainable equilibria, and what role do the actions of city administrations that we have hitherto examined play in this process?

ADOPTION AND THE ORGANIZATIONAL INFRASTRUCTURE

City administrations *do* play an important role in shaping local responses to climate change. They have set ambitious goals for carbon neutrality, launched resilience programs, and promoted green technologies related to transportation, energy, and waste management. But as we saw in chapters 3 and 4, administrative action related to climate change and sustainability is characterized by large geographic variation. The varying influence of city administrations is not simply a factor of whether or not they yield considerable power. State capacity, understood as the government's ability to pass and implement its policies, is often seen as a necessary ingredient for transformation or change, as is the strength of market-oriented actors seeking to pursue innovations.[5] For instance, when it comes to sustainable transportation, public policies providing incentives to buy EVs, the installation of a grid of EV chargers, and the development of new batteries is inarguably important. It is intuitive to believe that cities with greater state capacity can tackle climate change more swiftly and thoroughly. I will show, however, that state capacity alone is not sufficient and, in fact, often driven by an even more fundamental feature of cities: their civic capacity.

Civic capacity is the presence of prosocial organizations that seek to address social and environmental problems in the urban ecosystem.[6] The interplay between state and civic capacity is what enables cities to make meaningful moves toward becoming more sustainable. As the illustrative case of green construction will show, state and civic capacity together contribute to the catalysis, legitimation, and scaling of new ideas that can profoundly transform cities. The basic idea is simple but not quite as simple as social scientists of cities have hitherto imagined institutional change among cities through the *administrative adoption* of new practices or policies through a local government. Instead, I suggest that a process of *distributed adoption* best explains the relationship between the fundamental capacities of cities, the administrative adoption of strategies and policies of city climate action, and any city's sustainability transition.

This model of distributed adoption involves three steps, which draw from the innovation literature on diffusion. The first is catalysis:

Ordinary organizations get the ball rolling by creating proofs of concept that an innovative practice—that is, a practice that is new to the community, if not the entire world—works. The second is legitimation: City administrations pass public policies that legitimize innovative practices as desirable or even prescriptive. The third is scaling: The bulk of people and organizations—even conservative ones—widely adopt the innovative practice once the payoffs are clear. Understanding urban sustainability as a diffusion process triggered by catalyzing organizations puts the idea of starting with the scaling of social innovation on its head, as the case of green construction illustrates in this chapter. I'll provide a more complete picture of what city climate action looks like—because attributing city climate action to city administrations is meaningful only when it is paired with a sea change in individual actions that aggregate to urban transformation.

Cities are not only responsive to shared influences from above, they also are shaped by local influences.[7] For cities to learn from the bottom up, city administrations need to be able to do two things. First, they need to scope out whether ideas are present in the community. Second, they need to determine whether the city is open to listening to and responding to these ideas. This ability of cities to pay attention to their population is neither equally distributed nor self-evident. As discussed in chapter 4, some cities take listening to their citizens more seriously than others, and some prefer to listen to big business instead. The receptivity of novel ideas is a learned aptitude that requires interfaces that allow for the exchange of ideas between the administration and the organized citizenry—synapses that enable learning. Learning cities—those we encountered in chapter 3—are capable of catalyzing new ideas, amplifying the innovative potential in the population, and orchestrating diverse actors to work together on promising initiatives.

Does this simply mean that cities with greater punch—those with stocked coffers, officers educated at elite colleges, and widespread legislative and political powers—hold the best cards to green the city? State capacity surely does not hurt, as the success of European countries to simply regulate high expectations about energy efficiency shows. Instead, however, it is the cities that punch above their weight that make the greatest strides—and civic capacity is one feature that enables these cities to punch above their weight. As I will show, even after considering the city's state capacity and municipal resources, the cities that boasted greater civic capacity were dramatically more likely to see green buildings, saw them sooner; and ultimately saw more of them.

BUILDING GREEN AS DISTRIBUTED CITY ACTION

Green construction provides a powerful example of the distributed adoption of city climate action for multiple reasons. As any of the associations we encountered in chapter 3 will not get tired of emphasizing, cities are responsible for a massive percentage of global carbon emissions, and that number is climbing as more people flock from the countryside to opportunities in big cities. What is a city to do about its carbon footprint? The Chicago chief sustainability officer (CSO) reminded me before telling me about the city's new building decarbonization strategy to "just acknowledge that, you know, 70 percent of our greenhouse gas emissions come from our built environment." No topic or policy domain comes up as quickly as buildings.

Why do buildings have such a major impact? The answer is that both the construction and operation of the building are highly energy intensive. Most of the energy that you consume daily goes through buildings: Heating, cooling, ventilation, lighting, and electric appliances all require a significant amount of energy, not to speak of waste and water. Additionally, many buildings are poorly insulated and leak energy. Besides the high energy consumption of buildings, the manufacturing of concrete and steel is incredibly damaging to the planet; concrete alone is responsible for about 8 percent of global carbon emissions. Bill Gates's instruction manual for *How to Avoid a Climate Disaster* features a powerful image of the overall footprint of construction. In the next forty years, he estimates that the world will build new buildings roughly the size of all of Manhattan *every month.*[8] No wonder sustainability officers were so keen on tackling the built environment head-on.

Green construction is not just impactful from an environmental standpoint but also from an economic one. Construction makes up about $2 trillion, or somewhere between 5–10 percent of the gross domestic product (GDP) depending on which estimate one trusts. So even though the market for green buildings is still burgeoning, it is valued at more than $120 billion in the United States, and the industry has grown at an average of around 12 percent every year since the early 2000s.[9] This number was stable even during the Great Recession, when the construction industry as a whole plummeted. A study by the City University of New York showed that within the state, the construction industry has the highest percentage of "green jobs" of any industry they studied.[10]

The importance of green building is now widely acknowledged. The Intergovernmental Panel for Climate Change (IPCC), a high-caliber United Nations working group of climate experts that assembles evidence-based opinions on climate change and strategies to address it, identified energy-efficient buildings as the primary strategy in the category of settlements and infrastructure. Consider what the IPCC has to say about cities and their built environment:

> Urban systems are critical for achieving deep emissions reductions and advancing climate resilient development (high confidence). Key adaptation and mitigation elements in cities include considering climate change impacts and risks (e.g., through climate services) in the design and planning of settlements and infrastructure; land use planning to achieve compact urban form, co-location of jobs and housing; supporting public transport and active mobility (e.g., walking and cycling); the efficient design, construction, retrofit, and use of buildings; reducing and changing energy and material consumption; sufficiency; material substitution; and electrification in combination with low emissions sources (high confidence). Urban transitions that offer benefits for mitigation, adaptation, human health and wellbeing, ecosystem services, and vulnerability reduction for low-income communities are fostered by inclusive long-term planning that takes an integrated approach to physical, natural and social infrastructure (high confidence).[11]

The promise of reforming the urban environmental footprint should not mean that we can pause building new wind turbines and hydropower plants and buying solar panels and EVs. These things still matter a great deal, and oftentimes more than the built environment, but they are often beyond the reach of cities. Green buildings matter for two reasons: They only require urbanites to apply existing technologies, and they are within the control of cities.

Perhaps the most attractive feature of improving the energy efficiency of buildings is that most of the technology that is required to reduce carbon emissions to a level that would meet the Paris Agreement's 1.5°C goals already exists.[12] The problem is not so much the technical availability of better insulation, construction materials, or green concrete. Rather, the widespread adoption of such existing technologies for energy-efficient construction is a social and organizational problem. As with many other voluntary practices intended to address environmental issues, the spread of these behaviors is inherently skewed toward innovators—early adopters that have the means and desire to try out something new, even if such adoption comes at a potential cost. As a result, green construction is not only important but also analytically ideal for understanding the unequal turn to sustainability, in the transformational sense of the term.

Green construction is particularly relevant to understanding variation in city climate action. Greening the built environment is the most common strategy by which city administrations can influence their city's carbon footprint—passing policies related to green building is not only a form of distributed city action but also an *administrative* city action.

Cities address environmental concerns rather uniformly in building codes. "Where local governments have adopted green building programs, the vast majority have employed the LEED program," legal scholar Katherine Trisolini has argued.[13] As political scientist Paul Peterson's "stifled cities" argument about the legal and political limits of municipal policies suggests, many aspects of climate mitigation and adaptation are outside the control of local policymakers—but building codes are staunchly within their domain.[14] Building codes are almost exclusively the responsibility of municipal governments, and there is little legal preemption by states, which occurs routinely in the context of taxation or the regulation of emissions. Building codes—a rationalized practice that spread across US cities only relatively recently and is now ubiquitous—are also highly standardized and follow model codes, such as those created by the International Code Council. Many CSOs brought up the massive carbon footprint of construction when I spoke to them. In other words, the question that occupies this chapter is, what enables cities to meaningfully shape both the regulation and construction of the buildings?

Do not let the refreshing optimism of green construction proponents betray the fact that in most places, building green is tough as concrete. Although construction experts report that green construction has become a new default, *most* new buildings in the United States do not qualify as green buildings. Although there are many success stories of places that have entirely reformed how they do construction, there are many in which sustainability considerations come far behind the cost and ease of doing business. One problem is that greater energy efficiency costs money, at least upfront. In other words, the incentives for project owners to choose thicker insulation, more sustainable materials, or more sophisticated heating and cooling systems are weak. Unless you are keen to do the right thing without any pecuniary motive, why spend money if nobody can see the benefits? This is where building certifications enter the stage.

CERTIFYING GREEN CONSTRUCTION

What does green construction mean, exactly? If one of the leading associations to advance energy efficient design and construction is to be believed, it very much is whatever you want it to be. The US Green Building Council's (USGBC's) framework for LEED is the world's leading certification system for green buildings. LEED has become the standard for new construction (and increasingly for the operation and retrofitting of existing buildings) since the late 1990s. The idea of LEED was to consolidate the existing, disparate frameworks for green buildings and to provide a single guideline for project owners who wanted to construct an even greener building.

In practice, this means that when project owners apply for a building certification, they first select the credits they want to pursue on various dimensions ranging from indoor environmental quality, the use of sustainable materials, water efficiency and whether the building is located in a sustainable and accessible site. As the USGBC claims, the majority of credits are directly related to lowering carbon emissions: "Of all LEED credits, 35 percent relate to climate change, 20 percent directly impact human health, 15 percent impact water resources, 10 percent affect biodiversity, 10 percent relate to the green economy, and 5 percent impact community and natural resources. In LEED v4.1, most LEED credits are related to operational and embodied carbon."[15]

After project owners choose their credits, the GBCI, a sibling organization of the USGBC, then examines the extent to which the building meets their standards.[16] Depending on the number of points achieved in the evaluation, the building gets a beautiful plaque with its certification status, which ranges from simple certification as LEED Certified (40–49 points) to Silver (50–59 points) to Gold (60–79 points) to the coveted LEED Platinum (80+ points) certification. If you have recently been to a US city, you have almost certainly passed dozens of these plaques on the doors of government offices, apartment complexes, and corporate headquarters. The point system is designed not only to facilitate a low level of certification at a low cost but also to ensure that Gold- or Platinum-certified buildings are cutting edge.

Although there are some minimum requirements to both the site and the environmental performance of the building, there are no guidelines as to how exactly a building achieves the points necessary for certification, as one USGBC employee explained to me: "You need to choose your own adventure. You need to find the most economical pathway that gets you to your best value, and so there's a lot that's happening within there, and so that process itself is reinforcing and pushes outward to peers and competitors, right? So that's happening in the private market space, and as that happens, we get to a place where expectations are evolving."

LEED's primary purpose is to overcome knowledge gaps about how to build green. It is not a determinative framework but encourages innovation. For instance, in addition to combining various credits, project owners can get "innovation credit" by experimenting with new ways of building green, and then share the results with other members of the ecosystem through the LEED pilot credit library.[17] A building expert at the US Green Building Council told me about one of the most popular pilot credits: bird collision. Stickers designed to avoid bird collision are a success story because the practice is now included in federal guidelines for new construction, and this innovation has made it into the main credit library for new LEED projects. The idea behind these credits is to introduce new opportunities for building a more sustainable building that are relatively easy to achieve and have the potential of widespread adoption.

Is piloting bird credits a radical change in how energy-efficient buildings are? No, but it is one of many small steps toward making buildings more integrated with their natural environment, and the definitive improvement over bird graveyards.[18] Other examples include the inclusion of affordable housing units; locating buildings in particularly poor Census tracts; and environmental, social, and governance (ESG) reporting—all of these practices were pioneered by LEED users and later integrated in the ever-evolving library of building standards. As of 2024, there are 197,000 buildings in 186 countries and a total of 29 billion square feet that are LEED certified—and the number keeps climbing.

LEED is in ample use, but is it useful? Policy and management scholars have written tomes about the power and shortcomings of LEED because it is a prime example of a market-oriented certification scheme developed by an intermediary organization with a substantial impact on the behavior of private market actors. It is not the only such scheme—its cousins include the B Corps certification for responsible businesses, the Fair Trade label for ethical working conditions and wages in international trade, and the Rainforest Alliance certification for biodiversity in ecosystems.[19] According to critics, liberal market economies have come to rely on these certification schemes to a considerable extent over the past decades precisely because these countries have had weaker regulatory power and, therefore, have to rely on the goodwill of market actors. Let's assume that you run a company that has decided that it wants to do better—or at least look better in the eyes of the young, potential employees who care about the construction of the planet. Understanding how to be or to build green is precisely the question that these certifications seek to answer by codifying what they see as the best and most practical suggestions in a certification system. In addition, certifications provide a visible badge of honor for those who have managed to meet the certification scheme's standards.[20]

It should be clear from the onset that certification schemes are prone to error and manipulation. What some perceive as green, others may see as greenwashing. Certification schemes can be gamed, as a Midwestern

CSO's assessment of the small city's ordinance to favor LEED-certified buildings argued: "I'm skeptical that there's been a meaningful difference than if there wasn't an ordinance. I think we probably have more green roofs and more bike parking. These things are great and are important to have, but really weren't the initial impetus or intent of the green building ordinance, and I don't think the ordinance has been effective at changing some of these other things like greenhouse gas emissions or energy use intensity for buildings."

Many construction owners simply go for the easiest points and avoid more expensive improvements. One example of bad incentives that made it into the academic literature was a building by the US Environmental Protection Agency (EPA) in Kansas, which was greener by design but so removed from where employees lived that, arguably, the emission savings were dwarfed by additional commuting.[21] Even without any gaming or malfeasance, point systems are not always perfectly calibrated, so that some points are too easily acquired, whereas others are difficult to gain.

In other words, the fact that many cities use LEED should not be mistaken as an unconditional endorsement of LEED as the cutting edge of what is possible in green construction or sustainability more broadly. Arguably, *not* constructing a new building at all is sometimes the better choice when it comes to climate change mitigation. I will leave a nuanced critique of the certification scheme to architects and engineers, who are better suited at identifying and remedying the system's weak points.[22] What is clear, however, from the perspective of this social scientist, is that voluntary certification is an influential tool in weak states to change the behavior of market actors. And when cities act to curb carbon emissions—which we have established is one of the most meaningful things they can do if their goal is to be seriously sustainable—then they often turn to LEED.

The question we now need to ask for the purposes of understanding distributed city action is, who adopts LEED standards and when? Aggregated to the city-level, the uptake of green building shows whether a city's sustainability strategies are in fact associated with meaningful changes to the its ecosystem. As the leading sustainability planner of Washington, D.C., explained, the certification created a clear expectation and a measurement tool for the planning department: "The number of LEED buildings is a benchmark of the impact of the built environment on climate. Buildings are responsible for 70 percent of the greenhouse gas emissions in our city, so I think it's important. It's almost understood that if your project goes before the zoning commission, it's going to be minimum LEED Silver, and they're not that impressed by that. We use that as a benchmark, especially because we are near the top as for per capita green buildings. We like to *tout* that one."

TRACKING THE DIFFUSION OF
GREEN BUILDINGS AMONG US CITIES

The insight that US cities use LEED indicators as a measure of greening inspired me to dig deeper into what data were available to track green construction as an indicator of city climate action more broadly. To better understand how these credits work, I traveled to Washington, D.C., twice to speak to directors and employees of the USGBC about their work. I also interviewed representatives of statewide branches of the USGBC, some of which have grown to become independent of the national organization, and several vice presidents who after the pandemic were working all over the United States. I also spoke to several people involved in the initial decision to build green among the very first LEED projects in the San Francisco Bay Area and Chicagoland, two metropolitan areas that were early to adopt LEED standards. I have no doubt that what works and what doesn't with respect to LEED credits has much nuance that others have yet to explore, but I came away with a decent understanding of which theoretical accounts made the most sense for explaining the role of cities and their infrastructure in the spread of green construction.

To my delight, the USGBC kept good records of all buildings that ever applied for or received a LEED certification. The public directory not only lists where the buildings are located, and how large they are, but also who is behind the building effort and how well it performed in the assessment. In other words, I had what I needed to move from vision to action and to track the distributed adoption of green construction among cities. In the end, I examined the location, time, and sector of 130,000 buildings in more than five thousand municipalities across all 929 core-based statistical areas (metro- and micropolitan areas, hereafter metropolitan areas for simplicity). In addition, I had already encountered detailed records of which cities passed official policies to incentivize or require green construction in their jurisdiction—which helps track the administrative adoption of green construction policies. Because the US Census Bureau also keeps good track of its cities, I already had rich data on their organizational, social, and political ecosystems. This comprehensive dataset made it possible to examine whether the organizational infrastructure that we had seen to shape the administrative adoption of sustainability practices in chapter 4 also affects the turn to sustainability in the city's wider population. With the help of some insightful work that was published around the time, I figured out a way to measure civic capacity as a residual of the predicted nonprofit count—that means, the ability of the population to form nonprofits beyond what was to be expected.[23] My in-depth inquiry into the emergence and expansion of green construction among US cities was later published in the *American Journal of Sociology*, and the following discussion draws on findings and examples from that study and related writing. The methodological appendix offers further detail.

The most important explanation of differences in green construction turned out to be *civic capacity*, which is reflected in the presence and number of nonprofit organizations in a city and, indirectly, their relationship to their local environs. Many places feature nonprofits that have next to nothing to do with their city, except to be located there—think of NPR, a national radio broadcaster with its headquarters in Washington, D.C., or Kiva, a global crowdfunding platform based in San Francisco, California. Looking at the raw number of organizations is meaningful for a couple of reasons. As historian Claire Dunning and I have argued, *all* organizations are ultimately part of the local organizational infrastructure, even when the remit of their mission is international. For example, Stanford University is a global university, but all the groundskeepers are local, and even many of the administrators were born and trained locally. Many global nonprofit organizations have access to talent everywhere but nonetheless source much of their staff locally. Moreover, the number of nonprofit organizations is but a proxy for the actual network interactions among organizations. If more organizations are present, more frequent interactions are not guaranteed but probable. As the famous Austrian-American sociologist Peter Blau jested: "One cannot marry an Austrian when no Austrian is around."[24]

This inquiry generated a couple of insights—some expected and some surprising. The most important and thorough finding was that civic capacity matters greatly for how green construction diffused. The number of buildings was consistently and substantially higher in places with more civic capacity, controlling for many other things such as the city's politics, geography, and demographic makeup. Using a variety of econometric analyses to examine the timing and extent of LEED-certified buildings, I found that places with greater civic capacity adopt green construction sooner than others and that more nonprofit organizations per capita are associated with more green building certifications.

Figure 5.2 illustrates this relationship. Initially adopting cities included familiar leaders such as Boston and Cambridge, Massachusetts; Chicago,

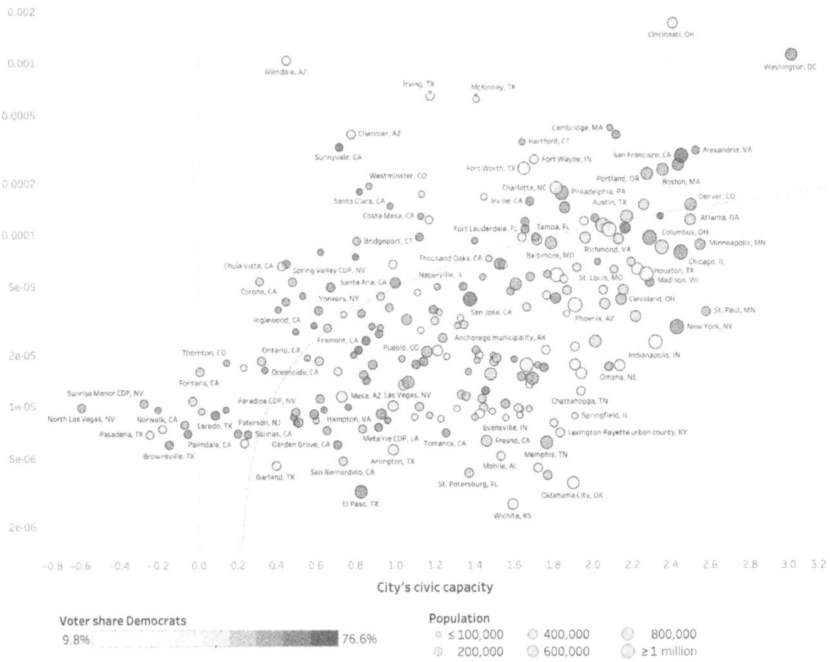

FIGURE 5.2 Relationship between green building count per capita and city's civic capacity.

Illinois; Los Angeles and San Francisco, California; Portland, Oregon; Seattle, Washington; and Washington, D.C.; but also less obvious trailblazers, such as Austin, Dallas, and Houston, Texas; Little Rock, Arkansas; Phoenix, Arizona; Pittsburgh, Pennsylvania; Raleigh, North Carolina; and Salem, Oregon. Many of the cities that had amassed the most LEED-certified buildings by 2016 were also among the earlier adopters, including Washington, D.C. (2,324 registrations), New York City (1,663 registrations), and Dallas (1,552 registrations). Cities that stood out in terms of green buildings per capita early on were often college towns, including Cambridge, Massachusetts; Fort Collins, Colorado; New Haven, Connecticut; Pittsburgh, Pennsylvania; Seattle, Washington; and Syracuse, New York. Although many leading cities are Democratic-leaning, political preferences were visibly not deterministic: several cities in relatively Republican-leaning counties were among the leaders. All these cities registered above-average civic capacity regardless of political affiliation. Only much later did places with large-scale prefabricated and luxury housing developments such as Irving and McKinney, Texas; Glendale and Chandler, Arizona; and the corporate campuses of Apple and Google in Sunnyvale, California, propel cities with lower civic capacity to the per capita top of green construction. Even sizable cities with lower levels of civic capacity registered a modest number of registrations, for instance Jacksonville, Florida; Mesa, Arizona; Las Vegas, Nevada; Philadelphia, Pennsylvania; and San Jose, California.

A comparison between Chicago, Illinois, and Cleveland, Ohio, illustrates the power of civic capacity. Chicago has a robust network of nonprofit organizations that are actively involved in addressing social and environmental issues. Its local chapter of the USGBC, the Illinois Green Alliance, has been working to promote the adoption of green construction practices and policies in the city. Chicago has seen constant growth in the adoption of green construction, with several new net-zero buildings dotting its skyline. The city of Cleveland, in contrast, has a weaker network of values-oriented organizations and has seen slower adoption of green construction practices. Despite early efforts to advance sustainable development by advocacy organizations like the Cleveland Green Building Coalition (a precursor of the local chapter of the USGBC for North Ohio), green construction did not take off in the early years. Cleveland now features roughly half as many green buildings per inhabitant as Chicago.[25]

One reason for the civic capacity effect was that cities with more nonprofit organizations were more likely to see green building early on because they make decisions not just based on financial considerations but also ethical ones. Early adoption is why the boost that civic capacity provided to the marginal number of new buildings registered for LEED was stronger in the first phase, before 2006. Just like when I examined survey data on urban sustainability in chapter 4, I found that it was not only *environmental* nonprofit organizations that were responsible for the effect. Environmental nonprofits alone had roughly the same effect as all other nonprofits, even when controlling for the share of environmental nonprofits in the overall nonprofit sector. Considering conventional measures of social capital also suggested that the effects of civic capacity are not just structural in that cities with greater civic capacity see more social capital—more trusting or denser networks among organizations.[26] What was it about civic capacity that made cities more prone to engaging in both administrative and distributed city action?

The statistical analyses of what explains differences in green construction that I am about to discuss are of course not beyond reproach. There are always remote possibilities that green buildings are behind the historical rise in the growth of nonprofit organizations and not the other way around, or that certain unaccounted factors like social capital or wealth could explain both variation in nonprofits and green building growth, or that a few outliers explain the strong relationship. But these were either accounted for or very unlikely. I controlled for many other potential explanations, including demographic factors, such as the race, income, and politics of the population; network spillovers from cities with similar politics or shared ties; and spatial spillovers from larger cities to smaller ones. Overall, I came away confident that the basic association between civic capacity and the adoption of green buildings was existent and not trivial.

The real question was *how* a certification primarily aimed at the market for office buildings landed on more fertile ground in cities with more civic capacity. The key to understanding this is to follow the diffusion process through all three stages: how it started, how it became popular, and how it became the default for new projects. As we will see, it would be reductive to say that nonprofits were the key, or that city administrations did the work, or that it was a market phenomenon, for that matter. Instead, it was the interplay among the public, nonprofit, and corporate spheres that was really responsible for the successful uptake of green construction practices in cities. In more theoretical terms, green building certification happened to be a perfect example of how cross-sector dynamics at the intersection of the institutional superstructure and the organizational infrastructure made a meaningful transition possible.

THE DIFFUSION OF GREEN BUILDING CERTIFICATION

The first step to answering this question is to look at where green buildings were constructed at different points in time. The map in figure 5.3 shows the number of places in the United States by the number of LEED buildings (dark gray) in 2006 and 2016—at the beginning and the end of the diffusion process, while figure 5.4 shows the timeline of the number of LEED buildings and city policies from 2000 to 2015. There are two crucial takeaways from looking at the geographic dispersion and the chronology of diffusion of green construction. The first is that, unsurprisingly, these green buildings are heavily clustered around population centers. Overlaying the map with the boundaries of metropolitan areas, and cities more specifically, shows that most of the geographic variation is indeed between cities—making the patterns a useful starting point for explaining variation among cities. The second insight is the geographic patterns of diffusion became locked in around 2006 and remained stable in the later stages, when the majority of adoptions occurred. At this stage, LEED was already widely disseminated. In other words, most businesses that have a good-looking plaque in their lobby joined the movement at a time when it was already clear that LEED offered a variety of benefits, both reputationally and in terms of cost saving.

Not all cities jumped onto the LEED bandwagon at the same time. The dark gray line in figure 5.4 shows that the first LEED building spread among US cities in an S-shaped curve that is familiar to anyone who has studied the diffusion of innovation. When any new behavior hits the market—from the new iPhone to TikTok dance moves—a few innovators get the practice started and early adopters are willing to try out a new and untested product or practice. At this stage, the S-curve is still a flat line. In the green building case, the first certifications spread in the

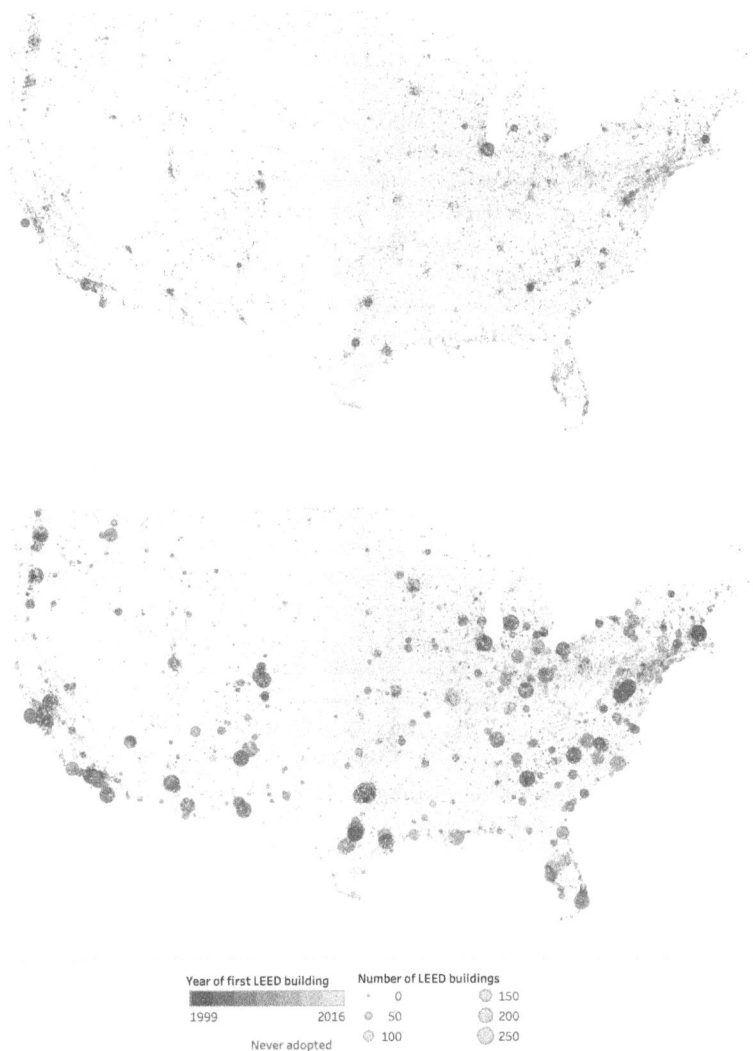

FIGURE 5.3 Map of cumulative LEED registrations in the contiguous United States, 2006 and 2015.

Note: Each dot represents one building registered for evaluation for LEED certification. The map shows registrations in all previous years.

Sources: Map uses map data from Mapbox, https://www.mapbox.com/about/maps; OpenStreetMap made available under the Open Database License, https://www.openstreetmap .org/copyright.

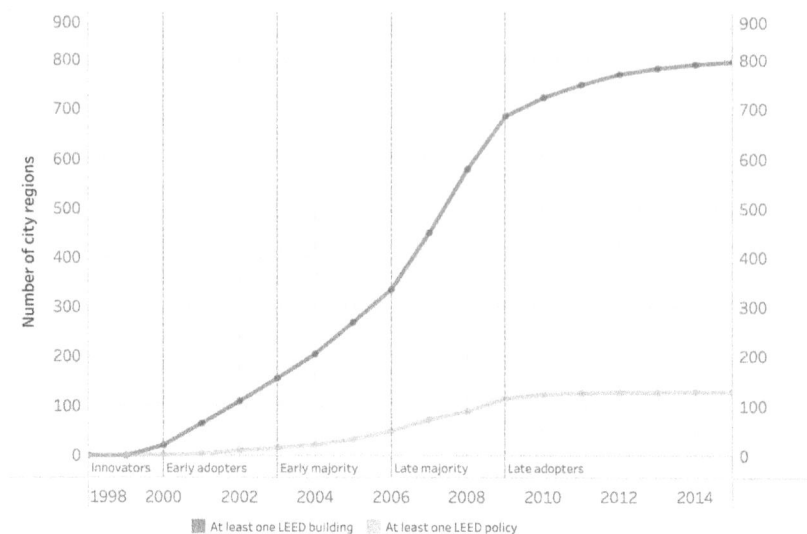

FIGURE 5.4 Diffusion of LEED green building certifications and policies by metropolitan area, 2000–2015.

early 2000s, and about 200 leading cities—from San Mateo to Washington, D.C.—saw their first buildings break ground during this period.

Then, most users see the benefits or trendiness of the product and jump on board. This adoption among a majority is responsible for S-curve's steep slope. We see this incline at the later end of the decade— paradoxically around the time of the Great Recession, which sent much of the construction industry reeling. As one executive of the USGBC explained in an interview, the financial crisis provided "a boost" for the certification protocol because, presumably, corporations publicly advertised their prosocial orientation: "During the economic downturn, 2008, 2009 through 2012 or so, the real estate industry tanked, right? . . . But the assets that carried green certifications like LEED pretty much the only investments that continued. As the construction market took a nose dive, you actually saw more of like a flat line [of new green construction]."

After a final period of lagged adoption, when even reluctant late bloomers adopt the practice, the S-curve plateaus. In 2015, only five metropolitan areas in the United States saw a first-time registration for LEED. Not much changed after that—all new buildings were in places that already had at least one LEED-certified building. At this point, the practice can be said to have successfully diffused. Making it to the end of the S-curve does not necessarily mean that everyone now has a new smartphone or the latest dance moves in their repertoire. In fact, a review of diffusion studies showed that among all organizational practices, almost half of practices organization scholars have studied only reach under 20 percent of potential adopters.[27] In other words, the fact that there are only one hundred or so—mostly very small—urban areas in the United States that do not have any certified green buildings is astonishing: By 2016, urban construction was thoroughly drenched in green construction principles.

I had three hints as to what explained the transformation: (1) the role of the USGBC and its associated professionals in creating a powerful brand, (2) the creation and sharing of proofs of concept on the part of ardent early adopters, and (3) the role of local governments in establishing LEED as an appropriate and attractive solution to the private sector. These three mechanisms roughly align with the *scaling, catalyzing*, and *legitimation* of green building. Although scaling to great numbers comes last in a typical diffusion process, I will start with the scaling stage because it shows that professionals and market forces offer a compelling partial account of how green construction became a new standard, albeit not the full story.

SCALING THROUGH PROFESSIONALS

One reason for LEED's success is due to the fact that it became *taken for granted*. LEED is not the luxury it may have once been—a label one slaps on a building to show off that one is on the bleeding edge—but it has become downright mundane in some parts of the construction industry. As the representative of the Illinois Green Alliance told me in 2022, "There's no requirement [that a new building has to be LEED certified] these days, but there are market forces that if it's a class A office building, that it should be. In the class A office building market, you don't stand out if you do LEED. It's just standard practice now. Now, if you want to stand out, you have to do something different, like a passive house." Although constructing a passive house that requires no active heating system at all or a net-zero building that meets all energy demand with onsite renewable energy may still require some mulling over, energy-efficient buildings have become second nature in the construction industry. Such automatism in decision-making, or taken-for-grantedness, is a core feature of what organization scholars think of as institutionalization—the process by which something becomes a norm or default whose adoption gives the adopter legitimacy in the eyes of others. This occurs when a tool or practice moves into the cognitive background, like a doctor putting on her white coat every morning or a professor standing at the front of the room rather than sitting down with the students.[28]

There is a certain appeal to these buildings. When I visited the USGBC headquarters in Washington, D.C., staff members pointed out renewable materials, modern window shades that automatically move with the sun, biophilic elements that create the impression of being in nature, and effective HVAC systems for heating, ventilation, and air-conditioning. It really was like an office building of the future.[29] Energy efficiency was not quite so attractive during the days of the 1973 oil crisis, when small windows and faceless concrete boxes were the norm. And the contrast to regular offices is visible for those already working, or better, being recruited to work, in fancy office buildings. A USGBC executive said that, "Once we start to see that some of those offices could offer a better office building and some of those are drab, and they're caught in the 1970s, it becomes a more likely scenario that you and I have evolved our expectations, and we say, well, I actually want to work in a nice office building." Unaware of the car brand's bleak future, he likened the protocol to "the Tesla of green building." He continued: "Until there was LEED, there was not a common language to define and explain and commit to sustainability in buildings. So prior to that you had, you know, let's make a building more efficient, or let's make a building more something, but LEED became a very simple and effective communications tool, which then drove a lot of action. The brand was very smartly built from the beginning. It was all about leadership."

Like the EV, the building protocol took some time to become a household name. "It took a few years for people to have heard about it, and for us to have enough projects to be able to say that here are a few, and they're fantastic, before people started just taking it and running with it, both private sector, public sector." As another USGBC staff member explained, the increased brand recognition has brought actors to the table who were more concerned with an explicit cost-benefit calculation of signing up for LEED. "The impact has scaled up now that the brand is more recognized. The folks who had heard of it obviously understood why it could be a mark of quality. But that only goes as far as the people who recognize the name and the brand."

The network of people and organizations that have contributed to the widespread adoption of green construction in cities and who continue, to this day, to proselytize the practice, has grown dramatically since the early days of the USGBC. Some members of the USGBC even came to the organization through the professional accreditation. A leader for the Pacific Coast chapter explained: "The company I was working for at the time was an environmental engineering company and they had one or two LEED projects and I was interested in the field, so just the concept of green design and sustainable design and construction was interesting to me. So you know, they were offering the exam and so I took the exam." As of 2024, more than 200,000 architects, engineers, construction managers, and urban planners worldwide—albeit mostly in the United States—have been certified as LEED Accredited Professionals (AP).[30] In some cases, sustainability is a career-defining specialization.

Many of the architecture and construction companies that are involved in individual green buildings develop expertise and even credentials that allow them to specialize in these activities later on, and that gives them a competitive advantage over other companies. As a USGBC fellow explained: "There are one-person architecture studios, and like they've built their entire tiny business around the fact that they are the sustainable guy or gal in the whole state, and so they have been poking, and prodding, and making sure that sustainability happens." These professionals are not just skilled at navigating the certification process; they also take on a normative role—as professionals from management consultants to diversity experts tend to do in other domains.[31] These professionals become a driving force behind new accreditation, as I learned from my USGBC source: "At every scale, when you have companies that have invested in a new way to deliver their services, and they see that they can be lucrative [with green construction], they are some of our biggest advocates."

Architecture studios specialized in sustainable construction have played, as one USGBC officer stressed, a "huge role." This activity has not been limited only to the large, international design power houses like of Gensler, HOK, or DPR Architecture. "Of course they're involved. They're the ones that go to their clients and say we can give you the best building for your money, and we can guarantee you that you'll earn at least LEED Silver certification because we know how to do that. . . . We built our business around it." These companies are persuasive, as in the case of one company that worked on a building owned by Donald Trump, whose disdain for climate action was evident. "He was not interested in any of this stuff, but guess what, we did it for him anyway. . . . We sneaked green stuff into his buildings, right?" Even when sustainability is not a primary consideration of project owners, these days, architects and construction companies may make the building green just by default. These firms, my informant explained, "are weaving it in."

The taken-for-granted culture of sustainability in construction was so profound in some places, that it spilled over to the least likely sites of green buildings. For instance, a policy expert at the USGBC told me about a forest products and paper manufacturing company in Atlanta, Georgia: "They're not the company that you would expect to pursue green building certification or LEED. But they decided for their headquarters building, for their own reasons, they would pursue LEED certification. . . . They felt like it was the right thing to do to be able to tell their shareholders and their community that they are an environmental company. . . . You know, what they do is chop down trees, right?"

Today, the branches of banks, hotels, and big-box retailers are among the top adopters of green construction, often applying for dozens of certifications at a time. The program that allows for mass certification is called LEED Volume and was started in collaboration with Starbucks, which is headquartered in one of the leading LEED cities: Seattle. This is not a coincidence, as the majority of these "batch adopters" have their headquarters located in cities that were on the cutting edge of green construction. This is precisely why the focus on post-2020 buildings with a LEED plaque—looking at a past process through the lens of the present—does not tell the full story.

Professionals put green construction on the map, and they helped make it a staple of contemporary construction practices. But it would be a mistake to rest our theoretical laurels on this very well-established explanation. The USGBC's ambition was not to ignite green construction on its own, but to use the "initial sparks" from local initiatives and to kindle a flame that could be seen far and wide: "USGBC has been founded on identifying, celebrating, and pushing leaders forward. . . . What we've discovered in our experience is that there are leaders on any side of the table or any side of this equation that are oftentimes acting collaboratively who are some of the *initial sparks* to make this happen."

At a later stage, the availability of environmental consultants, entrepreneurs, and industry specialists is equally important in explaining the rise of green building.[32] Yet the focus on market forces and its institutional backbone does not explain why some cities embrace green construction quicker and more thoroughly than others. Professionals and major corporations work nationally, and their influence increased over time. What *does* explain initial uptake?

For answers, we need to take a closer look at which buildings were built and when. The right-hand panel of figure 5.5 shows the cumulative numbers of all green buildings by sector of their project owners. These patterns suggest that it is primarily corporations and, to some extent, homes that have driven the success of LEED from 2006 onward.[33] Education, nonprofit, and government buildings made up only less than one-third of the total number of buildings by 2015. The number of buildings constructed directly by municipal governments pales in comparison to the number of retail banks, grocery stores, and corporate offices that have been certified by the end of the diffusion process. Residences certified with the LEED Home label account for a significant part of the wave, although the protocol was introduced only in the late 2000s. By then, cities had already baked LEED into building codes, architecture studios and construction companies had specialized in the practice, and thousands of LEED APs around the country had become cheerleaders who were paying attention to the carbon footprint of new construction well beyond the regulatory caps established by state and federal governments. In trying to understand the origins of LEED, however, this big

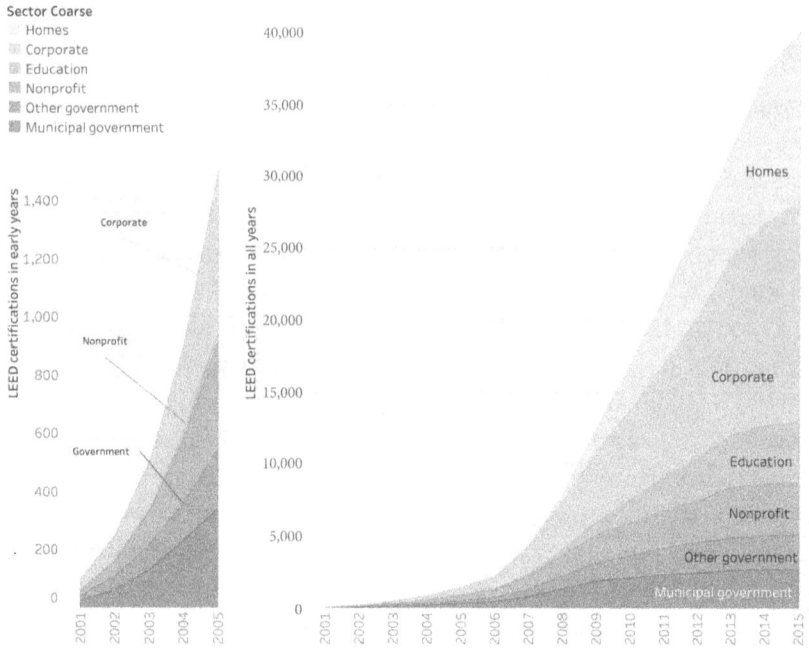

Sector Coarse
Homes
Corporate
Education
Nonprofit
Other government
Municipal government

(Left chart, LEED certifications in early years)
1,400
Corporate
1,200
1,000
Nonprofit
800
600
Government
400
200
0
2001 2002 2003 2004 2005

(Right chart, LEED certifications in all years)
40,000
35,000
30,000 — Homes
25,000
20,000
Corporate
15,000
Education
10,000
Nonprofit
5,000 — Other government
Municipal government
0
2001 2002 2003 2004 2005 2006 2007 2008 2009 2010 2011 2012 2013 2014 2015

FIGURE 5.5 Cumulative LEED registrations by year and sector, 2000–2015.
Source: Data from USGBC Project Directory, excluding buildings that did not list a sector.

picture is misleading. Looking at early patterns is essential because, in some important respects, they were similar to patterns that appeared in later periods and, in others, they were different.

First, in the early periods of LEED, it was not so clear that there would be clear payoffs to adopting LEED. For one, LEED was not widely known, as we learned with respect to the organization's continuously strengthening brand. Moreover, there was an ongoing debate about the actual cost and energy saving of LEED buildings. Economists have used regression discontinuity designs to examine whether buildings that are just above the point cutoff for high-level certification have a meaningfully different performance than those that fall just below that threshold.[34] Unsurprisingly, there was no big difference. Nevertheless, constructing a building that is eligible for LEED in the first place is a meaningful accomplishment that contributes to the sustainability transition. After two decades of evaluation, there is no doubt that LEED buildings overall perform at an environmentally superior level than buildings that are not certified; LEED-certified buildings use one-third less energy than noncertified buildings.[35] At the time, the question of whether certification would have a meaningful impact on green building was entirely unresolved. Although early adopters were important markers of "initial sparks," they were not sure what they were getting into. Adopters who came onboard after the financial crisis but before the big rush on green construction had to have other good reasons to pursue certification, especially considering that it cost money, and more rigorous construction standards also could increase the costs of a building quite substantially.[36]

Especially among early adopters, normative considerations played a crucial role in motivating certification, as one of my informants claimed: "What maybe drove it was really a statement of commitment to addressing environmental issues, but then close behind that was the benefits that accrue to the owner, both in terms of constructing the building and in terms of reducing costs of operation." These reasons are consistent with market research. According to a survey by the construction analytics company Dodge Data and Analytics, about three in four construction professionals said that both cost savings and encouraging sustainable business practices were key motives for green construction.[37] In other words, in addition to financial incentives, organizations can signal a concern for sustainability by building green.

Second, the first five years of LEED have a prognostic quality for the long-term trajectory even though who exactly adopted these standards was quite different from what emerged in the overall pattern. As the map in figure 5.3 shows, these early patterns solidified and became institutionalized, explaining geographic patterns down the road. For instance, cities with early LEED buildings by universities also saw a significant uptake of corporate LEED buildings later on. To understand how LEED became a legitimate solution in the construction business, it is important to wind back the clock and take a good look at what happened in the first years of the protocol.

The organizations that were responsible for the emergence of green construction in US cities were quite different from the ones we observed as being the most common adopters in the late stage of diffusion. The left-hand panel of figure 5.5 shows that nonprofits and municipal governments played a crucial role in getting LEED off the ground in many places. Environmental leaders such as the California Academy of Sciences were among the first adopters in the San Francisco Bay Area. This effect was not solely due to environmental organizations advocating for change. As we learned at the start of this chapter, librarians were among the early sustainability champions in San Mateo County. In Chicago, I learned about the interesting case of Access Living, a disability rights organization that sought to demonstrate that a high level of accessibility was possible in a modern building; sustainability became part of the package. Organizations in all domains were involved in doing the right thing, thus creating powerful proofs of concept that a better built environment was possible.

The reason that nonprofits and foundations were among the early adopters was that green buildings reflected their missions, not unlike the initial leadership of city administrations. A USGBC director confirmed the important role of mission-driven organizations in the early days of green construction: "Most of the early adopters of LEED were local government, or nonprofits, or foundations. First of all, this is their statement of their commitment to addressing environmental impacts." According to this director, the ethical consideration related to sustainability was particularly salient among nonprofit organizations:

> Many nonprofit organizations are founded around aspirational goals and a mission, right? . . . This is their statement of their commitment to addressing environmental impacts. Nonprofits have that mission orientation, and they need to attract the kinds of people that would be potentially more interested in working in those environments. Because of those missions, it's not just about the people or the mission itself, but it's about the fact that at the end of the year, I have to hand out this report about my sustainability leadership. A green building is a pretty simple way to do that. There's a lot of complex ways you can build a greener company, but a green building is probably the simplest way to add to that credential. LEED has become such an efficient way of communicating it that it is kind of is a no-brainer for them.

Differences in organizations with such environmental values are among the reasons for intercity differences, for instance, in the case of Washington, D.C. "That certainly is a piece of why the District of Columbia has more LEED-certified buildings per resident than any other place in the world." Together with stiff competition among commercial real estate companies, the presence of nonprofit and public organizations makes Washington, D.C., the "LEED capital of the world."

The building records back up this hunch that nonprofits were on the bleeding edge of green construction. Figure 5.6 shows the relative share of new registrations by sector: To what extent were public, nonprofit, and corporate projects responsible for the origin and uptick of green buildings? The downward trajectory of organizations in the nonprofit sector—particularly among leading cities that already had LEED buildings in the initial period—suggests that nonprofit organizations played an important role as pioneers that waned over time. Nonprofit organizations are the only project owners whose relative number of registrations declined over time; most of the growth in later years can be attributed to homes and corporations.

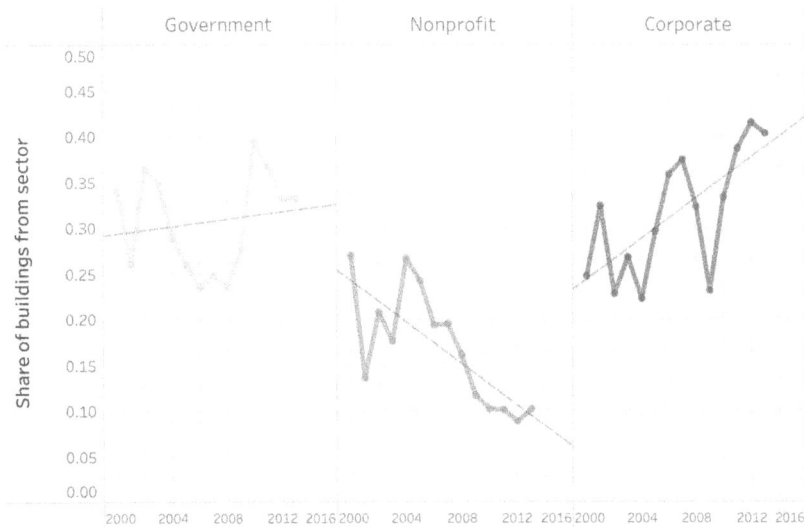

FIGURE 5.6 Relative share of new LEED registrations by sector, 2001–2015.
Note: Trendlines using ordinal least squares (OLS) regression.

This early onset of green construction in the nonprofit sector is one reason for the overall effect of civic capacity on the timing and extent of green building shown in the statistical analyses. In major cities such as Chicago, Cincinnati, and San Francisco, the initial push toward green buildings in the early 2000s was led by museums, laboratories, and foundations. Many governments buildings—local, state, and federal—are included among the initial adopters of LEED certifications. The green building stock later expanded to include shared office buildings, apartment complexes, and retail outlets, motivated by the dual benefits of cost savings and legitimacy associated with energy-efficient buildings. The critical role of nonprofit organizations in advancing green building practices was evident across the country, even amidst the adoption of state and local regulations aimed at enhancing construction standards.

The world's first LEED Platinum building, which opened in 2001, was located on the East Coast: the Philip Merrill Environmental Center, which houses the Chesapeake Bay Foundation. Foundations have also played an important role elsewhere. For example, the Hewlett Foundation, located in Palo Alto, California, spearheaded LEED in the private foundation world to be "consistent with the environmental grant portfolio at the time," as a board member at the time recalled. One thing these buildings had in common with the San Mateo Public Library was that both buildings had an exhibition to show off the materials and technologies used in the construction. In other words, the buildings were not only early successes that reflected well on the project owners but also turned the owners into champions of green construction. And as we learned, others followed suit.

These early buildings mattered because, through them, mission-oriented organizations and local governments created "demonstration projects" for others. The limited extent of new construction indicated that although leading by example did not flip the norms of construction, it did show the feasibility and desirability of a new norm, as one USGBC employee explained: "The idea of a city or a county or a state having a demonstration project is similar to why a company would want to certify its headquarters or certify its next building. I mean, there are very few companies that certify all of their buildings, right? That's very rare. It's very common for a company to say we have one and we're proud of it, right? Governments are another player that has a set of audiences that it wants to communicate that they're doing the right thing."

The idea of such demonstration projects is common. For instance, with funding from the Danish Industry Foundation, Copenhagen boasts a showroom for green solutions (the "House of Green") that according to their website has attracted 1,500 delegations since 2008.[38] In 2003, the Chicago Center for Green Technology was one of the first attempts to create a municipal building that was LEED Platinum–certified by the USGBC. The center featured a series of cutting-edge technologies, including an advanced heat pump system and automated HVAC systems. One local representative of the Illinois Green Alliance recounts:

> There were a set of these professionals who were really interested in LEED, and they actually went to the city and pitched the idea: Give us this building that you own, let us do something really cool with it. . . . It's an example of where there's a physical space that you could point to. People gave tours of it. Other professionals came to learn. There were education sessions held there in the space, so it became like a physical manifestation of being able to tell this story and get others interested wanting to do LEED.

As I discussed in an article for the *Stanford Social Innovation Review*, creating trailblazing green and net-zero buildings provides valuable evidence that new practices and policies can be successful, which can then be used to persuade others to adopt similar approaches.[39] Even when they are not immediately transformative, early initiatives can serve as prototypes for more significant systemic change. By focusing on achievable goals and demonstrating their effectiveness, small wins can build momentum and support for large-scale efforts.

Looking at different stages of the diffusion curve is important because the early stage of adoption is often different from later stages. A seminal study by institutionalists Pam Tolbert and Lynne Zucker showed that the initial adoption of a practice—in their case, the manager-council structure discussed in chapter 4—was determined by the properties of the city only at first.[40] Once the system had become institutionalized by consultants and think tanks, the characteristics that determined initial susceptibility to the innovative practice became less important. The example of civil service reform is often interpreted as the first documented example of a two-stage diffusion process in which initial uptake is driven by functional factors—like the need to curb corruption in large cities—and later replaced by legitimacy concerns—like nationwide associations beating the drum for reform and widespread copying of the leaders.[41] In this case, different organizations characterize different phases of adoption. In the first stage, in particular, the share of nonprofit and public organizations adopting LEED exceeded that of corporate adoptions. The full story became evident, however, only after I looked at how these pillars of the local organizational infrastructure interacted with the efforts of the city administrations.

LEGITIMATION THROUGH PUBLIC POLICY

The fact that municipal governments have considerable leeway to mitigate climate change through their building codes is not only a consensus among the IPCC climate scientists and legal scholars but also a major selling point for the USGBC. Many of these policies directly mention LEED in addition to, on occasion, other certification schemes such as the US Department of Energy and Environmental Protection Agency's EnergyStar, the British Building Research Establishment's BREEAM, or its North American counterpart, the Green Building Initiative's Green Globes.[42]

How do these policies come about? In some cases, the USGBC has a finger in the pie. "We've helped many of them make the case. That's . . . ," the policy director interrupted himself, looking at his office phone. "My phone here is blinking, so maybe that's somebody else calling for that kind of advice!" How often was the USGBC involved? "I'd say probably about 50 percent we were involved in in some way. But that also means that there were half of them that just happened, which is amazing, right? Because there was a very active company, a very active lawmaker, and people just wanted stuff to happen, which is really fantastic, and it's extraordinarily time consuming."

The number and complexity of city administrations posed a particular problem to the USGBC, but cities also had appeal as partners. Cities are "their own complex animal, but they are just more likely to act." For the policy director, this emphasis on cities is partially political: "At the state level, you have a majority of individuals who are representing very rural perspectives, which do not tend to prioritize density and sustainability in the built environment the way that we do. . . . At the city level, everybody's heard of it. So you have a big advantage."

In 2016, LEED pioneered a certification scheme called LEED Cities that is not just about single buildings but instead seeks to provide insight and create cultural change for entire cities. Based on the STAR Communities rating system, LEED Cities is intended to drive a movement in support of local sustainability directors who need help to develop sustainability strategies and attract funds to implement them. "Some of our directors just copy and paste text from the certification into their grant proposals," the USGBC's vice president for LEED Cities explained. This certification process helps urban sustainability directors generate resources for their work rather than just gain recognition for past achievements. Despite this admirable effort to bring many hitherto left-behind cities into the fold, the direct work with cities is relatively recent. A policy director at the USGBC explained why the organization prioritized working with states at first: "For us, it's been more efficient to work at the state level, because if you can get the state to enable a policy or a program, it's just a more efficient place . . . you can work with one state of Texas rather than 100 cities from Texas."

Public policy is not only a prerogative of local governments. As discussed in chapter 4, local governments are often preempted by state and federal governments that treat cities as mere "creatures of the state." Federal government agencies were also among the main adopters of LEED. By the end of the Obama era, the General Services Administration (GSA) had become the single largest project owner in the LEED system, and the US Department of Defense was also among the major users. Needless to say, climate action does not always have the favor of federal politicians, which means that the movement for green buildings has required some active framing and reframing. Just like local sustainability officers engaged in reframing of their sustainability strategies, Pentagon officials had to justify their investment in green construction in terms that resonated with their constituencies. A USGBC officer recalls that, around 2010, the US Secretary of the Navy explained why the Pentagon needed

LEED in a keynote speech: "He explained that every time a military convoy needs to move, there is a certain risk that they could lose a vehicle and lose soldiers. The main thing that militaries around the world move is oil. Just to be able to keep the bases powered, right? So for them, they felt like sustainability was about saving lives. And so, at that conference the Secretary of the Navy committed all US Naval facilities to LEED Gold certification."

The federal government's contributions to green construction are volatile for other political reasons. In another instance, the GSA announced that their facilities could use Green Globes—which, according to a USGBC representative who was reluctant to say anything bad about a competitor, is a less rigorous standard relative to the more widely disseminated LEED certification. Many government partners ended up switching to Green Globes, which is easier to acquire. The representative explained: "You could think about this much more critically, but if there's an endorsement, that carries a lot of weight." This endorsement, however, came late in the game. Rather than providing a regulatory guideline for using LEED or another energy-efficiency standard, as was common elsewhere, the US federal government simply led by example.

State governments also play an important role in advancing building standards. For instance, California's CALGreen (the Green Buildings Standards Code) was the first statewide green building code in the United States. The prescriptive policy became active in 2009 and was updated in 2011. The law is, in some respects, more stringent than LEED and was argued to set buildings up for rating success.[43] Without doubt state and federal regulations played an important role in mainstreaming LEED and other green building standards—but as discussed in chapter 4, state policies are of limited use for explaining variations among cities within states.

What do city administrations do to accelerate the distributed diffusion process? A USGBC policy director distinguished among three types of intervention: The first type is leading by example (as illustrated in chapter 4). The other two types of interventions are policies, including measures that compel the private sector to build green. Structural incentives require a change to the permitting rules either by speeding it up or by changing the rules, in his words, "to build higher or build wider. You can change something that doesn't cost you any money. It's very valuable to you the developer, but to me the city, it hasn't had an impact. Those have been popular for that reason!" Other incentives come with actual investments, such as grant programs or a reduction of property taxes. For instance, the City of Chicago's 2011 Green Permit Program states that "projects striving for higher levels of LEED certification will receive their permits within 30 days and are eligible to receive a partial permit waiver up to $25,000."[44] Another way that policies compel the private sector to build green is by passing mandatory requirements. In 2003, the City of Dallas was one of the first cities to pass a policy, Resolution 03-067, that required "all new municipal buildings larger than 10,000 square feet be constructed to meet LEED Silver [and as of 2006, LEED Gold] certification standards."

To determine whether these policies matter, I examined the influence of all 428 municipal policies listed in the USGBC's public policy library through a statistical analysis that examined the differences between cities with and without policies—understood as laws and ordinances to incentivize or require LEED certification of newly constructed buildings. My analysis showed that LEED policies had quite an impact; cities with such a policy saw approximately eighteen more certified buildings per year than cities without such a policy. Of course, certain cities could be more likely to adopt a policy, maybe because they are more liberal, larger, or have greater state and civic capacity than other cities that did not have a policy. When I took these factors into account through a technique called propensity score matching, I still found a ten building per year difference. Yes, cities' policies do matter in getting green building off the ground, even after considering that not all cities are equally disposed to pass a policy.

The policy effect was expected, but how it came about in the first place was more of a surprise. My intuition was that, of course, cities passed public policies before individuals took to certifying their buildings. As the light-gray line of the diffusion of public policies among metropolitan areas in figure 5.4 shows, the uptake of green building *policies* was far slower and less penetrative than the certification itself. Most places already had plenty of green buildings by the time the local government incentivized or mandated green building. I thus looked at what predicted the likelihood of having a public policy. I considered how existing green building practices affected the propensity to have a policy. The number of existing buildings was a strong predictor of policies, even considering other factors. This result showed that public policies were not in place at the beginning of the turn to green construction—it was the early buildings that were, in fact, transformative. Consider Cincinnati, for instance. In 2006, the City of Cincinnati passed a tax abatement program to incentivize new construction to replicate early achievements in green construction of the University of Cincinnati and the Cincinnati Zoo. I also found that city governments in cities with high civic capacity were more likely to pass policies that legitimized novel practices. This finding indicated that early adopters played a more important role than previously considered—and they were a primary reason for *why* civic capacity mattered so significantly.

Thus, policies did not only create an economic incentive for companies to build green. Some actors—like local and state governments—have little choice to comply with requirements. More important, municipal policies also legitimated LEED. They put LEED on the radar of every single person interested in obtaining a building permit. This process of LEED becoming universally known and providing a general solution to a problem—whether or not it was the best solution—is what organizational sociologists understand as institutionalization, a process by which something becomes a norm or default whose adoption gives the adopter greater legitimacy in the eyes of others. With few exceptions, for example, being a modern organization requires that one has a website; if basic information cannot be found on the internet, the organization may be perceived as nefarious or poorly run.[45] Conversely, a city with a green building policy—and general sustainability strategy for that matter—is doing the obvious thing, whereas a city that has no such policy may raise eyebrows. It is at this final stage, with proofs of concept and legitimating policies in place, that the forces of scaling discussed earlier in this chapter come to be in full swing.

WHAT WE SAW

This chapter adds depth to the finding that civic capacity enables city climate action. In accordance with the findings from chapter 4, places with a more vibrant civil society tend to adopt administrative action like green building policies more extensively—and they do it sooner. My analyses showed that civic capacity did not only help with the diffusion among the LEED green building policies from one city to another, as a rationalized practice. Often the role of catalysts was taken on by civil society organizations, although local governments prone to experimentation also contributed.[46] Nonprofit organizations were not solely responsible for legitimating and scaling green construction. Professionals like LEED AP-certified architects established an industry standard and lobbied for change, and local governments played an important role by passing legitimating public policies. But these policies were often inspired by the proofs of concept created by early adopters. Nonprofit organizations contributed to the institutionalization of green buildings from the get-go by advocating for and implementing changes and thereby catalyzing practices whose payoffs were not yet clear.

This observation is remarkable for two reasons. First, sociologists typically view public policies as triggers of institutional change among private actors. In this case, catalytic organizations were active before and enabled public policies. Second, transformative city climate action features administrative and distributed action. Both were enabled by civic capacity, which is a particularly important aspect of the local organizational infrastructure. Taking into consideration distributed behaviors is crucial, in part because urban sustainability, unlike sustainability plans as a rationalized practice or the sustainability actions of a city administration, cannot simply be adopted by a city. The question of how and when cities are sustainable adds the aggregate to the "attributive dimension" of city action discussed in chapter 1. City action cannot be understood solely as the rationalized practices of a monolithic decision-maker; reducing city action to the city administration alone would be limiting. In the cities I studied, the locus of innovation shifted over time—as the general model in figure 5.7 shows.

Existing theories of administrative adoption highlighted how central authorities (like city administrations, but also bosses in a company) learn from others and adopt policies that then affect decentral organizations (like companies with construction projects, but also employees in said company). My theory of distributed adoption emphasizes the important role of decentral catalysts (like nonprofit organizations as early adopters of green building, but also employees with initiative)

T₁ Initiation phase **T₂ Scaling phase**

$\mathbf{T_1}$ Initiation phase $\mathbf{T_2}$ Scaling phase

Administrative adoption

Distributed adoption

Learn & model

Teach & advocate

Learn

Incentivize & coerce

Proselytize

Mimic

◇ Central authority (a)   ○ Decentral adopters (b)   ▢ Decentral catalysts (c)   ○ Not adopted   ● Adopted

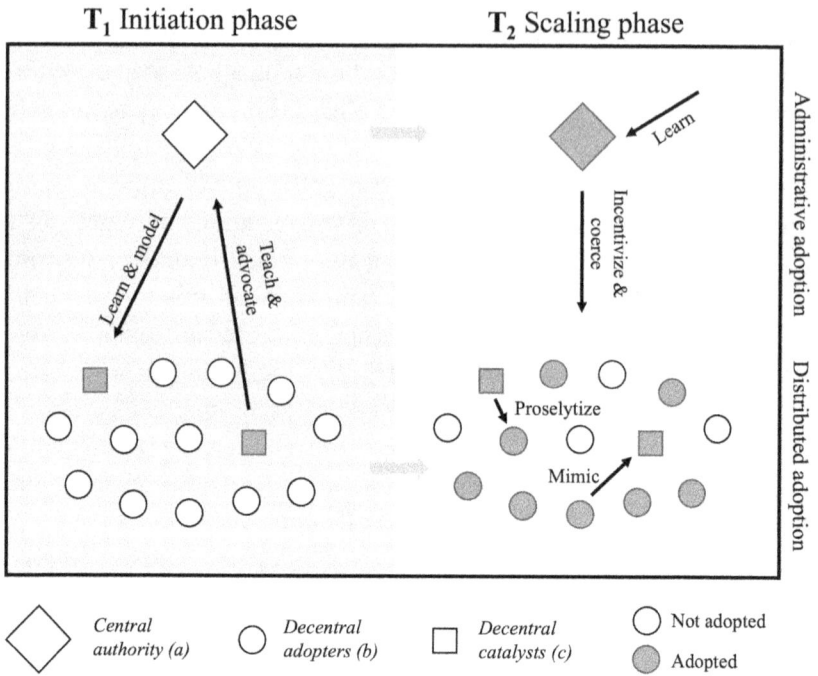

FIGURE 5.7 Schematic summarizing the general model of administrative and distributed adoption.
Source: Reprinted from Brandtner (2022).

that influence potential adopters and inspire central authorities to adopt formal policies by establishing proofs of concept. Following this process, nonprofit organizations played the most critical role in the initiation stage, and then receded in influence after public policy legitimated green construction and professionals helped scale it up.

Although corporations are responsible for the vast majority of total adoptions, most firms are followers.[47] City action involves urban governance more broadly, in which many stakeholders contribute to decision-making processes.[48] Although some have argued that local governments have a receding influence in these governance networks, they are crucial actors that legitimate private behaviors.[49] As I have shown, distributed adoption is an important component of city action in which decentral organizations inspire administrative action and ultimately aggregate to city action.

Conclusion

How can cities, as in the avant-garde architect Ron Herron's evocative image of the *Walking City*, move their inhabitants strategically toward sustainability and prosperity? I set out to explain why, over the past two decades, some cities have *acted* in response to the changing climate and others have *not acted*. Through my organizational lens, I explained how cities have engaged in strategic planning, have learned from and set examples in their peer city networks, have led local initiatives and demonstration projects, and have scaled individuals' and organizations' actions citywide. I also explained that cities act in response to social and environmental problems, such as climate change, based on whether they are enabled or inhibited by their organizational infrastructure and institutional superstructure. Figure C.1 summarizes my concluding arguments and their implications.

A city's ties to the *institutional superstructure* includes, in the case of climate change, a global network of associations that help cities collaborate with each other such as the Local Governments for Sustainability (ICLEI), the City Climate Leadership Group (C40), and the Resilient Cities Network (100RC). As I related in the first part of this book through a study of strategic planning and network memberships of the world's 360 largest cities, cities tied into the professional network of city associations more proactively create strategies to mitigate climate change,

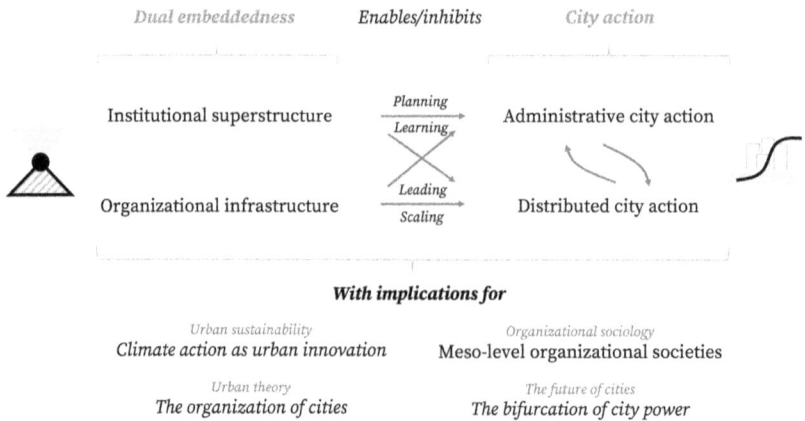

Dual embeddedness	Enables/inhibits	City action
Institutional superstructure	Planning / Learning	Administrative city action
Organizational infrastructure	Leading / Scaling	Distributed city action

With implications for

Urban sustainability	Organizational sociology
Climate action as urban innovation	Meso-level organizational societies
Urban theory	The future of cities
The organization of cities	The bifurcation of city power

FIGURE C.1 Visual summary of the conclusion.

adapt to it, and build resilience. These cities *plan* because the network has championed sustainability as a core professional responsibility of urban managers and planners, and because participating in the network has created collaborative ties among sustainability officers. Worldwide, cities that are more central in that professional network are substantially more likely to have a strategic plan that addresses climate change and that lays out a vision beyond what the city must do to include what it strives to do as well as steps to achieve those goals.

This is not to deny that some cities "punch above their weight" because of especially persuasive local advocates, or because they were lucky to receive funding from external supporters like the 100RC program or the American Cities Climate Challenge. Grit and luck, too, however, are never distributed quite at random.[1] Proactive cities like New York City and Copenhagen were founders of and early participants in these intercity associations, which later began to exert pressure on laggard cities like Nashville, Manila, or Warsaw. Moreover, the data show that a city's position in the intercity network shapes the boldness of their strategic planning. Generally, cities that emphasize climate change in their strategies tend to be part of a cluster of associations with dense social interactions that foster learning and promote competition for prestige.

Cities' membership in the global network is not just discursive or aspirational, nor does it merely virtue signaling or greenwashing. An analysis of sustainability practices among a cross-section of cities in the United States shows that cities that are members of a professional intercity association exhibit more sustainability practices, which frequently include but are never limited to strategic planning. This analysis also shows how institutions consistently affect the actions of individual people and organizations who step up to make cities greener. The institutional superstructure shapes norms among the elected officials and professionals who run contemporary cities and supercharges the diffusion of practices and policies, such as sustainability strategies and green building regulations. These practices and policies can be difficult to put in place when the roof is on fire, but they have meaningful consequences for the city's contributions to mitigation, capacity to adapt, and resilience to climate change now and down the road.

The city's *organizational infrastructure* is how local civil society organizations empower cities to act through aggregation. As I demonstrate in the second part of this book through an analysis of administrative practices and policies intended to advance green construction in the United States, city climate action is associated with greater civic capacity—the presence of nonprofit organizations above what is typical for the city's status, politics, and region. This relationship is both true for administrative actions (strategic plans and public policies) as well as for distributed action (individual organizations and people taking action that aggregates to cities). That is, civic capacity has a substantial effect on city action. An analysis of green construction demonstrates that nonprofits matter to cities both directly and indirectly: They are among the early adopters of green practices, they advocate for policy change, and they create positive spillover effects to cities that lack proactive administrations.

The focus on civil society does not mean we should gloss over the critical role of politics and a specific city's history and political economy. Based on a study of more than one thousand US city administrations, cities whose population leans progressive have a higher number of sustainability-related policies, administrative practices, and private sustainability actions. But the effect of politics is not explained solely by partisanship nor by an elected mayor's ambition. In fact, cities led by an appointed city manager rather than an elected mayor are more proactive. A strong organizational infrastructure attracts ideas to the city, and city administrations can encourage, amplify, and learn from local organizations' initiatives.

Together, the two sides of the *dual-embeddedness* model introduced in chapter 1 provide the underpinnings for my cohesive theoretical perspective on city action: that it is enabled or inhibited by a city's organizational infrastructure, by its embeddedness in the institutional superstructure, and by the interaction between these two structural features. My framework is not just theoretical but also substantiated by crunching the numbers on large comparative datasets of cities and interviewing remarkable experts on the ground about their experiences, strategies, and concerns. Thanks to this evidentiary basis, we now know how cities—from global metropoles to small places in rural areas, around the world and in the United States, with limited resources or supported by some of the largest foundations in the world—have acted on climate change as well as what features of their infra- and superstructures have enabled them to act.

What enables cities to act is a combination of three ingredients: First, this action is supported by the remarkable effort of urban sustainability officers, planners, and other professionals at planning and articulating the benefits of addressing climate change; second, a strong ecosystem of civil society organizations has been able to identify and mobilize support to resolve locally relevant social environmental problems; and third, receptivity to a globally extensive but locally oriented network of associations has filled knowledge gaps, provided scarce resources to cities, and created a cooperative network among bureaucrats. Even in places with strong political headwinds, cities have picked up the baton—and although political will and mayoral support matter, they are not sufficient conditions for success. Many cities fail to act because they are inhibited by a lack of administrative capacity, a crumbling local organizational infrastructure, and isolation from the intercity network that helps cities pay their due.

These findings offer insight and pose exciting new questions about city climate action related to four domains: urban innovation, the organizational constitution of cities, the meso-foundation of organizational society, and the social underpinnings of city power. I next lay out the implications of my findings for those interested in climate action, urban theory, organizational sociology, and the future of cities. Textbox C.1 summarizes the core insight for practitioners in a hurry.

TEXTBOX C.1: YOU RUN A CITY. WHAT ARE YOU TO DO?

What people who want to see their city "in action" should do depends on the city. A common barrier to city action is a lack of staff capacity and resources. If there is one thing that a city manager or mayor should do, it is to find the funds for one full-time staff person—for example, in the role of a sustainability officer—to tap into the knowledgeable network of associations that support local governments. Without someone who can apply for grants and programs, a city will be left out of the rich support network that promotes city action. Building capacity around collaboration and networking is crucial—one chief resilience officer (CRO) tongue-in-cheek called their role in the senior administrative team the "foreign minister." Sustainability officers also scout innovative solutions—within the administration, in the local ecosystem, and elsewhere—and can propose helpful legislation to amplify and scale those solutions. They serve as ambassadors from the broader networks to the city administration and the city's residents, and they reframe politically contentious issues for a variety of groups within the city, articulating the benefits city residents will enjoy by taking climate change issues seriously.

If a city already has adequate staff time devoted to climate change action, financially supporting organizations in the local ecosystem who otherwise have little hope of funding is a promising path to catalyze urban innovation. Do not require local grantees to write lengthy reports but allow them to showcase a solution that they wish others implemented and be prepared to replicate their successes in the city's operations. Organizing regular town halls where city officers and nonprofit leaders exchange their insight can build lasting "synapses" for city learning.[2] Investments in social infrastructure and civic capacity, such as offering small grants for climate solutions, providing general operating grants to nonprofit organizations that commit to contributing to local issues, and bolstering the local library—are likely more cost-effective in the long term than spending money on physical infrastructure. Although a sea wall or levy may keep disaster from striking, and is potentially necessary, *civic capacity* is the most effective insurance against all kinds of unanticipated shocks—it is what makes a city truly resilient.

CLIMATE ACTION AS URBAN INNOVATION

What role do cities play in addressing climate change? The most common answer—perhaps after a resigned "none"—is that they can adopt experimental technologies that will reduce or even remove carbon in the atmosphere. Tech-solutionist thinking is all around us, from cutting-edge urban planning consultancies to the airport bookstore. To take one popular version of this argument, philanthropist Bill Gates's *How to Avoid a Climate Disaster* argues that reducing carbon-emitting practices is not the primary path to carbon neutrality; new technologies like carbon capture or green concrete are the way to reduce the world's carbon footprint.[3] Urbanists have also toyed with the idea of engineering cities that are energy-efficient by design. These utopian models are darlings of the media: Eco-City Tianjin, a sustainable city built from scratch by the Singaporean government in China; the Line, a horizontal skyscraper with a hyperloop in the desert of Saudi Arabia; and Google's smart city designs for Toronto's waterfront.

It does not take the sharp eye of urban writer and activist Jane Jacobs to realize that grand technological designs for cities to become more sustainable are myopic at best.[4] The most germane reason is that cities are not just grids of roads with buildings in between, but rather are dynamic communities of organizations and people whose interactions create a compelling and complex experience for city dwellers. What I have shown is that city climate action is fueled not by the lofty aspirations of the wealthy and their dreams of technical solutions, but instead by the inspired activities of local and global civil society leaders, urban planners, and city strategists dedicated to connecting cities to each other. Like climate change itself, city climate action is not merely a technological problem, and we cannot simply engineer our way out of it without first attending to the social and organizational aspects. If scholars of cities and climate change take away one point from this book, it should be that city climate action—and urban sustainability broadly—is a social and organizational problem, which has three key implications for research.

The first implication is to take a broad view of what counts as city climate action rather than getting mired in debates about the nuances of adaptation, mitigation, smartness, or resilience. Actions furthering the sustainability of cities in a changing climate are not always directly motivated by climate change per se—environmental degradation and pollution, environmental injustice, and the threat of natural disasters are equally valid motivations. Discarding these facets as unrelated would be short-sighted. Sustainability is socially constructed as whatever city dwellers and managers understand to be the ability of cities to meet the needs of the current generation without compromising those of future generations. This is not a fixed definition of the sort we would need to optimize an engineering system.

Since I began working on this book, some of my respondents completely shifted how they discussed responsiveness to environmental change from a sustainability framework to a resilience framework. The transition from one framework to another, as I discovered in chapter 3, served to create a broader umbrella for climate change–related work so that the politicians and bureaucrats of conservative cities could frame their priorities and their actions as not directly related to climate change. A senior analyst who works at a think tank involved in the evaluation of urban sustainability programs suggested *not* measuring resilience because once money for resilience is on the table, "people are doing exactly the same work they did before, but now they are calling it resilience." Several other informants agreed that the conceptual boundaries of sustainability and resilience are permeable. One chief resilience officer (CRO) underscored this point effectively when he explained that, in the climate space, you just have to "deal with" buzzwords that come and go. "What is important is that we actually are doing holistic climate planning around both greenhouse gas emissions and adaptation. It's not one or the other, and it's not like you've given up on mitigation if you're doing adaptation."

Sustainability is complex and multifaceted, by my understanding and that of many of the sustainability officers with whom I spoke, and sometimes elements of this broad concept come into tension with each other. For example, the most effective decarbonization strategy may not be the most democratic or equitable one, like in certain forms of carbon pricing or low-emission zones. Contemporary strategic plans have focused more on equity than pure emissions reduction, covering topics such as the unequal impacts of environmental degradation, energy efficiency as a reason to build social housing, and the involvement of hard-to-reach groups in planning processes. The CRO of a university town noted that sustainability directors have recently "gotten far more training in racial inequality" and that the COVID-19 pandemic underscored that their work is "intrinsically tied to equity."

Because the concept encompasses emissions reduction, emergency preparedness, and equity, urban sustainability is a moving target whose meaning morphs over time—and that's the point.[5] Lasting systems that enable urban innovation are more important than redefining how we measure carbon emissions, solving whatever other short-sighted technical issue, or debating what counts toward urban sustainability and city climate action. Instead of getting caught in a debate about the nuances of sustainability, I advocate for a return to the intergenerational understanding of sustainability. The 1987 Brundtland report—in many ways, the origin of the modern sustainability movement—defines sustainability as the ability to serve the current population without compromising the ability to do so for future generations, with regard to population growth, economic growth, and environmental protection.[6] Cities continue to face persistent problems such as economic and racial inequality, which cut deep into the physical layout of cities and the challenge of planning, as well as political headwinds. Cities must dynamically respond to a changing world, but their capacity for doing so varies greatly; naming, understanding, and overcoming this disparity is the research frontier.

The second implication of city climate action's social and organizational dimensions is that researchers need to shift toward a more constructive debate about between-city disparities in urban innovation more generally. Sociologists have traditionally been concerned with inequality *in* cities, for instance, as a result of segregation or redlining, and environmental sociologists have rightly emphasized spatial inequalities within cities—segregated neighborhoods that lack food or medical care, an uptick in environmental gentrification, lack of access to parks and other natural amenities, disproportionate exposure to hazardous sites, and uneven vulnerability to climate disasters. There is no doubt that urban sustainability is centrally related to these issues, but inequality rarely came up in my interviews that took place before the murder of George Floyd. After the George Floyd protests of 2020 that shone a spotlight on the continued racial segregation in US cities and the lived experience of Black Americans, issues related to equality and equity came up more frequently, despite the fact that the majority of my respondents were elites. For instance, the sustainability officer of a liberal city in the US Midwest spoke of the opportunities to rectify institutional racism as a main motivation of his work. Another officer from the Great Lakes region mentioned the necessity of considering the interests of low-income residents in every decision. Issues of environmental inequality arose not only in US-based interviews but also in interviews in Lyon, Hong Kong, and Vienna, especially related to the affordability of sustainable cities and the unjust burden of climate change in the Global South. Concerns about intercity inequity notwithstanding, I want to draw attention to a form of inequality that is difficult to see from a single town square—inequality among continents, countries, and cities with respect to access to renewable energy, soft transportation, green and affordable housing, organic recycling, and other ingredients essential to a healthy and sustainable lifestyle. That is, inequalities *between* cities are another source of disparities between people within individual countries and, of course, the world. Low-income residents of El Paso or Manila will fare far worse than low-income residents of Pasadena or Toronto. In the social sciences of the West, such between-city inequalities have received dramatically less attention than within-city inequalities.[7]

The dual embeddedness of cities explains between-city inequalities: why some cities take action, develop their capacities to engage in innovative practices, and are at least somewhat responsive to emergent knowledge in their institutional environment. Private think tanks and transnational agencies dedicate significant amounts of time and energy to developing and accumulating knowledge about how to deal with problems that have not yet entered the public discourse. Cities, too, proselytize this insight to one another. It stands to reason that there is no shortage of solutions for the manifold problems that cities face. But there are serious limitations regarding the extent to which cities learn about, translate, and implement these solutions on an ongoing basis. Without thoughtful designs and deliberations for a "just transition," popular solutions to climate change may reify and reproduce environmental injustices. The unequal uptake of green innovation requires further study, for instance in the diffusion of environmental transition infrastructures such as photovoltaic panels or electric vehicle charging stations, access to which may not be available to all. The effect of markets of social innovation that are left to their own devices as well as the intended and unintended consequences of the "visible hand" that tries to usher adoption through place-based policies is an important frontier of scholarship. How to "unskew" climate action remains a major research frontier in policy and the social sciences alike.

I have shown that the uneven landscape of urban sustainability is the result of discrepancies in urban innovation—by which I mean new ways of doing things, regardless of their source and whether they are truly new to the world. Such innovations have included municipal identification cards for undocumented immigrants, opportunity zones to improve the life chances of youth in underprivileged neighborhoods, and regulations to protect employees and marginalized populations from displacement and precariousness in the sharing economy. Besides environmental impacts, such innovations are related to many outcomes that social scientists care deeply about: inclusion of immigrants, spatial segregation, the social mobility of the urban underclass, and the quality of jobs. Future work on the geography of urban innovation should investigate how organizational infrastructures affect the adoption and implementation of different forms of innovation, as developed and promoted by the public and private sectors. Moving knowledge forward may require researchers to collaborate in developing a city innovativeness index, mirroring the state innovativeness index that indicates early policy-adopting states.[8] A holistic city innovativeness index would provide insight into what enables urban innovation across the board, rather than treating different concepts such as smartness, competitiveness, inventiveness, and sustainability in isolation.[9]

The third implication of understanding city climate action as fundamentally a social and organizational problem is that cities' visions are worthy of consideration irrespective of whether these imagined futures always come to pass. Cities' strategies should be taken seriously not because of the steps they have already taken—which can be small—but because of their vision for the future—which is big.[10] From a climate science point of view, the elephant in the room is whether cities are meaningfully reducing carbon emissions to an extent that dodges the worst of climate change. Concerns about whether city plans are disconnected from real change are natural, considering that a generation of scholars has already shown how nations and firms decouple talk from action to comply with external expectations.[11] Similarly, urban planners are aware of the difficulties of actually implementing plans, many of which gather dust while on display in prominently located bookshelves. Yet efforts to create sustainability strategies have been crucial for identifying areas for bold action and structural interventions.[12] Even fads and fashions in city management can provide cities with role models for how to imagine and enact their futures. At their core, strategic plans assume that city leaders are capable of navigating the city and all its constituents to cope with highly complex problems. This agentic mindset is powerful. For example, work on corporate social responsibility has argued that the "aspirational talk" of firms can influence long-term attitudes so that initially decoupled corporate social responsibility policies (i.e., inconsistent talk and action because an organization placates external demands without changing its core activities) often become "recoupled," and organizations that employ euphemistic discourse without following through get into trouble.[13]

Whether, when, and how cities walk the talk is an important question. Cities with resilience plans and officers responsible for executing them create administrative capacities that enable more aspirational visions to mitigate climate change and the overall way in which city officials in all policy areas think and talk about climate change. For example, San Francisco's recurrent capital plan—which outlines the city's major investments for years to come—is likely more consequential than the city's resilience plan. But the CRO I interviewed *was* the person in charge of producing the capital plan, and thus wove resilience into the capital plan. Whereas climate and sustainability strategies are not always implemented, as I argued in chapter 2, the city's capital plan is legally binding. Although what urban planners refer to as an "implementation gap" is worthy of scholarly and public scrutiny, I suspect that future iterations of San Francisco's planning tool—which offers a blueprint for legally binding budget decisions—will continue to incorporate ideas and projects from the resilience plan. Understanding the pathways through which aspirational talk succeeds or fails to translate into meaningful outcomes remains a fruitful field of research. Researchers should pay less attention to the plans and more attention to the *planners*.[14]

One particular process that deserves scrutiny is how cities with solid process in place "ratchet up" their goals, as one sustainability officer in the Midwest called it. "The goals are so big—it's not like we're going to meet those 20 years ahead of schedule." But the next time the strategy gets revised, "a lot of things will feel possible that didn't feel possible in 2018. We'll be able to ratchet up our goals and make it much more ambitious." In other words, a commitment to a greener city just five years earlier had already given rise to a bolder and more ambitious set of actions by the time we spoke. This "ratcheting up" from minor promises to major ones, combined with the fact that other communities and organizations may be inspired to do more by one city's actions, is the reason that I refrain from judgments about whether city climate action has amounted to "enough" decarbonization.[15] Evaluating progress based on cities' twenty-year-old goals likely undervalues their potential and sometimes unexpected consequences.

Yet official goals also are not always where the action is. The civil servants working in cities are deeply committed bureaucrats with a heart, whose pursuit of the betterment of cities is, of course, heavily constrained in the absence of resources, limited staff, the lack of buy-in from mayors, and ever-more demanding reporting requirements requiring technical expertise.[16] Although the civil servants with whom I spoke were committed to upholding their promises, frequent turnover in mayors and political appointees has reduced their ability to follow through on the plans. As such, cities' *formal* commitments may be insufficient to reverse the worst-case climate scenarios. Alternatively, the impactful diffusion of green construction suggests a case for soft power and voluntary but audited certification as options when policy prescriptions and price adjustments are not possible. Even limited administrative action early on can amount to significant progress being made, both because of distributed action among individuals and organizations in the city, such as the commitment to build an energy-efficient building and historical decisions, endowments, and experiences that constrain and encourage later ones, such as the city's civic capacity.[17] Cities like Vienna, for example, as portrayed in chapter 2, have moved from cobbling together existing projects to developing strategic milestones. The planners chose the United Nations (UN) Sustainable Development Goals (SDGs) as an overarching framework and went beyond these externally defined, boilerplate goals to develop Vienna-specific goals. Understanding the creation and use of soft power to forward sustainability strategies is an important challenge for future scholars of climate action.

THE ORGANIZATION OF CITIES

Urban scholars, and urban sociologists in particular, are often preoccupied by drastic poverty, racial discrimination, and incapable or corrupt local governments. Despite cities' deep-seated and well-documented problems, this dismal view of cities' reality would not be complete without a more hopeful one. Cities are a great invention; some of what ails them, such as construction, traffic, and social inequality, is an undeniable part of the package of innovation, connection, and growth. A shift of perspective (and the unit of analysis) from people and their neighborhoods to organizations and their ecosystems allows us to understand not only the urban underbelly but also the creation, diffusion, and implementation of solutions to cities' greatest problems.

Social problems are often solved through government command or market interventions, and so it is no wonder that urbanists primarily look through analytical lenses concerned with the politics and economics of cities.[18] But policy prescriptions and price adjustments are costly and whip up resistance, often to such a degree that technically viable solutions become politically untenable.[19] The economic and political fiat of cities—their ability to exert power or "dominate"—are not the only keys to the city. I suggest that the crucial and often-overlooked "third key" to solving urban social problems is the organization of cities.[20] By this, I do not only mean how city administrations are organized but also how the organizational infrastructure—including nonprofits, public entities, and companies—shapes the city. I first pull together the evidence for this claim in the context of civil society organizations, which have emerged as particular champions of progress in cities' organizational infrastructure, and then lay out its broader implications for urban theory.

Whether cities act effectively is not just a matter of state capacity or market capacity but also of *civic capacity*. Market perspectives that emphasize the economic interests of developers and other members of "growth coalitions," and political perspectives that emphasize the role of elites and interested coalitions, have dominated understandings of urban governance. The perspective that emphasizes how cities are made up of organizations that create a place-based organizational infrastructure is rarely spelled out in the urban literature. The spotlight on business and elites is not wrong, but it does not acknowledge two of the most important players in urban innovation: the civil servants who make up the city's administration and the local nonprofit leaders. Both of these groups try to prepare and influence mayors, bring the private sector on their side, and create knowledge systems like ratings and guidelines. I am not the first to highlight the role of organizations in communities or the importance of communities in organizations. But even after a decade of working on this book, I remain concerned with the critical lack of integration of the urban and organizational perspectives.[21]

When I started my inquiry, I did not know or expect the "nonprofit effect" on city climate action to be so consistently strong and pervasive. According to Kent Portney, one of the pioneers of studying sustainable urban planning, greener communities usually also see higher levels of civic participation.[22] Following his work, I expected environmental advocacy organizations to be the linchpin of urban sustainability. Such a finding would be consistent with work by urban sociologists on the decline in crime in US cities, which has been attributed to the difficult work of specialized community organizations.[23] I do not doubt the direct substantial effects of engaged activists and citizens, but my research reveals that other, more indirect mechanisms are also at work. I found two major indirect channels through which civil society organizations shape urban sustainability.

One channel is *associational* and is closely linked to what eminent urban sociologist Robert Sampson calls "collective efficacy."[24] Nonprofit organizations are relays of the idea that locales should take action. Responsible corporations and universities, too, can play roles in establishing an expectation of sustainable behavior—in part because they employ high-status and highly educated individuals who support such interventions. Of course, not every organization is eager for change, and some are openly or covertly custodians of the status quo. The main point is that it is not only interested advocacy organizations, community-based interest groups, and urban social movements that exert pressure on city administrations.[25] Rather, some everyday organizations—

nonprofits chief among them—are attuned to global discourses of responsibility and virtue, which they incorporate into their goals and relay to administrators.

Civil society does not determine a city's climate action but rather it *coevolves with* the actions of city administrations. For instance, a great deal of urban gardening initiatives and innovative green businesses have sprung up in Fort Collins, Colorado, and my hometown of Innsbruck, Austria, simply because these cities have a commitment to sustainability grounded in an outdoorsy population that is enthusiastic about nature and willing to protect it. It is both a conceptual and a methodological challenge to establish a unidirectional causal link between the activity of civil society and administrative behaviors across a large number of cities. Yet there is mounting evidence that nonprofit organizations and, more broadly, what urban sociologist Eric Klinenberg calls "social infrastructures" activate behaviors related to sustainability, from planning to specific projects to positive outcomes, such as community cohesion.[26] Civic capacity adds an important factor that empowers the creation of such social infrastructures that should be incorporated in future research and local policymaking.

The other channel through which civil society organizations shape cities is *professional.* City administrators have a strong professional ethos and training in policy and planning from schools that have been teaching and preaching sustainability for some time. Cities with managers rather than mayors are particularly susceptible to normative influences in their wider institutional environment. Several professional associations also specifically target city managers and administrators, and offer them resources for both collaboration and learning. Some, such as Sustainable San Mateo and SPUR in the San Francisco Bay Area, can be seen as "shadow cabinets" advising local leaders with general and place-specific expertise. More global associations, including C40, ICLEI, and 100RC, are relatively limited in scope but tend to have specific requirements for members—for instance, requiring that a city have a sustainability or resilience plan with specific elements such as attention to citizen inclusion and a risk assessment. Interviews revealed the strength of the personal networks between city administrators who are members of such associations, including group social nights, common WhatsApp channels, and webinars that they both give and attend. How to best design and sustain these networks to amplify their effect on the planning practices of their members is a question ripe for further investigation.

Civil society organizations are only one of many aspects of the city's organizational infrastructure that shape the urban environment. Like urban sociologist Robert Park's relational idea of the city's intermingling "web of life," I suggest that the interplay among the organizations within a city shapes urban practices, governance, and innovation. In 1925, Park wrote that formal organizations were "utilities, adventitious devices which become part of the living city only when, and in so far as, through use and wont they connect themselves, like a tool in the hand of man, with the vital forces resident in individuals and in the community."[27] Although this instrumental understanding of organizations as tools seems to be in sync with contemporary organizational research on strategic alliances, his century-old insight invites revisiting how cities are shaped by modern organizations—universities, social movement groups, parties, news outlets, consulting firms, rating agencies, social enterprises, start-ups, and tech firms. All these organizations have a lasting imprint on the places in which they are located and are more than mere derivatives of the urban environment.[28] I crack open the door to a more organizational view of cities, according to which important decisions in urban governance are a consequence of the presence, practices, and interlinkages of associational, commercial, and public organizations in a place. This perspective, in my view, holds incredible promise that has only partially been realized, and only in areas that have emphasized the ecology of particular commercial organizations for economic growth while ignoring other types of organizations.[29] My findings suggest several promising opportunities for future research, two of which I highlight.

The first opportunity is to take a closer look at the variety of practices that organizations employ, and the consequences of these practices for urban outcomes. Although research has shown generally positive effects of nonprofits on outcomes such as the diffusion of urban innovations, cities also shelter nonprofit organizations (and firms) that hold them back. For example, Patrick Sharkey and his colleagues have shown that crime declined in areas with a greater concentration of nonprofit organizations. Yet as urban sociologist Robert Vargas's *Wounded City* demonstrated, some nonprofits funnel resources away from the communities that need them most.[30] In the context of green construction, the inclusion of both tenant organizations that fight skyrocketing rents, on the one hand, and astroturf organizations that advocate for free-market policies and fossil fuels, on the other, in the overall measure of civic capacity may in fact attenuate the results of my research: Nondiscriminatory nonprofit counts are an imperfect albeit conservative proxy of cities' civic capacity because they likely underestimate the effect of various organizations that are actively working toward climate change. In this case, taking a page out of the book of organizational researchers shows great promise for the study of cities. Research should urgently move on from counts and densities and toward the study and understanding of *how* different nonprofits actually interact with the communities in which they are embedded. Whereas recent research has shown that civic opportunities are scarce in some cities and neighborhoods, we know less about what causes organizations to be disconnected from their urban environs and, as a result, why they fail to become an active component of the organizational infrastructure. Research on the sources and consequences of such "civic disembeddedness" is much needed to develop an organizational understanding of urban dynamics and strategies to enable nonprofit ecosystems.[31]

A second research opportunity is to investigate how nonprofits interact with or substitute for other types of organizations, such as public universities, government agencies, or firms. Much ink has been dedicated to understanding why nonprofit organizations exist in the first place and, as a result, why there is variation among cities in the size of their nonprofit sectors.[32] Earlier research by organizational sociologists has shown that the corporate demography—including the size and age of local companies—has consequences for labor market sorting, income inequality, and innovation. We know much less about how the mix of organizations with different missions and organizing principles shapes places. Although the effect of civic capacity I have demonstrated underscores the finding that variation in organizational infrastructures is consequential, there is much work left to do. For instance, some cities are relatively dominated by corporations, whereas others are shaped by other organizations, such as nonprofits, federal government agencies, or research universities. This implies important and measurable differences in the local governance regimes. Washington, D.C., for instance, has a far greater relative share of nonprofit organizations and government agencies than companies; tech centers like San Jose, Boston, and San Francisco are more heavily dominated by firms; many suburban towns are defined by research universities that account for a large share of employment. Considering that urban governance is crucially shaped by the "cast of actors" who participate in it, the organizational infrastructure of places—including its dispositions, practices, and interlinkages—is a crucial and undertheorized characteristic of places that likely shapes their character in important ways.[33]

CITIES AS MESO-LEVEL ORGANIZATIONAL SOCIETIES

As much as urban theory can benefit from organizational perspectives, studying cities also offers valuable insight for scholars of organizations. At the end of the twentieth century, sociologists like James Coleman in his *Asymmetric Society* and Charles Perrow in his *A Society of Organizations* made the startling observation that, in a world of ever-larger corporations embedded in a pervasively capitalist society, people increasingly became enveloped in large organizations.[34] As the polymath Herbert Simon famously wrote, a Martian visiting Earth might not so much see a market society but a society of organizations.[35] Organizations remain important to many people's daily lives throughout the life course—from preschools to hospices—but the organizational makeup of the twenty-first century differs from that of the twentieth century in several key ways.[36] First, a few superstar firms such as Google, Meta, and NVIDIA in tech or Berkshire Hathaway and Visa in finance control large chunks of the new economy while employing a fraction of what corporate giants such as GM or Walmart once did.[37] Second, corporations publicly embrace their social responsibility and see themselves increasingly as "corporate citizens," while an ever-growing number of nonprofit organizations have looked at market revenues and managerial expertise.[38] Third, in many parts of the world, the public sector has faced privatization, downsizing, and encroachment by markets in the face of illiberal backlash against civil society and higher education. As a result, governments and their agencies have, in general, become one of many participants in network governance—although in some countries, the resurgence of nationalist parties and the turn toward authoritarianism is changing the organizational landscape once again.[39]

Despite, or rather because of these changes, organizational perspectives are more necessary than ever. Greater attention to place-based communities such as cities can breathe new life into macro-organizational inquiries that attend to the structure of fields and networks of organizations.[40] Cities are not just one significant part of organizational society, but in fact constitute thousands of individual organizational societies. Comparing these places means that organizational scholars can test how differences in organizational composition and practice shape societal outcomes.

Cities should be a level of analysis in the toolkit of anyone seeking to produce meso-level organizational theory, whether in sociology, management, or other disciplines. The meso-level emphasizes the behaviors of discrete organizations and organizational infrastructures—neither individuals nor groups within organizations, but within the organization itself.[41] Organizations relay institutional macro-pressures, such as culture and structure, the economy, and so on, to individuals; they often do so through soft influence and their mere presence by attracting educated people or setting standards for legitimate behavior. Research linking insight from organizational analysis to the local governance of social issues has been rare.[42] The future of organizational theory arguably lies in the contributions it can make to the understanding of important sociological outcomes, such as citizen participation, corruption, discrimination against low-status groups, technological change, legal entrenchment, and human impacts on the natural environment. Many of these outcomes are deeply local and indeed constitutive of place-based communities.

Place-based organizational sociology can provide the meso-foundations for understanding how individual judgments and prejudices are connected to broader societal shifts and rifts. This approach requires examining how organizations and their practices shape macro-level institutions, such as laws, modes of production, and models of organizing.[43] It implores scholars to think not only about the micro-foundations of institutional explanations but also about how individual dynamics affect such meso-level structures as occupational prestige, roles, and

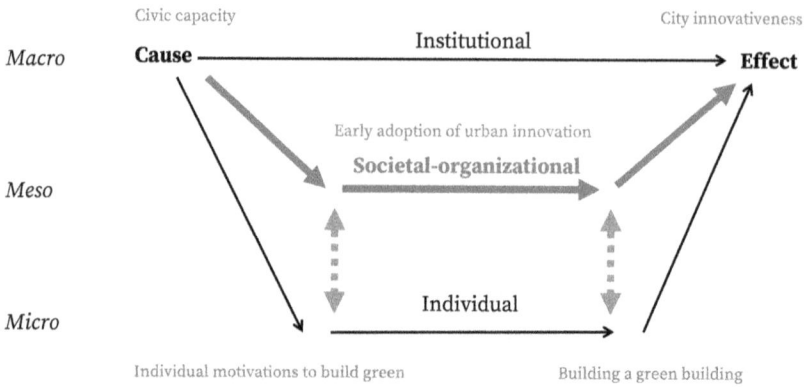

FIGURE C.2 Meso-foundations of social theory.
Source: Adapted from Jepperson and Meyer (2011) applied to the case of the diffusion of green building as a multilevel phenomenon developed in chapter 5 (gray text).

network formation. The dynamics connecting civic capacity to city action illustrate the societal-organizational connections between causes and effects, as figure C.2 illustrates. As organizational societies, cities fulfill Park's promise of showing "processes of civilization [. . .] under a microscope" and thereby offer a way to test theories about the consequences of organizations on institutions and individuals.[44]

Another affordance of studying place-based communities is to bring macro-level theories of organizational behavior onto the ground. Meso-level organizations are tied to macro-level social structures through institutions—taken-for-granted beliefs that establish society's rules of engagement. Institutional theory has lost traction as an explanation of social behavior in part because institutions are often located on such an abstract level that they are hard to track. Firms, for instance, respond to the expectations of industry-wide fields, and nation-states are responsive to global norms, as discussed in chapter 3. The study of cities offers another opportunity to revisit some of the most important aspects of institutions, including how they emerge, persist, and can be changed. Cities are where the rubber of institutions hits the literal and metaphorical road.

Yet urbanists have expressed some skepticism about the idea of applying institutional theory to urban outcomes—which outside of some creative analyses of global shopping streets and the study of diffusion of local government policies has rarely been done—because of the assumed expectation that shared institutions would lead to equal responses, and thus a kind of homogeneity among cities that empirically does not exist.[45] Despite urbanists' trepidation, I have found that organizational institutionalists have much to offer—and discuss with—scholars of urban governance. As my interviewees demonstrated, city professionals are not cultural dupes but are highly reflexive individuals who seek to actively shape their institutional environment.

My research shows what organization scholars call practice variation—variety in the depth and content of practices that have become universally seen as a gold standard. Strategic planning may be a common practice in contemporary cities, in part exactly because of the professional influences that institutionalists think of as normative pressures.[46] But the content and implementation of each plan depend heavily on the organization of each specific city, including its professionals and the "cast of actors" of urban governance. Furthermore, the wider context in which cities are embedded can change rapidly, both because new intermediaries enter the field (e.g., the Rockefeller Foundation's 100 Resilient Cities program inserting "resilience" into the urban sustainability debate) and because cities are embedded in the intercity network to different extents over time. As I have argued, the fact that organizational institutionalists have focused on the nation-state and international organizations to understand how global norms shape local outcomes is the result of data availability, especially around the time these arguments were first developed. But as I laid out in chapter 3, there is an important subnational dimension to global institutions that has evolved and is especially compelling as national support for liberal institutions crumbles in many parts of the world.[47] Urbanists and organizational scholars should work hand in hand to better understand how the newly emergent municipal world polity plays out on local levels of government.

THE BIFURCATION OF CITY POWER

Finally, the case of city climate action provides insight and raises new questions about the future of cities—insight and questions that cut across disciplinary literatures. The idea, perhaps even the ideal, of city action casts a new light on the relationship between cities and broader systems of local, regional, and national governments. Globalization, social inequality, and climate change have created mounting pressure to find solutions to tough problems like those undergirding the UN SDGs at a time when the concerted global effort to resolve them is under attack. With many countries experiencing political gridlock and polarization at the national level and the public's declining trust in liberal institutions, people from varied sectors—the media, the political and pundit class, the technology sector, and academia—have all called on cities to step up to the plate. So, can and should those who remain committed to climate action and social progress rely on cities to walk us through dire times?

Canonical scholarship on cities is skeptical about the prospect that cities can live up to these high expectations. For one, cities are starved for money and are constrained by state and national laws as well as the stifling influence of corporations.[48] Another problem is that how to address climate change in particular is not easily agreed on and resolved by *anyone*. Climate change is among a suite of social issues sometimes referred to as "grand challenges" or "wicked problems" because it requires the collective action of many public and private entities.[49] Although these problems can be overcome through collective action, it bears remembering that voters do not consider complex issues like climate change or segregation to be among the priorities of their city administrations, and nor do their elected officials. I center another problem, which is that cities are not monoliths but complex networks of organizations and people with conflicting interests.[50] Action often requires a higher level of coordination and unity among the many participants in urban governance than those involved can muster. Yet even though cities are stifled, starved, and split over strategic issues, I suggest that they are powerful agents of change—with three serious caveats that researchers and decision-makers alike should keep in mind.

The first caveat is that city action primarily operates through soft power. Discussions of "city power" in political science have emphasized municipal governments and the limits on their legislative hard power—of which cities do not have very much.[51] Yet regulatory constraints are secondary to how grassroots efforts create city power from below. In the context of climate action, governments have responded to proofs of concept that progressive adopters across all sectors have established in their communities. Future work on city power must recognize the question of civic capacity and the need for coordination across sectoral boundaries— such as collaborations with corporations or government agencies—and should investigate how cities interact through roundtables, contracts, or hybrid forms of organizing.[52] The carriers of soft power in my study include the city networks founded and funded by cities around the world, which are carriers of cultural expectations and therefore the hinge point at which cities' infra- and superstructures interact in important ways.[53] Studying this interaction in practice is an exciting opportunity.

I also show that political and professional networks intersect with cities' hard-wired positions in the global social order. Urbanists often think of cities as being more or less competitive, powerful, or high status, but what these dimensions *mean* for city action depends on cultural influences as well. Whether city networks will end up substituting for or complementing international politics, the emergence of a network among cities that affords learning, competition, and trust will have a lasting impact on the balance between nations and cities. The shifting rebalance between local, national, and global communities merits further research, for example with respect to public support for and resistance to community-based norms and local policies that may foster internal cohesion but also may exclude outsiders.[54] The formation of and domination exercised in city networks are also fruitful research areas because they are related to the rise and fall of political orders at the heart of political sociology.[55]

The second caveat is that the power of cities may stoke tension among political institutions at different levels of government. Cities' independent, bold actions may worsen the relationships between local, state, and national legislatures. In the United States, Republicans are giddy-eyed about the idea of sending national police forces into progressive cities to curb what they perceive as chaos, and the Trump administration followed suit. Governors of Southern states have bussed immigrants to progressive cities like New York, Chicago, and Denver, as a symbolic gesture that sanctuary cities, too, should share the burden of immigration. Examples from other parts of the world—like a Brexit vote against the interests of cosmopolitan Londoners or regions in Denmark or Northern Italy rejecting immigrants—abound. These tensions reflect the ideological differences in local and national politics but are not limited to mere party politics. They also result from a real lack of ideological alignment among federal, state, and local politics that political scientists have urged us to take seriously.[56]

To be sure, cities have not always been nor are they still hotbeds of freedom and equality. Earlier research on city power has accurately highlighted the many constraints that prevent cities from living up to idealized expectations. Many cities are coopted by interested companies or muzzled by states that have a historical tendency to preempt local autonomy. From a close-up view of the city's underbelly, segregation, discrimination, and corruption are so rampant that most cities are considered to be borderline dysfunctional. Studying elites and corporations, one cannot avoid the impression that a city's actions are but the extended arm of the rich and powerful. Outside influences on communities, whether globalization or state intervention, often exacerbate these inequalities. These issues are important, but resolving them may require looking beyond local power struggle.

The organizational lens affords a more optimistic take than one exclusively focused on city power in the sense of potentially coopted authorities, by understanding city power as *distributed* among a variety of organizations. Whether in the context of biotechnological innovation or crowd-sourced knowledge production, sociologists have pointed out how nonprofits and universities bring in important ideas. Companies collaborate to innovate and create new markets, foundations create new fields of philanthropy, and higher-level governments may empower organizations to act beyond the shackles of bureaucracy. In reality, institutional environments both constrain and enable organizational fields. The resource constraints, legal limitations, and urban power structures inherent to earlier work on city power shape the activities of cities as they assume their new role in society and the degree to which cities adopt voluntary practices, create nonbinding plans, seek legitimacy from sources other than institutionalized politics, and operate in a multistakeholder network. Thus, institutions empower cities to act and, through constraints, shape how this power is enacted. Because institutions empower organizations globally but constrain them in locally contingent ways, the outcomes are extraordinarily heterogeneous.

Whether, when, and how city action—whether it is administrative or distributed—invites backlash from within and outside the city borders is a frontier for future work.

The third and final caveat is that what empowers some cities isolates others—thus reinforcing existing disparities and creating new ones. This point requires putting the resurgence of city action into broader context. In his last, unpublished book, the great sociologist Charles Tilly had embarked on a project of tracking the historical development of the relationships between cities and states over five millennia.[57] Tilly argued that "with many a wiggle," cities lost dominance to nation-states and became increasingly integrated with them.[58] Figure C.3 shows this move from autonomous cities to centralized states.

Tilly may not have witnessed or foreseen the ties that cities have established with each other in the contemporary era. But the

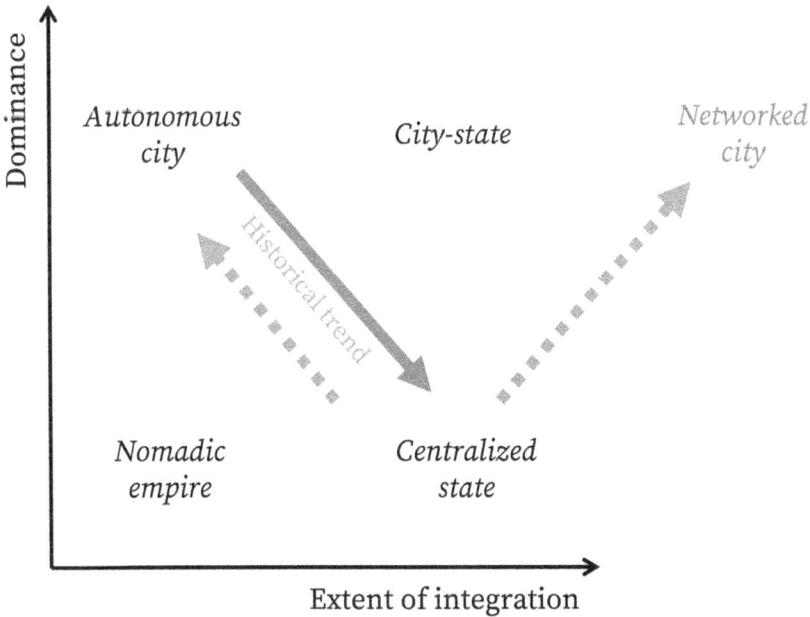

FIGURE C.3 Bifurcated paths for future cities, resulting from the shifting relations between cities and states in history.
Source: Adapted from Tilly (2011), with my additions in light gray.

collaborative network between cities documented in this book has significant implications for how we should interpret the relationships among cities in advanced capitalism. With the technological innovation of the late twentieth and early twenty-first centuries, coordination and communication between individuals in different places has become increasingly important, in contrast to the prior near-exclusive importance of the diplomatic corps or the red telephone between Moscow and Washington, D.C. As some cities have established direct lines of communication between one another, even more cities are being left out of the loop.[59] What determines the relative dominance and integration of cities and states is left to the next generation of sociologists, some of whom will hopefully help us understand the historical trends that put cities on and off the map. As Tilly writes, the value of studying the interaction among cities, states, and their networks "justifies the risk of stretching the evidence toward informed speculation."[60]

In closing, I thus dare speculate that the confluence of new insight about the dominance and integration of cities suggests new directions in the relationship between cities and states. Cities and their networks—with quick social media exchanges, regular conferences, and strategic partnerships—are now even more integrated than at the time of Tilly's writing. The regained dominance and the ever-tighter integration of cities will shift the map we have come to know toward *networked cities* in which strong city administrations and private intermediaries shape the accumulation of capital, coercive power, and soft commitments to address societal challenges. If this shift turns out to be true, there is a danger of elevating some cities to new heights of power while others become or remain segregated from the trust network, falling into a kind of isolation that society has not known since the existence of nomadic empires. To prevent this bifurcation, policymakers and social scientists must find ways to further the integration of cities that are already active with those that have yet to become activated.

Acknowledgments

What draws us to cities is the web of connections that imbues their physical presence with meaning. Cities thrive when people come together to exchange ideas, tricks of the trade, resources, and love. It is these connections that have given me the inspiration and aspiration to write this book, and so it only seems right to acknowledge each by city.

In my first home of Innsbruck, I am grateful to my parents, Petra and Josef, who have continued to be my trusted advisers and city guides ever since I left our village. A train ride to the East, in Vienna, Markus Höllerer, Dennis Jancsary, Stephan Leixnering, Florentine Maier, Michael Meyer, Renate Meyer, and Tobias Polzer, and many others at the Vienna University of Economics and Business kindled my interest in organizations and public governance and gave me a second home for reflection and renewed inspiration. The city kept me connected to my most trusted confidants, including Adriana Collini, Atila Kilic, Markus Kinschner, Tona Kollreider, Fedja Pivodic, Wolfgang Riedler, Vinzent Rest, Bettina Schauperl, Paul Schuierer-Aigner, Sebastian Schuster, and Fanny Shing, and many others. Like other cities, Vienna turned out to be more than a place—it's a bag of ideas that I took along on my journey.

At Stanford, these ideas fell on fertile ground and grew into the book's argument—from the first discussions about the role of cities in world society in John Meyer and Chiqui Ramirez's seminars to being grilled about civic capacity by Sarah Soule, Xueguang Zhou, Jacklyn Hwang, and Huggy Rao. In between, professors including Mark Granovetter, Justin Grimmer, Doug McAdam, Daniel McFarland, and Buzz Zelditch, and fellow students including Daniel Armanios, James Chu, Diana Dakhlallah, Priya Fielding-Singh, Aaron Horvath, Jennie Hill, Ece Kaynak, Liz McKenna, Hatim Rahman, and Ariela Schachter, taught me the tools I needed to pull off this work. Among all my mentors, Woody Powell became the most loyal companion on my academic journey and, with Marianne Broome Powell, a source of memorable Southwestern meals. Woody not only kept encouraging me to write this book but also taught me to study, challenge, and build institutions (whereas *being* one does not rub off); for that, I am immensely grateful. I have also benefited from the generosity of friends at the Center on Philanthropy and Civil Society (PACS), including Paul Brest, Tricia Bromley, Johanna Mair, Rob Reich, Priya Shanker, and Robb Willer. Teaching in the Urban Studies Program under the guidance of Michael Kahan and Fred Stout first exposed me to urbanists whose visions have become entangled with mine. Up on the shining hill where the Center for Advanced Study in the Behavioral Sciences (CASBS) towers over Silicon Valley, Bob Gibbons, Eric Klinenberg, Margaret Levi, and Woody Powell, and some of the brightest minds studying organizations today have given me so much to think about. Last but not least, the friendship of graduate school buddies, including Emily Carian, Jared Furuta, Max Hell, Molly King, Bethany Nichols, and Juan Pedroza, and Aaron Silverman has followed me around when I left Stanford and has brought me back.

Chicago became a third alma mater where I could apply and expand my understanding of organizations to the sociology of cities. At the University of Chicago's Mansueto Institute of Urban Innovation, I not only enjoyed the hospitality of Luís Bettencourt, Kate Cagney, Anne Dodge, and Paty Romero-Lankao, but I also found fellow travelers in Benji Fogarty-Valenzuela and Bia Cordeiro, Robert Manduca, Nicole Rosner, and Jared Schachner. My conversations with Nicole Marwell, Jennifer Mosley, and Lis Clemens shaped my thinking about the governance of cities profoundly. Chicago is also where, just before wrapping up my book proposal, Zenelia and I made each other another kind of proposal and got married. Zenelia has widened my horizons in many ways—by helping me prioritize experiences over things, be compassionate with myself, and explore her native Nicaragua. In Granada, Vivian and Rhina Sequeira have become trusted cheerleaders, manifested by many lovely days with their emissary in France, Viara Sequeira.

In Lyon, we built a new home of lavish meals and feathered friends that has so far proven to be a safe harbor from late-stage capitalism. My colleagues at EM Lyon Business School, under the stewardship of Bernard Forgues and Saulo Barbosa, have built an immense intellectual community of decent and humble people, for whose example I am grateful. Bulgarian poet Nevena Radoyn-voska has been the better half of my "dynamic duo"; Jean Clarke, Tom Elfring, Tao Han, Almantas Palubinskas, Celina Smith, Sonia Siraz, Farida Souiah, Adam Storer, and Thinley Tharchen, and other members of the InvEnt and STORM research centers helped me forget about work on Friday nights before getting back to this book on the weekend, or not doing so thanks to the hospitality of Marin and Mariyana Berket, Laura Callow and Pedro Zurbach, Karolina Korus and Paul Lenormand, and Maggie Norby-Adams and Adam Storer. Valerie Arnhold, Xavier Blot, Grégoire Croidieu, Federica Fusaro, JB Litrico, Jeroen Struben, Cris Taborda, and Yuhao Zhuang kindly read a rough version of this book so that you do not have to. Pauline Ade-not, Elisabeth Gelas, Géraldine Heurtel Tappaz, and Emma Ravinet, among countless others, have provided excellent work conditions. And I am grateful to Jean-Luc Arregle, Pia Arenius, Frédéric Delmar, Isabelle Huault, Tessa Melkonian, and Peter Wirtz for giving me the time and freedom to finish a sociological book in a business school—it is not a privilege I take for granted.

In Montréal, at the American Sociological Association, I was lucky to receive thoughtful guidance from some of my intellectual role models in the Community and Urban Sociology Section, including Hillary Angelo, Daniel Aldana Cohen, Jeremy Levine, Rich Ocejo, David Wachsmuth, and Jon Wynn. Somewhere between Milan and Chicago, at the European Group of Organization Studies and in the Academy of Management, I found in Laura Dupin, Santi Furnari, Markus Höllerer, Suntae Kim, and Silviya Svejenova treasured allies to work toward a burgeoning community of scholars interested in places, communities, and organizations.

Hong Kong was a fitting place to finalize the manuscript, thanks to the hospitality of Ioana Sendroiu, Tina Wu, Ron Chan, and many others at the University of Hong Kong, because this is where Frederic Lee first introduced me to a doorstopper about urban planning and governance. Reminding me of formative years during my earlier studies with Cheng Wing Sze, Jeffrey Leung, Raymond Luk, Eday Ng, and Yik Kai Lai, the city sent me on my way with the timely blessings of Man, the "civil god" of literature.

In New York, my wonderful editor Catherine Jampel helped me develop each chapter. Alyssa Napier, Eric Schwartz, and their colleagues including Rebecca Edwards, Meredith Howard, and Maureen O'Driscoll provided expert guidance at Columbia University Press. Evan Schofer played an important part in helping me pitch and develop this book project, and provided helpful insight, along with his editorial colleagues Dana Fisher and Lori Peek. I was also especially grateful for the helpful pointers from Ann Hironaka, Wes Longhofer, the Columbia University Press faculty publication committee, and unnamed others in the final stages of the project. The talented Oscar M Caballero helped me turn my black-and-white words into a colorful cover. From a little further north, Chris Rea and Caleb Scoville joined me virtually in workshopping our book manuscripts, which gave me hope that sociology has a future in the era of human-made climate change.

Visiting Washington, D.C., helped me understand the politics under-girding the green transition in the United States, initially thanks to the wonderful people at the Stanford Woods Institute for the Environment. Interviews at the US Green Building Council and the US Department of Housing and Urban Development were particularly insightful. Funding from the National Science Foundation got the work on this book started.

I have also found an important sounding board for my ideas in an intellectual traveling symphony at the Canadian Institute for Advanced Studies (CIFAR), where I have met and come to appreciate convers-ing with contemporary luminaries, including Yochai Benkler, Danny Breznitz, Maryann Feldman, Jane Gingrich, Sue Helper, Karen Levy, Neil Lee, Andrew Schrank, Olav Sorenson, Kate Zaloom, and Amos Zehavi, whose advice helped push this book across the finish line. The

CIFAR Azrieli Global Scholars program, masterfully orchestrated by Paula Driedger and Rachel Parker, provided support and encouragement without which this book would not have gotten done and introduced me to my friends Khalil Ramadi and Nicole Wu.

From the exceedingly long list of scholars who have inspired this work, none have had a greater imprint on my thinking than my collaborators: Parham Ashur, Khodr Badih, Patrick Bergemann, Tricia Bromley, Gordon Douglas, Claire Dunning, Santi Furnari, Aaron Horvath, Hokyu Hwang, Michelle Jackson, Krystal Laryea, Anna Lunn, Wei Luo, Florentine Maier, Axelle Miel, Michael Meyer, Jessica Sasso Lopes, Woody Powell, Dimitri Rodionov, Asef Salimi, Amanda Sharkey, Megan Tompkins-Stange, Katie Young, Cristobal Young, and Yuhao Zhuang. Our Civic Life of Cities Lab, which has also included Valerie Dao, Ling Han, Winnie Jiang, Jean Lin, Yan Long, Danielle Logue, Wei Luo, Gonnie Park, Nick Sherefkin, Berta Terzieva, Elizabeth Trinh, Qian Wei, Yi Zhao, and Wenjuan Zheng, has helped me articulate and deepen my understanding of nonprofit organizations in the urban context. Ana Gonzalez and David Suárez deserve special thanks for collaborating with me on some of the research reported in chapter 4. Traveling together is the best way of learning.

A multitude of bright research assistants has helped me gather the evidence for this book—chief among them Ana Gonzalez, Camille Le Brech, Olivia Rambo, and Paloma Hernandez—you met some of them and more throughout the book. Axelle Miel was indispensable in putting the finishing touches on the manuscript. If ever there was a good reason to write a book, it is so that they can read and, one day, teach or challenge it.

This book could not have happened without the occupants and informants of city halls around the world, and I appreciate the time and effort they have put into patiently explaining their world to me. I hope I did justice and gave voice to their important work. Bureaucrats are the backbone of our democracies; may many mayors and presidents serve under them.

These and countless other people have given me support and ideas along the way. I thank you all collectively for making me a better scholar and person while I finished this book; let's hope it sticks.

Methodological Appendix

I sought to explain and motivate my methodological choices, data, and measures in the chapters where the relevant research appears. Table A.1 complements this information in a compact format, and I direct readers interested in replicating the results to several companion papers that went through peer review and include much greater methodological detail.

Underlying chapters 2 and 3 is an empirical analysis of strategic plans among the 360 largest cities in the world as sampled from UN Population. In chapter 2, I discuss the content of plans and explain when a document qualifies as a strategy for the purposes of this book. As described in the chapter, several research assistants and I collected more than 650 documents from websites and online archives to examine each of the plans and identify whether they are holistic (describing the future of the city), visionary (describing the long-term plan), and purposive (connecting ends to means). Several cities had plans that did not qualify as city strategies because they either did not meet these criteria or because they were not published by the city (but rather an international

TABLE A.1

Overview of key variables and data sources for quantitative studies in this book

	Chapters 2 and 3	Chapter 4	Chapter 5
City climate action	Adoption of a strategic plan (presence, time)	Administrative sustainability practices	LEED municipal policies by city (administrative) and LEED certifications in city (distributed)
Data sources for dependent variables	Collected from city webpages and archives	International City/County Management Association (ICMA), 2021 Sustainability Survey	US Green Building Council
Explanatory variables related to institutional superstructure and organizational infrastructure	*Institutional:* Centrality in intercity network manually coded based on membership in twelve associations	*Organizational:* Civic capacity measured as presence and composition of nonprofit organizations; membership in Local Governments for Sustainability (ICLEI) *Institutional:* State policy innovativeness	*Organizational:* Civic capacity measured as presence and composition of nonprofit organizations *Institutional:* Spillovers from other cities, various state policy factors
Data sources for explanatory variables	Conferences and congresses hosted by INGOs, INGO headquarters Yearbook of International Organizations	*Organizational:* National Center for Charitable Statistics PC Core Files; scraped from ICLEI webpage *Institutional:* Institute for Public Policy and Social Research (IPPSR) Correlates of State Policy; National League of Cities (NLC)	*Organizational:* National Center for Charitable Statistics PC Core Files *Institutional:* IPPSR Correlates of State Policy

Other independent variables	Economic status in network of global service providers (competition)	Municipal form of government; firm activity	Environmental advocacy; firm activity
Data sources for other independent variables	Globalization and World Cities	ICMA Form of Government Survey; COMPUSTAT	Dynamics of Collective Action; COMPUSTAT; MSCI Environmental, Social, and Governance (ESG) Rating
Methods	Content analysis, term frequency–inverse document frequency (TF-IDF) weighted word scores; Cox models for event history analysis	Linear regressions	Poisson and zero-inflated negative binomial models for panel data; Cox models for event history analysis
Controls variables	Demographic, economic, social, political factors	Demographic, economic, social, political factors	Demographic, economic, social, political factors
Sample	360 global cities with more than 1.25 million inhabitants	1,540 respondent cities to ICMA survey	More than 5,000 US Census places

development agency, the national government, or the region). In chapter 3, I examine the diffusion of strategic plans with various content using a multivariate study that connects the adoption of a plan to the three social structural indicators of the city's position: economic status, approximated by the city's centrality in the global network of service firms as measured by the Globalization and World Cities (GaWC) as a measure of competition; integration in the world society, approximated by international organizations headquartered and meeting in the city as a measure of culture; and professional centrality, approximated by the city's membership in intercity associations as a measure of collaboration. The data are not currently publicly accessible, but they will become part of a repository of city strategies in the future. Further methodological detail can be found in a paper coauthored with Patricia Bromley.[1]

In chapter 4, I analyze secondary survey data from the International City and County Management Association (ICMA) about the presence of sustainability practices in a subset of 1,540 US cities. The study uses simple ordinary least squares (OLS) regressions to provide an overview of which city features—related to the city's organizational infrastructure and the structure of the city administration—are associated with sustainability practices. The survey is ideal for examining the distribution of urban sustainability at a particular time in history, when sustainability was becoming institutionalized. The empirical findings then served as a basis for subsequent qualitative work that further examined the mechanisms underlying the connections between membership in intercity associations—an important indicator of the institutional superstructure—and the work happening inside local sustainability offices in leading and lagging cities in the United States (see notes on qualitative methodology). Recent work has examined multiple such surveys and offered a deeper understanding of the covariates of urban sustainability beyond the factors considered in this book and providing further opportunities for future analysis. Further methodological detail can be found in a paper coauthored with David Suárez.[2]

Underlying chapter 5 is a complex diffusion study of the adoption of administrative and distributed action with respect to energy-efficient construction in US cities. To overcome the limitations of prior work and the limited sample of chapter 4, this study includes an expansive set of more than five thousand US Census-designated places—that is cities, towns, and villages with local governments—in its sample. This study's goal was to demonstrate the intertwined nature of institutional super-structure and organizational infrastructure. I combined event history models predicting the first incidence of administrative action (a LEED policy) and distributed action (a green building). I also used Poisson models to predict the count of buildings in a given city and an inter-rupted time series methods to examine the causal effect of municipal policies on subsequent green building adoptions. The data of the reg-istration for the LEED green building certification stems from the US Green Building Council's (USGBC's) LEED Project Directory and the presence of legislation advancing green construction was coded from the USGBC's LEED Public Policy Library. Both data sources are publicly accessible, and I share the dataset used for the analyses upon request. Further methodological detail can be found in an accompanying article.[3]

In addition to the statistical analyses performed in Stata, I also used Tableau to inspect and visualize the data. Data visualization is some-times seen as a last step to make the findings visible and accessible. But visual analysis of the data—be it the spread of strategic plans over time or the specific locations of buildings—brought many insights that inspired critical lines of inquiry once in conversation with theory, archi-val research, and interviews.

NOTES ON THE QUALITATIVE ANALYSIS

The sixty interviews reported in this book were initially intended as background interviews to be able to design and understand the com-parative quantitative studies; ultimately, the depth and reflectiveness of the responses was inevitable for being able to tell a comprehensive story. Respondents included representatives from intercity associations focused on sustainability, such as 100 Resilient Cities (100RC) and the Resilient Cities Network, the C40 City Climate Leadership Group, Local Governments for Sustainability (ICLEI), Resilience by Design, the Urban Sustainability Directors Network (USDN), the Southern Sustainability Directors Network (SSDN), and the National League of Cities (NLC). I also spoke to representatives of foundations and associations that have supported the work of these associations, including Bloomberg Philanthropies, the Rockefeller Foundation, the Hewlett Foundation, the Environmental Law and Policy Center (ELPC), the Great Plains Institute (GPI), and Clean Energy Resource Teams (CERT). I also interviewed policy experts in the US Department for Housing and Urban Development (HUD), the International City and County Management Association (ICMA), the Metropolitan Mayors Caucus (MMC), the Urban Institute, and, with a specific focus on green building certification, multiple respondents at the USGBC in the national and Northern California chapters and the Illinois Green Alliance. The respondents from city governments included representatives from US cities, including Ann Arbor, Chicago, Cleveland, Evanston, Milwaukee, Nashville, New York City, Palo Alto, Rochester, San Francisco, St. Louis, St. Paul, Toledo City, and Washington, D.C., as well as several cities outside the United States, including Copenhagen, Hong Kong, Vienna, and Sydney. All respondents were high-ranking city managers, urban planners, or sustainability officers. Most were career officials, with a few highly qualified political appointees among them. In Chicago, Palo Alto, San Francisco, Copenhagen, and Vienna, I conducted multiple interviews to get a fuller picture and to see changes over time. The interviews were conducted either by me, Ana Gonzalez, or the two of us together.

The work was deemed exempt from approval by the Stanford Institutional Review Board (IRB) for Human Subjects. Still, each respondent received an information sheet outlining the scope, significance, and goals of the research including independent contact details if they were concerned about research ethics (shown in textbox A.1). I promised respondents not to identify them by name, and in some cases, I also opted to anonymize the cities or organizations my respondents work for. This is because many of these officials may experience professional repercussions that are beyond my control and, possibly, understanding. Most interviews with city administrators were conducted in person in the city's offices, whereas most interviews with representatives of intercity associations were conducted on the phone or on a video call because of their varied locations and the COVID-19 pandemic. We typically scheduled interviews for thirty to sixty minutes to avoid overburdening our respondents, although several took more than ninety minutes. All interviews followed an interview guideline, shown in textbox A.2 and were recorded and transcribed by the wonderful Kendra Kline of Matchless Transcription (matchlesstranscription.com).

TEXTBOX A.1: INTERVIEW PROTOCOL

Background

Q 1. How did you end up in your position working for the City of X?

Q 2. What does your department do for the City of X?

Q 3. What professional background do the people have who work on this?
    What credentials do they have?
    Did this change over time?

History of strategic plans

Q 4. What's the background of current composition of plans that the city acts upon?

Q 5. What is the history of strategic planning in your city?
    When did this strategic approach start?
    How do recent strategy plans differ from more dated plans?
    What are the changes between the two most recent iterations of strategy plans?

Production and consumption of strategic plans

Q 6. Who would you say reads these plans?
    Who are the plans addressed to?
    Do the plans have an international audience?

Q 7. Who wrote the most recent plan?
    What role do politicians and administrators have respectively?
    Who are "actors inside and outside the city administration" that participated?

Q 8. Where do ideas for the plans come from?
    What is the role of civil society?
    What is the role of private consultants?
    What is the role of corporations?
    What is the role of other cities?

Q 9. How are the plans implemented, used, and monitored?

Q 10. Governance: Who do you collaborate with in these activities?
    What is the role of other levels of government (state, regional, national)?
    What is the role of the universities?
    What is the role of private business?
    What is the role of civil society organizations?
    What is the role of international organizations and audiences (e.g., United Nations)?

(*continued on next page*)

(*continued from previous page*)

Q 11. How do different plans interact? What about federal and supraregional plans?

International context of strategic planning

Q 12. [If a non-English-speaking city] Why was the plan published in English?

Q 13. Is the City of X ever compared to other cities?
What is the role of city rankings?

Q 14. Who are your reference groups/peer cities?
How do you determine them?

Q 15. Do you share your ideas and strategies with other cities?
How do you broadcast your strategies?
What does it mean to be a best practice city?

Q 16. What role do international certifications and accreditations play for the city?

Q 17. Have you or your colleagues interacted with international city associations?
If so, when did you join the association?
Were you involved in the founding of the association?

Q 18. Who do you have in mind when you speak of "international" audiences/collaborators?

Specific questions for program managers at intercity associations

Q 19. Could you describe how your organization seeks to influence city administrations?

Q 20. What is the motivation for your endeavors?
Do you have a theory of change? If so, what is it?

Q 21. How many cities participate in your organization's program?
How have the membership numbers changed over time?

Q 22. What activities do you implement in order to facilitate city learning?
Do you think they work?
Can you tell me of one recent example of an event and how participated?

Q 22. How do you fund your work?

TEXTBOX A.2: INFORMATION SHEET
FOR INTERVIEWEES

You are invited to participate in a research study on the ways in which intercity associations and foundations shape the organizational capacity of city administration's ability to respond to challenges to resilience, such as climate change. The interview will involve questions about your experiences as a program officer or city manager related to your training and background as well as the organization's practices and external relations to city administrations and other associations. The interview will also ask you about your perceptions of the landscape of associations that seek to further learning and collaboration in and between cities in general. The interview is part of a larger study and the purpose of the interview is to compare your responses with those given by other study participants. Your participation requires no preparation. You will be asked to answer open-ended questions related to the subjects outlined above. The interview will be audio recorded. The audio recording will be transcribed and transcriptions, along with transcriptions from interviews with other study participants, will be analyzed for content. The interview recordings and transcripts will be kept in a password-protected folder on Mr. Brandtner's computer. Only Mr. Brandtner and his research affiliates will have access to this file. Quotes from the interview may be used in public presentations, reports, journal articles, and books generated from this research. However, your name and identifying characteristics will never be used unless you explicitly waive your confidentiality.

Your participation will take approximately thirty to sixty minutes. If there are time constraints, please say so and the duration of the interview can be adjusted accordingly. The interview will not be paid. The primary risk associated with this study is a potential loss of confidentiality and/or anonymity. However, your answers will not be discussed with other study participants. Quotations from this interview may be used in written publications, but under no circumstances will your name or identifying characteristics be included unless you explicitly waive your confidentiality.

While there is no guarantee or promise that you will receive any benefits from this study, there may be a direct or indirect benefit to you from these procedures. In most cases, participants enjoy the interview process. It gives them an opportunity to reflect on their own lives and on how they experience the world around them. Furthermore, findings from this study will be publicly available. These findings may benefit you by providing a broader understanding of how your own experiences and perceptions compare to other study participants.

If you have read this form and have decided to participate in this project, please understand that your participation is voluntary and you have the right to withdraw your consent or discontinue participation at any time without penalty or loss of benefits to which you are otherwise entitled. The alternative is not to participate. You have the right to refuse to answer particular questions. Your individual privacy will be maintained in all published and written data resulting from the study.

All interviews were coded for themes in the same way that the interviews for chapter 4 were coded. Emerging categories of codes included actors (e.g., city associations, chief sustainability officers and chief resilience officers, mayors, nonprofits, foundations, companies), features that enable or undermine city action (e.g., size, urbanicity, co-benefits, external shocks including COVID-19, knowledge gaps, obstacles and challenges, political backlash, and history), governance (e.g., relationship between administration and politics, global scope, hard and soft law, interorganizational collaboration, relationship to the state and federal governments, relationship to other cities, the structure of city government), mechanisms (e.g., city learning, city collaboration, certification, grantmaking, implementation, innovation, metrics and measurement, personal relationships, public consultations, strategic management, technical planning), and sustainability concepts (e.g., sustainability, resilience, greenness, equity, environmental leadership, adaptation and mitigation). Ana Gonzalez helped me code the entirety of interviews in NVivo; I read through the coded excerpts multiple times before assigning quotes to particular chapters. For further detail about the sampling and coding process, specific to the findings presented in chapter 4, please see a published paper coauthored with Ana Gonzalez (especially the online supplementary material).[4]

Readers should note that, although I recommend the experience, any interview study comes with caveats. One is my positionality. As a European sociologist trained at a recognized US university interviewing elite decision-makers in different US institutions, my background may have influenced the emphasis placed on different topics that emerged in the conversation, such as the theme of international collaboration or professional expertise. Some of the interviews were conducted conjointly with or alone by Ana Gonzalez, however, who as a Latina would have triggered different responses; we noted no such differences. All interviews followed a well-laid-out interview guideline that we adapted to the interviewee considering whatever background information we could find. We are not aware of significant interviewer biases with respect to what we learned, but they may nonetheless exist. Other biases are certain. What we learned certainly was shaped by the fact that we spoke with planners and high-level bureaucrats and, with a few exceptions, neither lower-ranking employees within their offices nor elected politicians. There is no doubt that a detailed ethnographic approach (in the tradition of important ethnographic work by urban sociologists like Nicole Marwell, Jeremy Levine, or John Krinsky) could have revealed aspects of the urban governance that I did not foreground. Such an ethnographic approach would be particularly useful to understand changes in the meaning of urban sustainability over time, as well as how urban sustainability officers motivate their work. Such an approach could follow the idea of "following the planners, not the plans" expressed in the conclusion. Despite the value of this book's quantitative, comparative nature, qualitative and especially ethnographic research has much to add to understanding the embeddedness of cities.

The fact that I prioritized in-person interviews also meant that there is a geographic bias to who I was able to interview because I was located in the United States and Europe. Many of the organizations with whose leaders I spoke are actively global and gave examples from work around the world or, if they were focused on North America, a wide variety of cities that they have worked with. As a result, I approached saturation in what I could learn from intercity associations related to urban sustainability. In contrast, the interviews with city leaders are certainly not representative of the experience of all cities. In particular, I could not systematically interview respondents in many cities that are ranking low in terms of their propensity to take action about sustainability and the climate. The comparison of cities that face high barriers to sustainability differ from those that face lower barriers in chapter 4 helped get a better understanding of the experience of these cities (see table A.2 for the case selection). In contrast, interviewing representatives of more cities in the Global South, including the populous countries of India and China that together account for 108 cities compared with the 37 cities in the United States, was simply not within the realm of what was possible or affordable. Dear reader, you have your work cut out for yourself.

TABLE A.2

Summary of city- and state-level characteristics of cities in interview sample (chapter 4)

City	State	Population (as of July 2022)	Share Democrat (percentile)	Sustainability practices (percentile)	State energy efficiency score (%)
Ann Arbor	MI	119,875	0.67 (97)	31.3 (76)	27
Chicago	IL	2,665,039	0.64 (95)	22.7 (48)*	27
Cleveland	OH	361,607	0.62 (92)	39.0 (89)	14.5
Evanston	IL	75,544	0.64 (95)	22.7 (48)	27
Milwaukee	WI	563,305	0.54 (81)	21.9 (44)*	17
Nashville	TN	683,622	0.42 (45)	16.6 (24)*	15.5
New York City	NY	8,335,897	0.67 (97)	26.5 (64)	34.5
Palo Alto	CA	66,010	0.7 (99)	52.0 (99)	42
Rochester	MN	121,878	0.49 (69)	31.0 (74)	33
San Francisco	CA	808,437	0.75 (100)	43.0 (93)*	42
St. Louis	MO	286,578	0.53 (76)	18.2 (31)*	12.5
St. Paul	MN	303,176	0.55 (82)	27.8 (66)*	33
Toledo	OH	266,301	0.61 (91)	18.0 (28)*	14.5
Washington, D.C.	—	671,803	0.68 (97)	24.6 (55)*	25.5

*Aggregated at the core-based statistical area (CBSA) level.

Notes

PREFACE

1. Ebenezer Howard, *Garden Cities of To-Morrow*, 11th ed. (MIT Press, 2001); Ernest Callenbach, *Ecotopia* (Banyan Tree, 1975). For a discussion, see Christof Brandtner et al., "Where Relational Commons Take Place: The City and Its Social Infrastructure as Sites of Commoning," *Journal of Business Ethics* 184, no. 4 (2023): 917–32.
2. Herron's work *Walking City on the Ocean* inspired the book cover, designed by Nicaraguan architect and visual artist Oscar M Caballero. Herron was part of a group of avant-garde architects who sought to challenge establishment modernism, among others by questioning the permanence of cities. Herron thought the existing proposals were not radical enough. The *British Independent* writes in an obituary in 1994 that "the missing dimension was indeterminacy of place and what could be more obvious, to Herron at least, than a city which moved?" Caballero's *Automata* envisions a twenty-first-century version of the *Walking City*, in which cities inhabit the changing climate differently. Instead of depicting a single pod in the wide ocean, Caballero, like I, considers how the interconnectedness and isolation of some pods shape their relationship to nature.
3. Martin Kornberger and Stewart Clegg, "Strategy as Performative Practice: The Case of Sydney 2030," *Strategic Organization* 9, no. 2 (2011): 136–62; Kent E. Portney and Jeffrey M. Berry, "Participation and the Pursuit of Sustainability in U.S. Cities," *Urban Affairs Review* 46, no. 1 (2010): 119–39; Richard Florida, "The Economic Geography of Talent," *Annals of the Association of American*

Geographers 92, no. 4 (2002): 2; Saskia Sassen, *Cities in a World Economy*, 4th ed. (Pine Forge, 2012); Ronan Paddison, "City Marketing, Image Reconstruction and Urban Regeneration," *Urban Studies* 30, no. 2 (1993): 339–49; Kevin Morgan, "The Learning Region: Institutions, Innovation and Regional Renewal," *Regional Studies* 31, no. 5 (2010): 491–503.

4. Lewis Mumford, "What Is a City?," *Architectural Record* 82, no. 5 (1937): 59–62.
5. James D. Thompson, *Organizations in Action: Social Science Bases of Administrative Theory* (Transaction, 1967).
6. Herbert A. Simon, *Administrative Behavior*, 4th ed. (Simon and Schuster, 2013); Philip Selznick, *TVA and the Grass Roots: A Study of Politics and Organization*, vol. 3 (University of California Press, 1949); Royston Greenwood, *Patterns of Management in Local Government* (Martin Robertson, 1980); Pamela S. Tolbert and Lynne G. Zucker, "Institutional Sources of Change in the Formal Structure of Organizations: The Diffusion of Civil Service Reform, 1880–1935," *Administrative Science Quarterly* 28, no. 1 (1983): 22–39; John W. Meyer and W. Richard Scott, *Organizational Environments: Ritual and Rationality* (SAGE, 1983).
7. Organization studies have been stuck on a broad interpretation of what this open system constitutes and rarely have looked at the local context of what is going on where the rubber hits the road. Paying greater attention to place, space, and community in organizing promises to rectify this shortcoming. See M. Tina Dacin et al., "Navigating Place: Extending Perspectives on Place in Organization Studies," *Organization Studies* 45, no. 8 (2024): 1191–212; Christopher Marquis et al., "Introduction: Community as an Institutional Order and a Type of Organizing," in *Research in the Sociology of Organizations*, ed. Christopher Marquis, Michael Lounsbury, and Royston Greenwood, vol. 33, Communities and Organizations (Emerald Group, 2011), ix–xxvii; Christof Brandtner and Walter W. Powell, "Capturing the Civic Lives of Cities: An Organizational, Place-Based Perspective on Civil Society in Global Cities," *Global Perspectives* 3, no. 1 (2022): 36408.
8. For a more expansive definition of "the urban," see Henri Lefebvre, *The Urban Revolution* (University of Minnesota Press, 2003); Sue Ruddick et al., "Planetary Urbanization: An Urban Theory for Our Time?," *Environment and Planning D: Society and Space* 36, no. 3 (2018): 387–404; Neil Brenner and Christian Schmid, "The 'Urban Age' in Question," *International Journal of Urban and Regional Research* 38, no. 3 (2014): 731–55.

1. ACTING

1. C40 Cities, "Global Network of Mayors React to President Trump Withdrawing the US from the Paris Climate Agreement," press release, January 21, 2025, https://

www.c40.org/news/global-network-of-mayors-react-to-president-trump-paris
-agreement/

2. City climate *in*action is an often noted but unresolved problem in research and policy. In an authoritative textbook about cities and climate change, political scientist Harriet Bulkeley points out that the burgeoning of urban sustainability and the industry that surrounds it *"mask a significant diversity* between and within cities in terms of how climate change is being addressed. . . . [F]or the vast majority of the world's cities, climate change is far from being a significant issue. . . . Climate change is taking place, and processes of urbanization are a significant contributor to rising levels of GHG emissions, but municipal authorities and other urban actors remain either *unaware or unwilling to act"* (emphasis added). Harriet Bulkeley, *Cities and Climate Change* (Routledge, 2013), 104.

3. In 2019, a program manager at one of the largest philanthropic supporters of efforts to help cities reach their climate goals told me of their back-of-the -envelope calculation of how many cities that made commitments to the Paris Agreement are "actually on track to reach those goals or have a plan for getting there. And the results are kind of scary." Among the fifty largest cities in the United States, they estimated, "only about 10 of them seem to be on track to their goals and have a really solidified road map and sort of next steps for how to get there."

4. Some scholars have argued that cities' official commitments to climate change mitigation are small relative to the global output of carbon emissions. Environmental scholar Angel Hsu estimates that all proposed annual reductions in carbon dioxide emissions will add up to 2.2 billion tons by 2030—which pales in comparison to the global total output of 59 billion tons projected for 2030; see "California Leads Subnational Efforts to Curb Climate Change," *The Economist*, September 15, 2018, https://www.economist.com/international/2018/09/15/california-leads -subnational-efforts-to-curb-climate-change.

5. Like other scholars of city climate action, I understand cities' sustainability and resilience strategies as relevant to human responses to climate change. There is good reason *not* to make a strict distinction among the different types of climate action, because one thing often leads to another. Rather than treating city climate action as being specific to a particular type of environmental protection, conservation, or intervention, I understand it as a specific case of *city action*—a general phenomenon relevant beyond the climate (see chapter 6).

6. Hannah Ritchie et al., "Urbanization," *Our World in Data*, February 21, 2024, https://ourworldindata.org/urbanization.

7. Jonathan Rothwell et al., *Patenting Prosperity: Invention and Economic Performance in the United States and Its Metropolitan Areas* (Brookings Institution, 2013),

https://www.brookings.edu/wp-content/uploads/2016/06/patenting-prosperity
-rothwell.pdf; Bruce Katz and Jennifer Bradley, *The Metropolitan Revolution*
(Brookings Institution, 2014).

8. Air pollution, water quality, and green space have been local concerns long
before climate change became a global one. As environmental sociologist Hillary
Angelo argues in her history of urban greening in the German Ruhrtal over the
twentieth century, there is an "enduring conviction that green space will transform us into ideal inhabitants of ideal cities." Nature has been on the municipal
docket for a long time. Hillary Angelo, *How Green Became Good: Urbanized
Nature and the Making of Cities and Citizens* (University of Chicago Press, 2021).

9. In an issue of *Wired* guest-edited by former US President Barack Obama, Eric
Klinenberg demands that "cities, nations, and international agencies face up to
the challenge of global warming" and that the day of reckoning is "inevitable;
the only question is when it will begin." Eric Klinenberg, *Heat Wave: A Social
Autopsy of Disaster in Chicago*, 2nd ed. (University of Chicago Press, 2015), 192;
Eric Klinenberg, "The Key to Surviving Climate Change? Build Tight-Knit Communities," *Wired*, October 25, 2016, https://www.wired.com/2016/10/klinenberg
-transforming-communities-to-survive-climate-change.

10. Hokyu Hwang, "Planning Development: Globalization and the Shifting Locus
of Planning," in *Globalization and Organization: World Society and Organizational Change*, ed. Gili S. Drori, John W. Meyer, and Hokyu Hwang (Oxford
University Press, 2006), 69–90; Neil Brenner, *New State Spaces: Urban Governance and the Rescaling of Statehood* (Oxford University Press, 2004).

11. Matthew J. Hoffmann, *Climate Governance at the Crossroads: Experimenting
with a Global Response After Kyoto* (Oxford University Press, 2011); Bulkeley,
Cities and Climate Change; Cynthia Rosenzweig et al., "Cities Lead the Way in
Climate–Change Action," *Nature* 467, no. 7318 (2010): 909–11.

12. Scott Frickel and James R. Elliott, *Sites Unseen: Uncovering Hidden Hazards in
American Cities* (Russell Sage, 2018); Robert Manduca and Robert J. Sampson,
"Punishing and Toxic Neighborhood Environments Independently Predict the
Intergenerational Social Mobility of Black and White Children," *Proceedings of
the National Academy of Sciences* 116, no. 16 (2019): 7772–77.

13. Bulkeley, *Cities and Climate Change*; Hoffmann, *Climate Governance at the
Crossroads*.

14. Laura Tozer et al., "Transnational Governance and the Urban Politics of Nature-
Based Solutions for Climate Change," *Global Environmental Politics* 22, no. 3
(2022): 81–103. For critiques of ecomodernization, see Dana R. Fisher and William R. Freudenburg, "Ecological Modernization and Its Critics: Assessing the
Past and Looking Toward the Future," *Society and Natural Resources* 14, no. 8
(2001): 701–9.

15. Ion Bogdan Vasi and David Strang, "Civil Liberty in America: The Diffusion of Municipal Bill of Rights Resolutions After the Passage of the USA PATRIOT Act," *American Journal of Sociology* 114, no. 6 (2009): 1716–64; Justin Peter Steil and Ion Bogdan Vasi, "The New Immigration Contestation: Social Movements and Local Immigration Policy Making in the United States, 2000–2011," *American Journal of Sociology* 119, no. 4 (2014): 1104–55; Els De Graauw, *Making Immigrant Rights Real: Nonprofits and the Politics of Integration in San Francisco* (Cornell University Press, 2016); Christof Brandtner et al., "Creatures of the State? Metropolitan Counties Compensated for State Inaction in Initial U.S. Response to COVID-19 Pandemic," *PLOS ONE* 16, no. 2 (2021): e0246249.

16. For instance, the United Nations dedicated a specific sustainable development goal, SDG 17, to partnerships. Also see Rod A. W. Rhodes, "Understanding Governance: Ten Years On," *Organization Studies* 28, no. 8 (2007): 1243–64; Chris Ansell and Alison Gash, "Collaborative Governance in Theory and Practice," *Journal of Public Administration Research and Theory* 18, no. 4 (2008): 543–71. In light of the resulting complexity, city administrations have stopped publishing official organizational charts because of the increasing complexity of the city's bureaucracy and agencies at arm's length, and because the boundaries between public and private have blurred.

17. Charles Perrow, "Organisations and Global Warming," in *Routledge Handbook of Climate Change and Society*, ed. Steven Brechin and Seungyun Lee (Routledge, 2010), 59–77; Don Grant et al., *Super Polluters: Tackling the World's Largest Sites of Climate-Disrupting Emissions* (Columbia University Press, 2020).

18. For a detailed treatment of this point, see Christof Brandtner and Walter W. Powell, "Capturing the Civic Lives of Cities: An Organizational, Place-Based Perspective on Civil Society in Global Cities," *Global Perspectives* 3, no. 1 (2022): 36408. Place is also gaining a more central role in organization studies and entrepreneurship: see, for instance, M. Tina Dacin et al., "Navigating Place: Extending Perspectives on Place in Organization Studies," *Organization Studies* 45, no. 8 (2024): 1191–212; Tristan L. Botelho et al., "The Sociology of Entrepreneurship Revisited," *Annual Review of Sociology* 50 (2024): 341–64.

19. Saskia Sassen, *The Global City: New York, London, Tokyo* (Princeton University Press, 2001); Saskia Sassen, *Cities in a World Economy*, 4th ed. (Pine Forge, 2012); Richard Florida, "The Economic Geography of Talent," *Annals of the Association of American Geographers* 92, no. 4 (2002): 743–55; Richard Florida, *Who's Your City? How the Creative Economy Is Making Where to Live the Most Important Decision of Your Life* (Vintage Canada, 2010).

20. Nicole P. Marwell, "Privatizing the Welfare State: Nonprofit Community-Based Organizations as Political Actors," *American Sociological Review* 69, no. 2 (2004): 265–91.

21. It may seem obvious to consider organizations as similarly important determinants of urban outcomes as space, socioeconomic groups to which people
belong, or politics, but urban scholars have long neglected organizations. As
urban sociologists Michael McQuarrie and Nicole Marwell argue: "Marxian
and Chicagoan urban sociologists are becoming increasingly attuned to the
importance of formal organizations in urban processes." Michael McQuarrie and Nicole P. Marwell, "The Missing Organizational Dimension in Urban
Sociology," *City and Community* 8, no. 3 (2009): 262.

22. Christopher Marquis and Julie Battilana, "Acting Globally but Thinking Locally?
The Enduring Influence of Local Communities on Organizations," *Research in
Organizational Behavior* 29 (2009): 283–302; Christopher Marquis et al., Introduction: Community as an Institutional Order and a Type of Organizing," in
Communities and Organizations, Research in the Sociology of Organizations,
ed. Christopher Marquis, Michael Lounsbury, and Royston Greenwood (Emerald Group, 2011), 33:ix–xxvii; Christopher Marquis et al., "Community Isomorphism and Corporate Social Action," *Academy of Management Review* 32, no. 3
(2007): 925–45.

23. András Tilcsik and Christopher Marquis, "Punctuated Generosity: How Mega-
Events and Natural Disasters Affect Corporate Philanthropy in U.S. Communities," *Administrative Science Quarterly* 58, no. 1 (2013): 111–48; Henrich R. Greve
and Hayagreeva Rao, "Echoes of the Past: Organizational Foundings as Sources
of an Institutional Legacy of Mutualism," *American Journal of Sociology* 118, no.
3 (2012): 635–75; Sunasir Dutta, "Creating in the Crucibles of Nature's Fury:
Associational Diversity and Local Social Entrepreneurship After Natural Disasters in California, 1991–2010," *Administrative Science Quarterly* 62, no. 3 (2017):
443–83; Hayagreeva Rao and Henrich R. Greve, "Disasters and Community
Resilience: Spanish Flu and the Formation of Retail Cooperatives in Norway,"
Academy of Management Journal 61, no. 1 (2018): 5–25, https://doi.org/10.5465
/amj.2016.0054.

24. John F. Padgett and Christopher K. Ansell, "Robust Action and the Rise of the
Medici, 1400–1434," *American Journal of Sociology* 98, no. 6 (1993): 1259–1319;
John F. Padgett and Paul D. McLean, "Organizational Invention and Elite
Transformation: The Birth of Partnership Systems in Renaissance Florence,"
American Journal of Sociology 111, no. 5 (2006): 1463–1568; John F. Padgett and
Walter W. Powell, *The Emergence of Organizations and Markets* (Princeton
University Press, 2012).

25. Joseph Galaskiewicz pioneered this insight about network dynamics in cit-
ies with his studies of the social system of philanthropy and corporations in
Minnesota's Twin Cities. Joseph Galaskiewicz, "Interorganizational Relations,"
Annual Review of Sociology 11 (1985): 281–304; Joseph Galaskiewicz, "An Urban

Grants Economy Revisited: Corporate Charitable Contributions in the Twin Cities, 1979–81, 1987–89," *Administrative Science Quarterly* 42 (1997): 445–71; Joseph Galaskiewicz and Stanley Wasserman, "Mimetic Processes Within an Interorganizational Field: An Empirical Test," *Administrative Science Quarterly* 34, no. 3 (1989): 454–79. Similarly, geographic spillovers between proximate organizations influence organizational decision-making and the adoption of practices through various mechanisms, including diffusion migration of skilled workers, and collaboration. On diffusion, see David Strang and Sarah A. Soule, "Diffusion in Organizations and Social Movements: From Hybrid Corn to Poison Pills," *Annual Review of Sociology* 24 (1998): 265–90; Barbara Czarniawska and Guje Sevón, eds., *Translating Organizational Change* (Walter de Gruyter, 1996); on migration, see AnnaLee Saxenian, "Inside-Out: Regional Networks and Industrial Adaptation in Silicon Valley and Route 128," *Cityscape: A Journal of Policy Development and Research* 2, no. 2 (1994); Lee Fleming et al., "Small Worlds and Regional Innovation," *Organization Science* 18, no. 6 (2007): 938–54; and on collaboration, see Kjersten B. Whittington et al., "Networks, Propinquity, and Innovation in Knowledge-Intensive Industries." *Administrative Science Quarterly* 54, no. 1 (2009): 90–122.

26. Michael E. Bratman, *Shared Agency: A Planning Theory of Acting Together* (Oxford University Press, 2013); Max Weber, *Economy and Society* (University of California Press, 1978). This understanding of social action follows the standard Weberian definition of an act that takes others into consideration. Philosophers associate action with self-awareness stemming from an engagement with the outside world. Sociologists and philosophers of action attribute action to groups of people, but the idea of collective or group agency is contentious.

27. As organizational economist Bob Gibbons reminds us, the idea of "the firm" as the unitary decision-maker in markets is problematic because this simplification denies the fact that firms are in themselves political coalitions with "unresolved political conflicts." See James G. March and Herbert A. Simon, *Organizations* (Wiley, 1993); Robert Gibbons, "March-ing Toward Organizational Economics," *Industrial and Corporate Change* 29, no. 1 (2020): 89–94; James G March, "The Business Firm as a Political Coalition," *Journal of Politics* 24, no. 4 (1962): 662–78.

28. Hillary Angelo and David Wachsmuth, "Urbanizing Urban Political Ecology: A Critique of Methodological Cityism," *International Journal of Urban and Regional Research* 39, no. 1 (2015): 16–27. The epistemological problem of city agency also resonates with a critique of methodological cityism.

29. James Coleman, *Foundations of Social Theory* (Harvard University Press, 1990); Peter Hedström, *Dissecting the Social: Social Mechanisms and the Principles of Analytical Sociology* (Cambridge University Press, 2005).

30. Jesper Strandgaard Pedersen and Frank Dobbin, "The Social Invention of Collective Actors on the Rise of the Organization," *American Behavioral Scientist* 40, no. 4 (1997): 431–43; Brayden G. King et al., "Perspective-Finding the Organization in Organizational Theory: A Meta-Theory of the Organization as a Social Actor," *Organization Science* 21, no. 1 (2010): 290–305; Amanda Sharkey and Patricia Bromley, "Can Ratings Have Indirect Effects? Evidence from the Organizational Response to Peers' Environmental Ratings," *American Sociological Review* 80, no. 1 (2014): 63–91.

31. It is from this institutionalist view on actorhood that I arrived at the definition of city action as discretionary, purposive, and relational. Institutionalists John Meyer and Ron Jepperson note the importance of not taking the actorhood of any individual or organization for granted but asking how it is constituted and shaped: "Modern culture depicts society as made up of 'actors'—individuals and nation-states, together with the organizations derived from them. Much social science takes this depiction at face value, and takes for granted that analysis must start with these actors and their perspectives and actions." John W. Meyer and R. Jepperson, "The 'Actors' of Modern Society: The Cultural Construction of Social Agency," *Sociological Theory* 18, no. 1 (2000): 100–120; Hokyu Hwang and Jeannette A Colyvas, "Ontology, Levels of Society, and Degrees of Generality: Theorizing Actors as Abstractions in Institutional Theory," *Academy of Management Review* 45, no. 3 (2020): 570–95.

32. Ronald Jepperson and John W. Meyer, "Multiple Levels of Analysis and the Limitations of Methodological Individualisms," *Sociological Theory* 29, no. 1 (2011): 9.

33. James Coleman views the organization as "a system of action in which relations among actors are highly constrained by the social structure." James Coleman, *Foundations of Social Theory* (Harvard University Press, 1990), 426; John Levi Martin, *The Explanation of Social Action* (Oxford University Press, 2011).

34. Mark Granovetter, "Economic Action and Social Structure: The Problem of Embeddedness," *American Journal of Sociology* 91, no. 3 (1985): 481–510.

35. Patricia Bromley and Amanda Sharkey, "Casting Call: The Expanding Nature of Actorhood in U.S. Firms, 1960–2010," *Accounting, Organizations and Society* 59 (2017): 3–20; Peter L. Berger and Thomas Luckmann, *The Social Construction of Reality: A Treatise in the Sociology of Knowledge* (Anchor, 1966); John W. Meyer and Brian Rowan, "Institutionalized Organizations: Formal Structure as Myth and Ceremony," *American Journal of Sociology* 83, no. 2 (1977): 340–63.

36. Harvey Molotch et al., "History Repeats Itself, but How? City Character, Urban Tradition, and the Accomplishment of Place," *American Sociological Review* 65, no. 6 (2000): 791–823; David Harvey, *A Brief History of Neoliberalism* (Oxford University Press, 2007); David Harvey, *Rebel Cities: From the Right to the City to*

the Urban Revolution (Verso, 2012), http://www.worldcat.org/title/rebel-cities
-from-the-right-to-the-city-to-the-urban-revolution/oclc/943969433; Sassen,
Global City.

37. Harvey Molotch, "The City as a Growth Machine: Toward a Political Economy
of Place," *American Journal of Sociology* 82, no. 2 (1976): 309–32.

38. Floyd Hunter, *Community Power Structure: A Study of Decision Makers* (University of North Carolina Press, 1953).

39. Cristobal Young, *The Myth of Millionaire Tax Flight: How Place Still Matters
for the Rich* (Stanford University Press, 2017).

40. Berger and Luckmann, *Social Construction of Reality*; Amy Binder, "For Love
and Money: Organizations' Creative Responses to Multiple Environmental
Logics," *Theory and Society* 36, no. 6 (2007): 547–71.

41. Pamela Tolbert and Lynne G. Zucker, "Institutional Sources of Change in the
Formal Structure of Organizations: The Diffusion of Civil Service Reform,
1880–1935," *Administrative Science Quarterly* 28, no. 1 (1983): 22–39.

42. Frank Dobbin, *Inventing Equal Opportunity* (Princeton University Press,
2009), 16.

43. Evan Schofer and Ann Hironaka, "The Effects of World Society on Environmental Protection Outcomes," *Social Forces* 84, no. 1 (2005): 25–47; Wesley
Longhofer and Evan Schofer, "National and Global Origins of Environmental
Association," *American Sociological Review* 75, no. 4 (2010): 505–33.

44. McQuarrie and Marwell, "Missing Organizational Dimension in Urban
Sociology."

45. My model also seeks to transcend classic perspectives in environmental sociology, which are either concerned with the idea that the treadmill of economic
production necessarily leads to ecological degradation, or that ecological modernization lowers carbon emissions. What these two frameworks have in common when applied to issues of urban sustainability is that they leave no room
for cities to appear as organizational actors in their own right, but they see cities as sites of the actions of greater powers. My organizational approach shifts
away from debating whether cities can balance economic growth with sustainability to examining how and why they engage in purposive, discretionary, and
relational actions to respond to environmental challenges.

46. Weber, *Economy and Society.*

47. Park was born in the same year as Max Weber but was a late convert to sociology and promoted to full professor of sociology at age fifty-nine. Edward
Shils, "The Sociology of Robert E. Park," *American Sociologist* 27, no. 4 (1996):
88–106.

48. To illustrate this point, Park cites Darwin, who explained that the red clover
surprisingly owes its existence to the cat: Humble-bees alone visit red clover, as

other bees cannot reach the nectar. The inference is that if "the whole genus of humble-bees became extinct or very rare in England, the heartsease and red clover would become very rare, or wholly disappear. [However, t]he number of humblebees in any district depends in a great degree on the number of field-mice, which destroy their combs and nests[.]" Thus, next year's crop of red clover in certain parts of England depends on the number of humble-bees in the district; the number of humble-bees depends on the number of field mice, and the number of field mice depend on the number and the enterprise of the cats. Charles Darwin, *On the Origin of Species by Means of Natural Selection, or the Preservation of Favoured Races in the Struggle for Life* (Murray, 1859), 73–74.

49. This idea belonged to the "same intellectual family," Shils writes, as "Tönnies's idea of Gesellschaft, Simmel's ideas about the differentiated individualistic urban society, in which money plays a great part as a measure of value, and finally Max Weber's ideas about modern capitalistic and bureaucratized society." Shils, "Sociology of Robert E. Park," 90.

50. Robert E. Park, "Succession, an Ecological Concept," *American Sociological Review* 1, no. 2 (1936): 139. Robert J. Sampson, "Neighbourhood Effects and Beyond: Explaining the Paradoxes of Inequality in the Changing American Metropolis," *Urban Studies* 56, no. 1 (2019): 3–32. Robert Park explains the distinction as follows: "Human society, as distinguished from plant and animal society, is organized on two levels, the biotic and the cultural. There is a symbiotic society based on *competition* and a cultural society based on *communication and consensus* . . . The two societies are merely different aspects of one society, which, in the vicissitudes and changes to which they are subject remain, nevertheless, in some sort of mutual dependence each upon the other. The cultural superstructure rests on the basis of the symbiotic substructure, and the emergent energies that manifest themselves on the biotic level in movements and actions reveal themselves on the higher social level in more subtle and sublimated forms."

51. This view aligns with dominant theories of organizations as being interconnected through organizational networks and embedded in an institutional environment. Arthur L. Stinchcombe, "Social Structure and Organizations," in *Handbook of Organizations*, ed. James G. March (Rand McNally, 1965), 142–93; Walter W. Powell and Paul DiMaggio, eds., *The New Institutionalism in Organizational Analysis* (University of Chicago Press, 1991).

52. I refer to city administrations rather than local government: Although local governments may give direction to what city administration does, a multitude of others factors motivate city administrations to act in contemporary systems of urban governance, and I am agnostic to this aspect of urban politics—this means that local government deliberations, including the activities of

mayors and city councils are "black-boxed," for the time being, but it does not mean that they do not matter, as we will see in chapter 5.

53. Corporations are often considered as parts of the "growth coalition" that has an inhibiting effect on progressive policies—especially including tax increases, welfare policies, and environmental protection. Because corporations have many political orientations, it may make sense to distinguish between the mere presence of firms and the degree to which they embrace their roles as corporate citizens. Sharkey and Bromley, "Can Ratings Have Indirect Effects?"; Marquis et al., "Community Isomorphism and Corporate Social Action."

54. A focus on organizations should not conceal the people working in those organizations. The question of who actually implements urban transitions—be it recharging electric vehicles at night, building sea walls, or mounting clean-up efforts after an extreme weather event—is an important question that calls attention to the "human infrastructure" of labor and laborers. As sociologists John Krinsky and Maud Simonet remind us in their analysis of urban governance of New York parks, labor is a frequently forgotten dimension of this organizational infrastructure. In one striking example, the politics of garbage in Dakar, Senegal, urban geographer Rosalind Fredericks highlights that some of the most fundamental urban services like waste disposal are underpinned by "infrastructures of labor." Rosalind Fredericks, *Garbage Citizenship: Vital Infrastructures of Labor in Dakar, Senegal* (Duke University Press, 2018); John Krinsky and Maud Simonet, *Who Cleans the Park? Public Work and Urban Governance in New York City* (University of Chicago Press, 2017).

55. Although appropriate for understanding the mechanisms undergirding city action, these interviews are not a full representation of what all cities experience or plan—I wish I could have spoken to many more people in Africa, Latin America, Eastern Europe, and the Asia-Pacific region to better understand not only the barriers to inaction but also potential motivations for city inaction. Extending the qualitative component of this book to other places remains a major area for future research.

56. In other words, the precise meaning of city climate action is socially constructed. Consider, in analogy, the many often euphemistic meanings of "floating signifiers" like community or public participation discussed by Jeremy R. Levine, "The Paradox of Community Power: Cultural Processes and Elite Authority in Participatory Governance," *Social Forces* 95, no. 3 (2017): 1155–79. Broad concepts like sustainability often persist over time not because their meaning is solidified, but rather because their meaning adapts to new and often unpredictable contexts. For example, the idea of public housing persisted over a century in Vienna not because it was institutionalized but rather because it took on new meanings as the institutional context shifted. Christof Brandtner et al.,

"Dynamic Persistence of Institutions: Modeling the Historical Endurance of Red Vienna's Public Housing Utopia," *Organization Studies* (2025), https://doi .org/10.1177/01708406251317258.

2. PLANNING

1. Christof Brandtner et al., "Enacting Governance Through Strategy: A Comparative Study of Governance Configurations in Sydney and Vienna," *Urban Studies* 54, no. 5 (2016): 1075–91. The paper contains methodological considerations for the document analysis presented in this chapter. Further details are discussed or referenced in the methodological appendix. See also Martin Kornberger, "Governing the City: From Planning to Urban Strategy," *Theory, Culture, and Society* 29, no. 2 (2012): 84–106.

2. Richard Florida, *The Rise of the Creative Class* (Basic, 2002).

3. The City of New York, *PlaNYC: A Greener, Greater New York*, 2007, https:// www.nyc.gov/html/planyc/downloads/pdf/publications/full_report_2007.pdf.

4. Mercer surveys expatriates to arrive at its rankings, which has been a regular source of contention with respect to the validity and meaning of these rankings. In reality, there is a wide variety of different city rankings, discussed in chapter 3, which city leaders cherry pick; see Christof Brandtner, "Putting the World in Orders: Plurality in Organizational Evaluation," *Sociological Theory* 35, no. 3 (2017): 200–227.

5. When I returned to Hong Kong thirteen years later, the bypass was there, as was a new art district in West Kowloon, and a mind-boggling "Hong Kong 2030+" strategy preparing to integrate the city with Shenzhen, Guangzhou, and Macau into a Greater Bay Area of eighty-six million people. All three initiatives were following a strategy of "Smart, Green, and Resilient" (SGR). The city was also planning to become carbon neutral before 2050. I came away from my conversation with the chief city planner with a stack of plans and realized that this goal of urban sustainability was not unique to North American and European cities—many "overseas" cities were, in fact, ahead of the curve.

6. A master frame is an umbrella for a variety of different goals, grievances, and guidelines for action. For example, the idea that "all humans have equal rights" is a possible master frame underlying various specific concerns about class, gender, and race inequality. As Rao, Monin, and Durand discuss the idea that haute cuisine should be fresh and creative as a master frame of the nouvelle cuisine movement, the idea that life in a city should be sustainable—with a variety of more nuanced meanings and implications—is a common master frame underpinning cities' strategic turn. Hayagreeva Rao et al., "Institutional Change

in Toque Ville: Nouvelle Cuisine as an Identity Movement in French Gastron-omy," *American Journal of Sociology* 108, no. 4 (2003): 795–843.

7. This trend toward publishing strategies—even before climate change became its primary frame—was initially seen as a drive toward "competitiveness" brought about by neoliberal ideology seeking to curb government interven-tions to the benefit of free markets. I will turn to this potential explanation in chapter 3. See David Wachsmuth, "City as Ideology: Reconciling the Explosion of the City Form with the Tenacity of the City Concept," *Environment and Plan-ning D: Society and Space* 32, no. 1 (2014): 75–90.

8. "Vienna in Figures 2022," Statistics Vienna, 2022, https://www.wien.gv.at/statistik/pdf/viennainfigures-2022.pdf. In 2022, the City of Vienna employed roughly one hundred thousand people, compared with some sixty thousand employed by the European Union.

9. Joseph W. Kane et al., "Not According to Plan: Exploring Gaps in City Climate Planning and the Need for Regional Action" (Brookings Institution, 2022), https://www.brookings.edu/articles/not-according-to-plan-exploring-gaps-in-city-climate-planning-and-the-need-for-regional-action. Policy think tanks like the Brookings Institution regularly produce white papers showing that the pledges of most cities do not translate into "action," encouraging cities to move from inaction to action.

10. There is significant variation also within countries that have a national plan-ning regime generally unfavorable to city action. Consider, for example, the difference between Mumbai and Surat. Mumbai, India, is the "second most-at-risk coastal city in the world" and is "expected to see economic damages to $162.2 billion by 2050." According to Singh et al., the city not only lacks a strategy plan but also any "sectoral or city-scale planned adaptation projects or policies." Contrast this dearth of action with Surat, whose early strategic dis-cussions of flood management, energy consumption, and natural resource con-versation in 2011 that lead Singh et al. to cite Surat as an example of "successful adaptation." The authors state that "Surat provides an example of initiating city resilience through project-based interventions and using them to implement longer-term, sectoral projects." Chandni Singh et al., "Climate Change Adapta-tion in Indian Cities: A Review of Existing Actions and Spaces for Triple Wins," *Urban Climate* 36 (2021): 100783, 4, 14; Xuefei Ren, *Governing the Urban in China and India: Land Grabs, Slum Clearance, and the War on Air Pollution* (Princeton University Press, 2020).

11. David Strang and Sarah A. Soule, "Diffusion in Organizations and Social Movements: From Hybrid Corn to Poison Pills," *Annual Review of Sociology* 24 (1998): 265–90.

12. Everett Rogers, *Diffusion of Innovations*, 5th ed. (Free Press, 2003).

13. Based on the number of cities that published a plan that we coded as purposive, visionary, and holistic between 2000 and 2023. The remaining 170 cities had no such strategic plan, or did not make it publicly accessible.

14. In a review article, Ivana Naumovska, Vibha Gaba, and Henrich R. Greve estimate that among more than 80 percent of practices studied by diffusion researchers, this practice diffused to less than 40 percent of potential adopters (in 45 percent of cases, to less than 20 percent). For a detailed discussion of organizational research on innovation diffusion, see Ivana Naumovska et al., "The Diffusion of Differences: A Review and Reorientation of 20 Years of Diffusion Research," *Academy of Management Annals* 15, no. 2 (2021): 377–405; Strang and Soule, "Diffusion in Organizations and Social Movements."

15. NYC Mayor's Office of Climate & Environmental Justice, *PlaNYC: Getting Sustainability Done*, 2023, https://www.nyc.gov/content/climate/pages/reports-and-publications/planyc.

16. The City of Amsterdam, *New Amsterdam Climate Roadmap: Amsterdam Climate Neutral 2050*, 2020, https://assets.amsterdam.nl/publish/pages/943415/roadmap_amsterdam_climate_neutral_2050_2.pdf.

17. The City of Brussels, *City of Brussels Climate Plan*, 2022, https://www.brussels.be/sites/default/files/bxl/221130%20%20Plan%20Climat%20Version%20finale%20EN.pdf.

18. The City of Brussels, *City of Brussels Climate Plan.*

19. Kuala Lumpur City Hall, *Kuala Lumpur Climate Action Plan 2050*, 2021, https://aipalync.org/storage/documents/main/kuala-lumpur-climate-action-plan-2025_1713867229.pdf.

20. Although sustainability, resilience, or smartness can have specific meanings with respect to the environment, disasters, or technology, when they are used as master frames, such concerns take on a broader meaning. Sustainability, for instance, often follows the tripartite Brundtland definition of acting to meet current economic, social, and environmental needs without compromising the ability of future generations to do the same. These so-called open or floating signifiers defy narrow definition, meaning what you want them to. What terms like sustainability or resilience really mean in a particular context can depend on the specific needs of the city. For a more thorough discussion, see chapter 6.

21. Gemeente Rotterdam, *Rotterdam Resilience Strategy: Ready for the 21st Century*, 2016, https://resilientcitiesnetwork.org/downloadable_resources/Network/Rotterdam-Resilience-Strategy-English.pdf.

22. The City of Hiroshima, *International Peace Culture City: Connecting Hiroshima to the Future (6th Hiroshima City Basic Plan)*, 2020, https://www.city.hiroshima.lg.jp/_res/projects/default_project/_page_/001/009/818/220813_339228_misc.pdf.

23. The City of Hiroshima, *International Peace Culture City.*
24. The City of Hiroshima, *International Peace Culture City.*
25. Brandtner et al., "Enacting Governance Through Strategy."
26. Barbara Czarniawska, *A Tale of Three Cities: Or the Glocalization of City Management* (Oxford University Press, 2002); Paul J. DiMaggio and Walter W. Powell, "The Iron Cage Revisited: Institutional Isomorphism and Collective Rationality in Organizational Fields," *American Sociological Review* 48, no. 2 (1983): 147–60.
27. Patricia Bromley and Amanda Sharkey, "Casting Call: The Expanding Nature of Actorhood in U.S. Firms, 1960–2010," *Accounting, Organizations and Society* 59 (2017): 3–20.
28. "Text as data" refers to a set of computational methods used to describe, classify, and analyze large corpora of text. See Justin Grimmer et al., *Text as Data: A New Framework for Machine Learning and the Social Sciences* (Princeton University Press, 2022).
29. I used Google's OCR software Tesseract, to make all PDFs machine readable and translated the text in all fifteen languages into English using the Google Translate API. I then used a so-called TF-IDF weighted word scores to estimate the prevalence of certain terms, considering how common they were. TF-IDF measures a word's term frequency (TF) over the inversed document frequency (IDF), thus ranking highest terms that are common in one document while uncommon in other documents.
30. The District of Columbia, *Sustainable DC Plan*, 2013, https://sustainable.dc .gov/sites/default/files/dc/sites/sustainable/page_content/attachments /SustainableDCPlan_web.pdf.
31. Cynthia Rosenzweig et al., "Cities Lead the Way in Climate–Change Action," *Nature* 467, no. 7318 (2010): 909–11.
32. Harvey Molotch, "The City as a Growth Machine: Toward a Political Economy of Place," *American Journal of Sociology* 82, no. 2 (1976): 309–32; Jeremy R. Levine, *Constructing Community: Urban Governance, Development, and Inequality in Boston* (Princeton University Press, 2021).
33. Robert Kunzig, "The World's Most Improbable Green City," *National Geographic*, April 4, 2017, https://www.nationalgeographic.com/environment /article/dubai-ecological-footprint-sustainable-urban-city.
34. Singapore Ministry of National Development, "About the Project," https:// www.mnd.gov.sg/tianjinecocity/who-we-are.
35. John W. Meyer and Brian Rowan, "Institutionalized Organizations: Formal Structure as Myth and Ceremony," *American Journal of Sociology* 83, no. 2 (1977): 340–63.; Patricia Bromley and Walter W Powell, "From Smoke and Mirrors to Walking the Talk: Decoupling in the Contemporary World," *Academy of Management Annals* 6, no. 1 (2012): 483–530.

36. Christof Brandtner and Patricia Bromley, "Neoliberal Governance, Evaluations, and the Rise of Win–Win Ideology in Corporate Responsibility Discourse, 1960–2010," *Socio-Economic Review* 20, no. 4 (2021): 1933–60.

37. Dan Honig, *Mission Driven Bureaucrats: Empowering People to Help Government Do Better* (Oxford University Press, 2024).

38. Linda Shi et al., "Explaining Progress in Climate Adaptation Planning across 156 US Municipalities," *Journal of the American Planning Association* 81, no. 3 (2015): 191–202.

39. Michael Lounsbury et al., "Social Movements, Field Frames and Industry Emergence: A Cultural–Political Perspective on US Recycling," *Socio-Economic Review* 1, no. 1 (2003): 71–104.

40. Christof Brandtner, "Decoupling Under Scrutiny: Consistency of Managerial Talk and Action in the Age of Nonprofit Accountability," *Nonprofit and Voluntary Sector Quarterly* 50, no. 5 (2021): 1053–78.

41. Mariana Mazzucato, *Mission Economy: A Moonshot Guide to Changing Capitalism* (Penguin UK, 2021).

42. The view of organizations as orchestrators contrasts with their role as innovators and agitators. This debate also mirrors the broader discussion of governments as mere coordinators in a system of network governance, which has criticized the encroachment of private "social innovators" on the public sector. See Brandtner and Bromley, "Neoliberal Governance, Evaluations, and the Rise of Win–Win Ideology"; Chris Ansell and Alison Gash, "Collaborative Governance in Theory and Practice," *Journal of Public Administration Research and Theory* 18, no. 4 (2008): 543–71; Julie Battilana and Marissa Kimsey, "Should You Agitate, Innovate, or Orchestrate?," *Sanford Social Innovation Review*, September 18, 2017 https://doi.org/10.48558/3YGB-3M56; Jason Spicer et al., "Social Entrepreneurship as Field Encroachment: How a Neoliberal Social Movement Constructed a New Field," *Socio-Economic Review* 17, no. 1 (2019): 195–227.

43. The City of Amsterdam, *New Amsterdam Climate Roadmap*.

44. Michael Burawoy, *Manufacturing Consent: Changes in the Labor Process Under Monopoly Capitalism* (University of Chicago Press, 2012). The analogy to Burawoy's work—which focuses on labor relations in capitalist systems—is that the public may participate in its own subordination by elites. Although strategic planning is a discursive device that helps get different constituencies on the same page about a unified vision for the city, I did not find any support for the Marxian version of this argument. See chapter 4 for a deeper discussion of the participatory production of city strategies for sustainability.

45. The city's Smart City Strategy is deeply integrated with the UN's Sustainable Development Goals (SDGs) and mentions citizen participation. As the city's

sustainability coordinator told me, gathering existing projects can be a legitimate first step, but ultimately only if it is in the interest of involving the public: "It's legitimate way to start—as long as you move forward from there."

46. Martin Kornberger et al., "Exploring the Long-Term Effect of Strategy Work: The Case of Sustainable Sydney 2030," *Urban Studies* 58, no. 16 (2021): 3316–34.

47. Brandtner et al., "Enacting Governance Through Strategy."

48. The chief of staff of the city councilor responsible for housing in a major European city underscored how relevant some of these priorities are for people's most private decisions. Imagine you just bought an apartment and are now thinking about how to furnish it. Whether you put in a gas stove or cook electric, whether you will choose to install an air-conditioning unit, a heat pump, or no technology to regulate temperature, and even the thickness of your windows will be directly informed by economic incentives and regulatory frameworks there are directly derived from the strategic ambitions of municipal governments.

49. For one of the best attempts at doing so, see Peter Hall, *Cities of Tomorrow: An Intellectual History of Urban Planning and Design Since 1880* (Wiley-Blackwell, 2014). For understanding how the more recent transformation fits into the historical arc, there are interesting provocations from both the perspective of strategy scholars with an urban bend and critical urban scholars. On the former, see Kornberger, "Governing the City"; on the latter, see Hillary Angelo and Gianpaolo Baiocchi, "The Moral Work of Participation: Disillusio, Expertise, and Urban Planning Under Neoliberalism," *Qualitative Sociology* 47, no. 3 (2024): 493–515. Both find some blame in the emergence of the neoliberal form of governance that inspired my initial inquiries into the topic noted at the beginning of the chapter.

50. The absence of plans in the Chinese context illustrates well that it is possible to see a large-scale ecological transition in cities without city action being responsible for it. Beijing, China, for example, has no publicly accessible strategy plan but nevertheless become a global leader in electrification and recorded significant progress in terms of its air pollution. Nevertheless, the integration of the city's planning into the national planning framework indicates a lack of autonomy documented by comparative urbanist Xuefei Ren. This is not so say that civil society and objections to environmental degradation are absent, but rather that another process underpins the activities than in other cities, as Jean Lin's research on middle-class neighborhood resistance to incinerator projects chronicles. See Jean Yen-chun Lin, *A Spark in the Smokestacks: Environmental Organizing in Beijing Middle-Class Communities* (Columbia University Press, 2023); Lin Zhang et al., "Heterogeneity of Public Participation in Urban Redevelopment in Chinese Cities: Beijing Versus Guangzhou," *Urban Studies* 57, no. 9 (2020): 1903–19; Ren, *Governing the Urban in China and India*.

51. Meyer and Rowan, "Institutionalized Organizations."

52. John W. Meyer et al., "World Society and the Nation-State," *American Journal of Sociology* 103, no. 1 (1997): 144–81; Gili S. Drori et al., "Sources of Rationalized Governance: Cross-National Longitudinal Analyses, 1985–2002," *Administrative Science Quarterly* 51, no. 2 (2006): 205–29.

53. Georg Krücken and Frank Meier, "Turning the University into an Organizational Actor," in *Globalization and Organization: World Society and Organizational Change*, ed., Gili S. Drori, John W. Meyer, and Hokyu Hwang (Oxford University Press, 2006), 241–57; Gili S Drori et al., *Globalization and Organization: World Society and Organizational Change* (Oxford University Press, 2006); Patricia Bromley and John W. Meyer, *Hyper-Organization: Global Organizational Expansion* (Oxford University Press, 2015); Frank Dobbin, *Inventing Equal Opportunity* (Princeton University Press, 2009).

54. Christof Brandtner et al., "From Iron Cage to Glass House: Repurposing of Bureaucratic Management and the Turn to Openness," *Organization Studies* 45, no. 2 (2024): 193–221; Krystal Laryea and Christof Brandtner, "Organizations as Drivers of Social and Systemic Integration: Contradiction and Reconciliation Through Loose Demographic Coupling and Community Anchoring," *Research in the Sociology of Organizations* 90 (2024): 177–200; Christof Brandtner and Walter W. Powell, "Capturing the Civic Lives of Cities: An Organizational, Place-Based Perspective on Civil Society in Global Cities," *Global Perspectives* 3, no. 1 (2022): 36408.

55. John W. Meyer and Ronald Jepperson, "The 'Actors' of Modern Society: The Cultural Construction of Social Agency," *Sociological Theory* 18, no. 1 (2000): 100–120.

3. LEARNING

1. Jonathon Kass, "How Copenhagen Can Inspire Bay Area Cities to Go Big on Bikes," SPUR Urban Center, August 31, 2022, https://www.spur.org/news/2022-08-31/how-copenhagen-can-inspire-bay-area-cities-go-big-bikes.

2. John Padgett and Woody Powell, *The Emergence of Organizations and Markets* (Princeton University Press, 2012). In their book, sociologists John Padgett and Woody Powell argue that ripple effects to fields that are relatively far removed from municipal governments in a narrow sense—in national governments, universities, corporations, foundations, and professional associations—indicate innovation.

3. John L. Campbell, *Institutional Change and Globalization* (Princeton University Press, 2004).

4. José Luis Alvarez and Silviya Svejenova, *The Changing C-Suite: Executive Power in Transformation* (Oxford University Press, 2022); also see Patricia Bromley

and John W. Meyer, *Hyper-Organization: Global Organizational Expansion* (Oxford University Press, 2015).

5. Rakesh Khurana, *From Higher Aims to Hired Hands: The Social Transformation of American Business Schools and the Unfulfilled Promise of Management as a Profession* (Princeton University Press, 2010); Michel Anteby, *Manufacturing Morals: The Values of Silence in Business School Education* (University of Chicago Press, 2013).

6. Other design consultancies with urban strategy divisions include London-based Arup, Rotterdam-based KCAP Architects and Planners, and San Francisco-based Gensler. For example, the latter writes on its webpage: "We believe that thoughtful design connects nature, technology, and people while planning for resilience in a changing world. The places we create are rooted in local culture and act as a catalyst for socially sustainable urban transformation." Gensler, "Cities and Urban Design," https://www.gensler.com/expertise/cities -urban-design; Gehl Architects, "Home—Gehl," https://www.gehlpeople.com.

7. Saskia Sassen, *Cities in a World Economy*, 4th ed. (Pine Forge, 2012).

8. For organizations shown, one of the terms was listed as a keyword of the organization, either because it is in the title or a mission statement. This likely produces a conservative estimate; as of March 2025, the Yearbook of International Organizations lists more than four thousand organizations related to Sustainable Development Goal 11 (SDG 11), Sustainable Cities and Communities. Since findings are often reported in retrospect, the data can show right censoring, or time-lag bias. It is, however, possible that the number of newly founded international nongovernmental organizations (INGOs) has generally declined. See Sarah Sunn Bush and Jennifer Hadden, "Density and Decline in the Founding of International NGOs in the United States," *International Studies Quarterly* 63, no. 4 (2019): 1133–46.

9. Harriet Bulkeley and Vanesa Castán Broto, "Government by Experiment? Global Cities and the Governing of Climate Change," *Transactions of the Institute of British Geographers* 38, no. 3 (2013): 361–75; Laura Tozer et al., "Transnational Governance and the Urban Politics of Nature-Based Solutions for Climate Change," *Global Environmental Politics* 22, no. 3 (2022): 81–103.

10. The Rockefeller Foundation created 100RC as a nonprofit organization to implement one of the major grant programs that were part of its one-hundredth anniversary. In 2012, "we were barreling towards our centennial year in 2013. . . . But the centennial shouldn't be just looking back and reflecting on our history. We didn't want to just be black-and-white photographs of men in top hats. There was an opportunity to do something big that we should take advantage of. . . . Around the same time, Sandy hit. We mobilized to do a lot of work here in response to that. . . . A lot of that engagement was driven by

learnings from the work that we had done in New Orleans post-Katrina, which when that started in 2005 we weren't calling resilience but had I think quite a resilience approach to it in terms of thinking about physical rebuilding but also social rebuilding."

11. Resilient Cities Network, "Homepage," https://resilientcitiesnetwork.org; Brandt-ner et al., "From Iron Cage to Glass House."

12. Wade M. Cole, "World Polity or World Society? Delineating the Statist and Societal Dimensions of the Global Institutional System," *International Sociology* 32, no. 1 (2017): 86–104; John W. Meyer et al., "World Society and the Nation-State," *American Journal of Sociology* 103, no. 1 (1997): 144–81; Sassen, *Cities in a World Economy*; Richard Florida, *The Rise of the Creative Class* (Basic Books, 2002).

13. David John Frank et al., "The Nation-State and the Natural Environment over the Twentieth Century," *American Sociological Review* 65, no. 1 (2000): 96–116; Sassen, *Cities in a World Economy*; Florida, *Rise of the Creative Class*.

14. Urban sociologist Harvey Molotch defines a city as "the areal expression of the interests of some land-based elite, [which] competes with other land-based elites in an effort to have growth-inducing resources invested within its own area as opposed to that of another." In other words, real estate developers, business owners, and other members of a "growth coalition" promote specific policies that pursue economic growth at the expense of other community needs. In this elitist system, local authorities are not the authors of the city's future, but they are rather "utilized to assist in achieving this growth at the expense of competing localities." In such a world, cities dominated by a growth machine would see strategies as an opportunity to amass land and enrich themselves, even if that comes at the cost of marginalized communities. Harvey Molotch, "The City as a Growth Machine: Toward a Political Economy of Place," *American Journal of Sociology* 82, no. 2 (1976): 309.

15. Bob Jessop, "The Entrepreneurial City: Re-Imaging Localities, Redesigning Economic Governance, or Restructuring Capital," in *Transforming Cities: Contested Governance and New Spatial Divisions*, ed. Nick Jewson and Susanne MacGregor, vol. 46 (Routledge, 1997), 28–41; Bob Jessop and Ngai-Ling Sum, "An Entrepreneurial City in Action: Hong Kong's Emerging Strategies in and for (Inter) Urban Competition," *Urban Studies* 37, no. 12 (2000): 2287–313.

16. Another concept used by sociologists is power in a world city system, but this term is conceptually close to what I mean by economic status. For a nuanced discussion, see Arthur S. Alderson and Jason Beckfield, "Power and Position in the World City System," *American Journal of Sociology* 109, no. 4 (2004): 811–51; Peter J. Taylor, "Comment: Parallel Paths to Understanding Global Intercity Relations," *American Journal of Sociology* 112, no. 3 (2006): 881–94.

17. The actual ranking is more nuanced, differentiating, for example, between alpha++, alpha+, alpha, and alpha– cities. This ranking is Illustrated well at GaWC, "Globalization and World Cities," https://gawc.lboro.ac.uk.

18. Wendy Nelson Espeland and Michael Sauder, *Engines of Anxiety: Academic Rankings, Reputation, and Accountability* (Russell Sage, 2016).

19. Norman D. Ford, *America's 50 Best Cities in Which to Live, Work, and Retire* (Harian, 1958); Rudy Koshar, *German Travel Cultures* (Berg, 2000). The evaluative landscape of cities has become more rugged in recent years, combining a variety of "soft" rankings that examine livability or greenness and "hard" factors that measure competitiveness or the city's credit worthiness. For a discussion of city rankings, see Christof Brandtner, "Putting the World in Orders: Plurality in Organizational Evaluation," *Sociological Theory* 35, no. 3 (2017): 200–227. On the political dynamics of city credit ratings, also see Davon Norris, "The Illusion of Transparency: The Political Double Standard in City Credit Ratings," *Socio-Economic Review* 21, no. 2 (2023): 1125–50.

20. Michèle Lamont, "Toward a Comparative Sociology of Valuation and Evaluation," *Annual Review of Sociology* 38, no. 1 (2012): 201–21; Wendy N. Espeland and Mitchell L. Stevens, "A Sociology of Quantification," *European Journal of Sociology* 49, no. 3 (2008): 401–36; Gili S. Drori et al., *Science in the Modern World Polity: Institutionalization and Globalization* (Stanford University Press, 2003); Gili S. Drori et al., *Globalization and Organization: World Society and Organizational Change* (Oxford University Press, 2006).

21. Martin Kornberger and Chris Carter, "Manufacturing Competition: How Accounting Practices Shape Strategy Making in Cities," *Accounting, Auditing, and Accountability Journal* 23, no. 3 (2010): 325–49; Rudolf Giffinger et al., "The Role of Rankings in Growing City Competition," *Urban Research and Practice* 3, no. 3 (2010): 299–312; Brandtner, "Putting the World in Orders."

22. Paul L. Knox and Peter J. Taylor, *World Cities in a World-System* (Cambridge University Press, 1995); Sassen, *Cities in a World Economy.*

23. Patricia Bromley and Amanda Sharkey, "Casting Call: The Expanding Nature of Actorhood in US Firms, 1960–2010," *Accounting, Organizations and Society* 59 (2017): 3–20; Amanda Sharkey and Patricia Bromley, "Can Ratings Have Indirect Effects? Evidence from the Organizational Response to Peers' Environmental Ratings," *American Sociological Review* 80, no. 1 (2014): 63–91; Espeland and Sauder, *Engines of Anxiety.*

24. Economist Impact, Resilient Cities Index, 2023 https://impact.economist.com/projects/resilient-cities/assets/documents/Resilient-Cities_Report.pdf.

25. Kornberger and Carter, "Manufacturing Competition."

26. Brandtner, "Putting the World in Orders."

27. A policy director in the city clerk's office told me about another intriguing instance of city learning. The city clerk's office was working on a so-called municipal identification (ID), now CityKey, allowing undocumented immigrants to partake in the city's social and economic life. You cannot found a business or rent a car without government-issued ID, and in cities trying to provide sanctuary for immigrants whose fortune did not come with papers, integrating them was a major political, social, and even economic factor. Chicago's mayors—then Rahm Emanuel and later Lori Lightfoot—were trying to change this. And they were not the first. New York had a similar municipal ID policy and had begun to offer workshops for other cities to learn about the intricacies of the policy. As it turns out, the municipal ID New York had developed created a significant risk in light of a requirement to store the addresses of applicants. The federal government, then on an anti-immigrant crusade that was becoming manifest in a literal wall at the country's southern border, could now from the comfort of its offices request a list of many undocumented immigrants in the city and start deporting them, one by one. Chicago was eager to learn how they could overcome these problems. Els De Graauw, *Making Immigrant Rights Real: Nonprofits and the Politics of Integration in San Francisco* (Cornell University Press, 2016); Ross Barkan, "What Happens to New York's Municipal ID Card Under the Trump Administration?," *New York Magazine*, December 4, 2016, https://nymag.com/intelligencer/2016/12/what-happens-to-idncy-under-the-trump-administration.html.

28. Urban Sustainability Directors Network, "History," https://www.usdn.org/history.html.

29. Sadhu Aufochs Johnston et al., *The Guide to Greening Cities* (Springer, 2013).

30. Sara McTarnaghan et al., *Urban Resilience: From Global Vision to Local Practice* (Urban Institute, 2022).

31. This trust stems from repeated interactions. See Mark Granovetter, *Getting a Job: A Study of Contacts and Careers*, 2nd ed. (University of Chicago Press, 1995); Stewart Macaulay, "Non-Contractual Relations in Business: A Preliminary Study," in *The Sociology of Economic Life* (Routledge, 2018), 198–212; Mark Granovetter, *Society and Economy: Framework and Principles* (Harvard University Press, 2017).

32. Peter Michael Blau, "Consultation Among Colleagues," in *The Dynamics of Bureaucracy: Study of Interpersonal Relations in Two Government Agencies*, rev. ed. (University of Chicago Press, 1963), 157–69.

33. In social network analysis, multiplexity refers to the fact that organizations—including city administrations—are embedded in multiple, coexisting social orders. It is important to consider an organization's embeddedness because an organization's cultural and structural positions are more receptive to adopting

a practice. David Strang and John W. Meyer, "Institutional Conditions for Diffusion," *Theory and Society* 22, no. 4 (1993): 487–511. Social and cultural embeddedness are therefore of consequence for the likelihood of adoption, in part because social norms to adopt are stronger for some entities.

34. I call these relationships "fundamental" because they are not based on periodic interactions between specific people in the organization but on stable relations derived from joint membership (e.g., in a nation, or an elite college like in Charles Kadushin's seminal study of grandes écoles), reputations (e.g., of underwriters in Joel Podolny's study of security markets), or highly aggregate forms of interaction (e.g., trade in Saskia Sassen and Paul Ingram). In contrast, the city network gives rise to *particular* ties between staff members of city administrations and governments. See Charles Kadushin, "Friendship Among the French Financial Elite," *American Sociological Review* 60, no. 2 (1995): 202–21; Joel M. Podolny, "Networks as the Pipes and Prisms of the Market," *American Journal of Sociology* 107, no. 1 (2001): 33–60; Saskia Sassen, *The Global City: New York, London, Tokyo* (Princeton University Press, 2001); Paul Ingram et al., "The Intergovernmental Network of World Trade: IGO Connectedness, Governance, and Embeddedness," *American Journal of Sociology* 111, no. 3 (2005): 824–58.

35. This way of thinking about cities being connected by associations gives us a so-called bipartite network: one that connects cities to each other through their membership in or association with intercity associations.

36. My approach to constructing the network could be missing some associations and, thus, connections. One concern is that a bipartite network based on a subset of world cities conceals certain ties between associations mediated by smaller cities or that we missed a particular association that created "bridging ties" among cities. Some urban scholars, like Michele Acuto and Benjamin Leffel, have specialized in tracking all memberships in transnational municipal networks as of 2018. Michele Acuto and Benjamin Leffel, "Understanding the Global Ecosystem of City Networks," *Urban Studies* 58, no. 9 (2021): 1758–74. They report a single cluster of highly central cities that are particularly well networked, as well as cities that stay outside of this central cluster because they have more selected memberships. Their analyses—which capture a more comprehensive snapshot at the cost of being able to give insight into the evolution of the network—suggest that even though cities are members of different clusters, the intercity network is also not truly "polycentric," so members of associations do not speak to each other. In other words, this massive hairball of cities and an even greater list of associations confirm that thinking of cities as more or less central in the global city network makes sense—much like sociologists had described the *centrality* of cities in the global power hierarchy of

cities earlier. Also see Acuto, Michele, Daniel Pejic, Sombol Mokhles, Benjamin Leffel, David Gordon, Ricardo Martinez, Sayel Cortes, and Cathy Oke. "What Three Decades of City Networks Tell Us About City Diplomacy's Potential for Climate Action." *Nature Cities* 1, no. 7 (2024): 451–56.

37. World polity theory posits that states integrated in a global institutional system converge around a set of legitimate, shared expectations, see Cole, "World Polity or World Society?"; John Boli and George M. Thomas, "World Culture in the World Polity: A Century of International Non-Governmental Organization," *American Sociological Review* 62, no. 2 (1997): 171–90.

38. For an alternative reading about the diplomatic actions of cities, see Michele Acuto, *Global Cities, Governance and Diplomacy: The Urban Link* (Routledge, 2013); Michele Acuto et al., "What Three Decades of City Networks Tell Us About City Diplomacy's Potential for Climate Action," *Nature Cities* 1, no. 7 (2024): 451–56; Sharon Zukin, "Reconstructing the Authenticity of Place," *Theory and Society* 40, no. 2 (2011): 161–65; Sharon Zukin et al., *Global Cities, Local Streets: Everyday Diversity from New York to Shanghai* (Routledge, 2015); Sassen, *Cities in a World Economy*.

39. "Our Cities, Ourselves," *Nature Cities* 1, no. 1 (2024), https://doi.org/10.1038/s44284-023-00030-4.

40. For a primer in event history models, see David Strang and Nancy Brandon Tuma, "Spatial and Temporal Heterogeneity in Diffusion," *American Journal of Sociology* 99, no. 3 (1993): 614–39; Marc Schneiberg and Elisabeth S. Clemens, "The Typical Tools for the Job: Research Strategies in Institutional Analysis," *Sociological Theory* 24, no. 3 (2006): 195–227.

41. World society is often understood as a normative and cultural system of liberal values that empowers individuals and organizations to act, and thus "the domain of non-state actors," including civil society, which play an important role in constituting the institutional environment of state actors like city administrations. Meyer understands the world polity as a "system of creating value through the collective conferral of authority. . . . It includes state action, as is conventional, but also other forms of collective action that might in the modern social scientific lexicon be dismissed as merely 'cultural.'" Cole, "World Polity or World Society?", 87; Meyer et al., "World Society and the Nation-State."

42. Hokyu Hwang and Walter W. Powell, "The Rationalization of Charity: The Influences of Professionalism in the Nonprofit Sector," *Administrative Science Quarterly* 54, no. 2 (2009): 268–98; Kerstin Sahlin-Andersson and Lars Engwall, *The Expansion of Management Knowledge: Carriers, Flows, and Sources* (Stanford University Press, 2002); Cole, "World Polity or World Society?"

43. The likelihood of having a strategic plan is 59 percent higher for cities in the C40 cluster and even 143 percent higher for cities in the North American

cluster containing USDN, compared with other cities. Compared with the 104 isolates, all cities save for those in the identity network Metropolis are at least twice as likely to adopt a plan.

44. The most daring plans, like New York's PlaNYC, combine both. Novel rhetorical emphases emerge at the interstices of multiple professional networks; see Balázs Vedres and David Stark, "Structural Folds: Generative Disruption in Overlapping Groups," *American Journal of Sociology* 115, no. 4 (2010): 1150–90; Santi Furnari, "Interstitial Spaces: Microinteraction Settings and the Genesis of New Practices Between Institutional Fields," *Academy of Management Review* 39, no. 4 (2014): 439–62; Padgett and Powell, *The Emergence of Organizations and Markets*. Remember that this plan was not the first, but nevertheless is credited for a sea change.

45. Christof Brandtner et al., "From Iron Cage to Glass House: Repurposing of Bureaucratic Management and the Turn to Openness," *Organization Studies* 45, no. 2 (2024): 193–221.

4. LEADING

1. The City of Palo Alto, 2022 Sustainability and Climate Action Plan, 2023, https:// www.paloalto.gov/files/assets/public/v/1/sustainability/reports/2022-scap -report_final.pdf

2. Katherine A. Trisolini, "All Hands on Deck: Local Governments and the Potential for Bidirectional Climate Change Regulation," *Stanford Law Review* 62 (2009): 698. The term comes from Sherry Arnstein, a US Housing and Urban Development employee who in 1969 published a punchy essay skeptical of attempts to dupe the public into feeling involved through participation charades. She was not overly optimistic. Sherry R. Arnstein, "A Ladder of Citizen Participation," *Journal of the American Institute of Planners* 35 (1969): 216–24.

3. Even the city's poorer cousin, East Palo Alto, had a Climate Action Plan to reduce GHG emissions since 2011. But the city is less outspoken about turning its plan into a holistic strategy for sustainable development and livability. Palo Alto was ahead of this curve.

4. Or at least where the cancer researchers have a pint at the end of the day, as Stanford is technically an independent Census-designated place. The university's centrality became quite visible during the town hall's next poll, about what Palo Alto means to attendees. Stanford appeared as the largest word on the screen until someone excluded it from the word cloud.

5. On planetary urbanization, see Neil Brenner, "The Hinterland Urbanised?," *Architectural Design* 86, no. 4 (2016): 118–27; Sue Ruddick et al., "Planetary Urbanization: An Urban Theory for Our Time?," *Environment and Planning*

D: Society and Space 36, no. 3 (2018): 387–404; Neil Brenner and Christian Schmid, "The 'Urban Age' in Question," *International Journal of Urban and Regional Research* 38, no. 3 (2014): 731–55.

6. See Rachel M. Krause and Christopher Hawkins, *Implementing City Sustainability: Overcoming Administrative Silos to Achieve Functional Collective Action* (Temple University Press, 2021).

7. Richard Feiock et al., "The Integrated City Sustainability Database," *Urban Affairs Review* 50, no. 4 (2014): 577–89.

8. Seymour Martin Lipset, *American Exceptionalism: A Double-Edged Sword* (Norton, 1996); John W. Meyer et al., "World Society and the Nation-State," *American Journal of Sociology* 103, no. 1 (1997): 144–81; Gili S. Drori, *Globalization and Organization: World Society and Organizational Change* (Oxford University Press, 2006).

9. There is, in fact, some interesting variation within the United States with respect to two legal doctrines that allow cities more or less leeway. Cities that operate in the so-called Home Rule doctrine have significant discretion. Cities that operate based on the default Dillon's rule are strictly confined by the legal boundaries of the states and, in many cases, preempted from taking autonomous action that has not previously been cleared by the state government by "nullifying municipal authority" Nicole DuPuis et al., "City Rights in an Era of Preemption: A State-by-State Analysis" (National League of Cities, 2018), https://www.nlc.org/resource/city-rights-in-an-era-of-preemption-a-state-by-state-analysis.

10. See Ana Gonzalez, "Making Do: How Cities Pursue Progressive Sustainability Solutions in Light of Empowering or Constraining State Contexts" (master's thesis, University of Chicago, 2023). See also evidence on the diffusion of state preemption against local decarbonization efforts by Edward T. Walker and Andrew Malmuth, "The Natural Gas Industry, the Republican Party, and State Preemption of Local Building Decarbonization," *npj Climate Action* 3, no. 1 (2024): 98.

11. International City/County Management Association (ICMA), "ICMA Survey Research: 2015 Local Government Sustainability Practices Survey Report" (ICMA, 2016), https://icma.org/documents/icma-survey-research-2015-local-government-sustainability-practices-survey-report. In the 2015 ICMA survey, 62 percent cited "lack of funding" and 25 percent cited lack of staff capacity as minor factors hindering sustainability, compared with the threat of lawsuits (6 percent), opposition of business (9 percent) or elected officials (23 percent), and lack of information (17 percent). Remember that these average values include smaller towns and villages, creating a distorted image of what motivates major cities.

12. The measure has a Cronbach's alpha of 0.83 and an Eigenvalue of 4.4. The composition of the factor as well as individual factor loadings are reported in Table 4.1.

13. Whereas scope 1 greenhouse gas emissions consider only emissions that physically occur within the boundaries of a city (e.g., industry in the city, cars driving in the city), scope 2 emissions also include emissions from purchased energy and scope 3 emissions also include all other indirect emissions from goods and services that are brought into the city. The debate of which emissions are relevant for cities is particularly important in the context of defining what it means to be net zero (i.e., having a carbon footprint of zero or even negative). Incorporating out-of-boundary emissions is no trivial task, as this discussion of Paris planners shows: "Paris Climate Action Plan: Plan Climat 2024–2030," C40, https://www.c40knowledgehub.org/s/article/In-Paris-we-re-implementing-accelerated-actions-now-to-halve-our-emissions-by-2030-and-achieve-carbon-neutrality-by-2050. Also see Giulia Ulpiani et al., "Towards the First Cohort of Climate-Neutral Cities: Expected Impact, Current Gaps, and Next Steps to Take to Establish Evidence-Based Zero-Emission Urban Futures," *Sustainable Cities and Society* 95 (2023): 104572.
14. Adam Millard-Ball, "The Limits to Planning: Causal Impacts of City Climate Action Plans," *Journal of Planning Education and Research* 33, no. 1 (2013): 5–19.
15. Kent E. Portney, *Taking Sustainable Cities Seriously: Economic Development, the Environment, and Quality of Life in American Cities*, vol. 67 (MIT Press, 2003); Kent. E. Portney and Jeffrey M. Berry, "Participation and the Pursuit of Sustainability in U.S. Cities," *Urban Affairs Review* 46, no. 1 (2010): 119–39.; Harriet Bulkeley, *Cities and Climate Change* (Routledge, 2013).
16. Although public administration scholars like Krause and Feiock have examined variations in these and other survey data of US local governments, they rarely pay attention to both issues about collaborative governance and meso-level features of the bureaucracy or super- and infrastructure.
17. Among a random sample of 8,569 local governments that received the survey, 2,176 (25.4 percent) responded. Because I focused on municipal governments (rather than county governments), the overall number of observations is 1,540 cities.
18. Even careful proponents of functionalist explanations, like Richard Feiock and his colleagues, note that climate change action defies models based on rationality: "The centrality of cities to achieving global sustainability as well as their seeming ability to overcome the problem of collective action defies simple rational choice explanations or parsimonious linear models." Feiock et al., "Integrated City Sustainability Database," 579.
19. Frank Dobbin, *Forging Industrial Policy: The United States, Britain, and France in the Railway Age* (Cambridge University Press, 1994).
20. John W. Meyer and Brian Rowan, "Institutionalized Organizations: Formal Structure as Myth and Ceremony." *American Journal of Sociology* 83, no. 2

(1977): 340–63; Paul J. DiMaggio and Walter W. Powell, "The Iron Cage Revisited: Institutional Isomorphism and Collective Rationality in Organizational Fields," *American Sociological Review* 48, no. 2 (1983): 147–60.

21. Pamela S. Tolbert and Lynne G. Zucker, "Institutional Sources of Change in the Formal Structure of Organizations: The Diffusion of Civil Service Reform, 1880–1935," *Administrative Science Quarterly* 28, no. 1 (1983): 22–39 (more in chapter 5).

22. Even when cities are fully committed to climate goals, how they will get there remains an open question. As Chicago's CSO reflected, these goals are intended not to push the boundaries of what is conceivable but to float all boats: "I think that one of the interesting things going forward is, particularly for smaller cities, how do they get there, right? So they know they want to commit to Paris. They know what they want to do, so what tools will be necessary to help them get there? How do they build a climate action plan? How do they create an emissions inventory, which is run-of-the-mill stuff for cities like us? New York has a staff that does it every year, but that isn't necessarily true in Peoria, Illinois, or other parts of the country."

23. Portney, *Taking Sustainable Cities Seriously*; Portney and Berry, "Participation and the Pursuit of Sustainability in U.S. Cities."

24. If anything, these priorities may differ from one city to another. Aaron Deslatte explains, reviewing Krause and Hawkins's 2021 book, the unsurprising but nevertheless important fact that "different cities define sustainability differently, giving greater weights to equity, environmental, and economic dimensions. The case studies reinforce the conclusion that sustainability as a normative call for 'balancing' these dimensions may miss the mark when you consider how the concept is operationalized in specific contexts. In cities such as Oakland, where historic racial and income inequalities persist, 'balancing' may not be needed because framing sustainability around addressing these equity problems is paramount. Ann Arbor, a highly educated college town, has oriented its efforts more toward environmental goals, while Gainesville has focused on cost-savings." Aaron Deslatte, "Implementing City Sustainability: Overcoming Administrative Silos to Achieve Functional Collective Action, by Rachel M. Krause and Christopher Hawkins," *Journal of Urban Affairs* (July 19, 2021): 279.

25. Hillary Angelo and David Wachsmuth distinguish between "two ideological conceptions of urban sustainability: green sustainability posits a natural aesthetic simplicity associated with conservation, while gray sustainability takes a technological approach, engineering a human-centered environment that is complex yet optimized." Hillary Angelo and David Wachsmuth, "Why Does Everyone Think Cities Can Save the Planet?," *Urban Studies* 57, no. 11 (2020): 2201–21; Hillary Angelo and David Wachsmuth, "Urbanizing Urban Political

Ecology: A Critique of Methodological Cityism," *International Journal of Urban and Regional Research* 39, no. 1 (2015): 16–27.

26. Neil Brenner, *New State Spaces: Urban Governance and the Rescaling of Statehood* (Oxford University Press, 2004).

27. Maryann P. Feldman, *The Geography of Innovation*, vol. 2 (Springer Science & Business Media, 2013); Neil Lee, *Innovation for the Masses: How to Share the Benefits of the High-Tech Economy* (University of California Press, 2024); Olav Sorenson and Pino G Audia, "The Social Structure of Entrepreneurial Activity: Geographic Concentration of Footwear Production in the United States, 1940–1989," *American Journal of Sociology* 106, no. 2 (2000): 424–62. Also see chapter 5 for a further discussion of urban innovation.

28. Mildred Warner and Amir Hefetz, "Managing Markets for Public Service: The Role of Mixed Public-Private Delivery of City Services," *Public Administration Review* 68, no. 1 (2008): 155–66; Martin Kornberger et al., "When Bureaucracy Meets the Crowd: Studying 'Open Government' in the Vienna City Administration," *Organization Studies* 38, no. 2 (2017): 179–200; Stephan G. Grimmelikhuijsen and Mary K. Feeney, "Developing and Testing an Integrative Framework for Open Government Adoption in Local Governments," *Public Administration Review* 77, no. 4 (2017): 579–90.

29. Els De Graauw, *Making Immigrant Rights Real: Nonprofits and the Politics of Integration in San Francisco* (Cornell University Press, 2016); Justin Peter Steil and Ion Bogdan Vasi, "The New Immigration Contestation: Social Movements and Local Immigration Policy Making in the United States, 2000–2011," *American Journal of Sociology* 119, no. 4 (2014): 1104–55; United Cities and Local Governments (UCLG), "The Lampedusa Charter for Dignified Human Mobility and Territorial Solidarity," UCLG, n.d., https://www.uclg.org/sites/default/files/lampedusa-_carta-eng.pdf. On the municipal ID, see note in chapter 3.

30. Edward Glaeser, *Triumph of the City* (Penguin, 2011); Bruce Katz and Jennifer Bradley, *The Metropolitan Revolution* (Brookings Institution, 2014); Ion Bogdan Vasi and David Strang, "Civil Liberty in America: The Diffusion of Municipal Bill of Rights Resolutions after the Passage of the USA PATRIOT Act," *American Journal of Sociology* 114, no. 6 (2009): 1716–64; Michael Storper, *Keys to the City* (Princeton University Press, 2013).

31. Doug McAdam, *Political Process and the Development of Black Insurgency, 1930–1970* (University of Chicago Press, 2010).

32. In France, for example, the *Agence de la Transition Écologique* (ADEME) funds and encourages local initiatives for ecological transition.

33. Kate Aronoff et al., *A Planet to Win: Why We Need a Green New Deal* (Verso, 2019).

34. The distinction among more and less metropolitan/urban counties follows a conventional CDC classification. US Centers for Disease Control and Prevention,

"NCHS Urban-Rural Classification Scheme for Counties," https://www.cdc
.gov/nchs/data-analysis-tools/urban-rural.html.

35. Today, analysts are of two minds concerning the health-protecting impact of shelter-in-place policies, mostly because people in counties with shelter-in-place policies stayed home anyway. See Christopher R. Berry et al., "Evaluating the Effects of Shelter-in-Place Policies During the COVID-19 Pandemic," *Proceedings of the National Academy of Sciences* 118, no. 15 (2021): e2019706118. Regardless, the main point is that some cities were taking action in a way that experts at the time saw as the correct approach from a public health standpoint.

36. Neil Brenner, *New State Spaces: Urban Governance and the Rescaling of Statehood* (Oxford University Press, 2004).

37. Dimitrios Gounaridis and Joshua P. Newell, "The Social Anatomy of Climate Change Denial in the United States," *Scientific Reports* 14, no. 1 (2024): 2097.

38. As of April 2019, the EPA climate change page features the following disclaimer: "We've made some changes to EPA.gov. If the information you are looking for is not here, you may be able to find it on the EPA Web Archive or the January 19, 2017, Web Snapshot. US Environmental Protection Agency, "Homepage," https://19january2017snapshot.epa.gov. More at Laignee Barron, "Here's What the EPA's Website Looks Like After a Year of Climate Change Censorship," *Time Magazine*, March 1, 2018, http://time.com/5075265/epa-website-climate-change-censorship.

39. Amy Green, "With Gov. Scott and Legislature in Denial, Tiny Town Adapts on Its Own to Climate Change," *The Miami Herald*, July 15, 2018, https://www.miamiherald.com/news/state/florida/article214355019.html.

40. Chuck DeVore, "Texas Town's Environmental Narcissism Makes Al Gore Happy While Sticking Its Citizens with the Bill," *Fox News*, January 29, 2019, https://www.foxnews.com/opinion/texas-towns-environmental-narcissism-makes-al-gore-happy-while-sticking-its-citizens-with-the-bill.

41. Liz Koslov, "The Case for Retreat," *Public Culture* 28, no. 2 (2016): 359–87.

42. This kind of preemptive activity specific to climate change-related municipal policies may also be on the uptick. For example, Walker and Malmuth found that more than half of US states passed laws preempting municipalities from restricting legacy utilities such as natural gas, which cities have done in an effort at decarbonization. Walker and Malmuth, "The Natural Gas Industry, the Republican Party, and State Preemption of Local Building Decarbonization."

43. Gerald E. Frug and David J. Barron, *City Bound: How States Stifle Urban Innovation* (Cornell University Press, 2013).

44. Paul E. Peterson, *City Limits* (University of Chicago Press, 1981).

45. Fritz W. Scharpf, "Community and Autonomy: Multi-Level Policy-Making in the European Union," *Journal of European Public Policy* 1, no. 2 (1994): 219–42;

B. Guy Peters and Jon Pierre, "Governance Without Government? Rethinking Public Administration," *Journal of Public Administration Research and Theory* 8, no. 2 (1998): 223–43; B. Guy Peters and Jon Pierre, "Developments in Inter-governmental Relations: Towards Multi-Level Governance," *Policy and Politics* 29, no. 2 (2001): 131–36.

46. Katz and Bradley, *Metropolitan Revolution.*

47. Patricia Bromley and John W. Meyer, *Hyper-Organization: Global Organizational Expansion* (Oxford University Press, 2015); Christopher Marquis et al., "Golfing Alone? Corporations, Elites, and Nonprofit Growth in 100 American Communities," *Organization Science* 24, no. 1 (2013): 39–57.

48. Jack L. Walker, "The Diffusion of Innovations Among the American States," *American Political Science Review* 63, no. 3 (1969): 880–99; Frederick J. Boehmke and Paul Skinner, "State Policy Innovativeness Revisited," *State Politics and Policy Quarterly* 12, no. 3 (2012): 303–29; Devin Caughey and Christopher Warshaw, "The Dynamics of State Policy Liberalism, 1936–2014," *American Journal of Political Science* 6, no. 4 (2015): 899–913.

49. Richard Schragger, *City Power: Urban Governance in a Global Age* (Oxford University Press, 2016); Frug and Barron, *City Bound.*

50. Kale Roberts, "Three Decades of Sustainability: ICLEI at 30 Enters Next 'Decade of Local Action,'" ICLEI USA, September 4, 2020, https://icleiusa.org /iclei-at-30.

51. Nayiri Mullinix, "Nearly 15 Percent of Americans Deny Climate Change Is Real, AI Study Finds," *Michigan News*, University of Michigan, February 14, 2024, https://news.umich.edu/nearly-15-of-americans-deny-climate-change-is-real -ai-study-finds.

52. Tiffany Hsu, "He Wanted to Unclog Cities. Now He's 'Public Enemy No. 1,'" *The New York Times*, March 28, 2023, https://www.nytimes.com/2023/03/28/technology /carlos-moreno-15-minute-cities-conspiracy-theories.html.

53. The political tension of urban development is not unique to the United States. In my current home of Lyon, France, a Green Party mayor faced severe criticism for raising the idea of serving vegetarian meals in school canteens and banning diesel cars from the city center. "A French City Announced It Would Serve Meatless School Lunches. The Backlash Was Swift," Vox, April 1, 2021, https://www.vox.com /future-perfect/22360062/meat-vegetarian-vegan-lyon-france-culture-identity.

54. Any spatial planning and growth policy is contentious. See, for example, the class conflict among Manila neighborhoods recorded in Marco Z. Garrido, *The Patchwork City: Class, Space, and Politics in Metro Manila* (University of Chicago Press, 2019); Harvey Molotch, "The City as a Growth Machine: Toward a Political Economy of Place," *American Journal of Sociology* 82, no. 2 (1976): 309–32.

55. Although some cities and towns share sustainability officers with others.

56. Martin Kornberger et al., "When Bureaucracy Meets the Crowd: Studying 'Open Government' in the Vienna City Administration," *Organization Studies* 38, no. 2 (2017): 179–200.

57. David Wachsmuth and Hillary Angelo, "Green and Gray: New Ideologies of Nature in Urban Sustainability Policy," *Annals of the American Association of Geographers* 108, no. 4 (2018): 1038–56.

58. Public intellectuals like Benjamin Barber saw "urban sovereignty" as a potential "fix for global warming." Barber was among the first to articulate the belief that activist mayors are at the forefront of city climate action, responding to the public's deafening demands to finally do something about climate change. In his successful book *If Mayors Ruled the World*, he makes a case that cities can indeed "save the world" if decision-makers band together in a Global Parliament of Mayors. This belief is as inspired as it has become a trope. Benjamin R. Barber, *If Mayors Ruled the World: Dysfunctional Nations, Rising Cities* (Yale University Press, 2013); Benjamin R. Barber, *Cool Cities: Urban Sovereignty and the Fix for Global Warming* (Yale University Press, 2017).

59. Sara McTarnaghan et al., *Urban Resilience: From Global Vision to Local Practice* (Urban Institute, 2022), xiii.

60. Forrest Briscoe and Sean Safford, "The Nixon-in-China Effect: Activism, Imitation, and the Institutionalization of Contentious Practices," *Administrative Science Quarterly* 53, no. 3 (September 1, 2008): 460–91.

61. International City/County Management Association (ICMA), "ICMA Survey Research: 2018 Municipal Form of Government Survey," ICMA, July 2, 2019, https://icma.org/2018-municipal-fog-survey.

62. About 72 percent of city managers are men, down from 84 percent in 1985.

63. Portney and Berry, "Participation and the Pursuit of Sustainability in U.S. Cities"; also see Dan Honig, *Mission Driven Bureaucrats: Empowering People to Help Government Do Better* (Oxford University Press, 2024); Christopher M. Rea and Scott Frickel, "The Environmental State: Nature and the Politics of Environmental Protection," *Sociological Theory* 41, no. 3 (2023): 255–81.

64. Rachel M. Krause, "An Assessment of the Greenhouse Gas Reducing Activities Being Implemented in US Cities," *Local Environment* 16, no. 2 (February 1, 2011): 193–211.

65. Christopher V. Hawkins et al., "Making Meaningful Commitments: Accounting for Variation in Cities' Investments of Staff and Fiscal Resources to Sustainability," *Urban Studies* 53, no. 9 (2016): 1902–24.

66. Richard C. Feiock and Hee Soun Jang, "Nonprofits as Local Government Service Contractors," *Public Administration Review* 69, no. 4 (2009): 675.

67. John W. Meyer and Ronald Jepperson, "The 'Actors' of Modern Society: The Cultural Construction of Social Agency," *Sociological Theory* 18, no. 1 (2000): 100–120; Bromley and Meyer, *Hyper-Organization: Global Organizational Expansion.*

68. Drori et al., *Globalization and Organization.*

69. Gili S. Drori et al., "Sources of Rationalized Governance: Cross-National Longitudinal Analyses, 1985–2002," *Administrative Science Quarterly* 51, no. 2 (2006): 205–29; Gili S. Drori et al., "Global Organization: Rationalization and Actorhood as Dominant Scripts," *Research in the Sociology of Organizations* 27 (2009): 17–43; Evan Schofer and Ann Hironaka, "The Effects of World Society on Environmental Protection Outcomes," *Social Forces* 84, no. 1 (2005): 25–47; Ann Hironaka, *Greening the Globe* (Cambridge University Press, 2014).

70. Christopher Hood, "A Public Management for All Seasons," *Public Administration* 69, no. 1 (1991): 3–19; Christopher Pollitt and Geert Bouckaert, *Public Management Reform: A Comparative Analysis—Into the Age of Austerity,* 4th ed. (Oxford University Press, 2017).

71. Patricia Bromley and John W. Meyer, "'They Are All Organizations': The Cultural Roots of Blurring Between the Nonprofit, Business, and Government Sectors," *Administration and Society* 49, no. 7 (2014), https://doi.org/10.1177/0095399714548268; Bromley and Meyer, *Hyper-Organization: Global Organizational Expansion.*

72. David John Frank et al., "The Nation-State and the Natural Environment Over the Twentieth Century," *American Sociological Review* 65, no. 1 (2000): 96–116.

73. Vasi and Strang, "Civil Liberty in America."

74. The positive association between state and city holds despite a series of city- and state-level control variables, which align with previous studies on city policy diffusion and indicate the sources of intrastate variation. The models show a consistent effect even upon including a range of city- and state-level control variables. Unsurprisingly, larger cities show a much larger capacity for sustainability action.

75. As the Chicago chief sustainability officer told me: "We recognize that people can have a better quality of life if we invest in systems that prioritize people and health, right? There's so much great potential for us to reframe how we thought about our environment and sustainability agenda historically." See framing, Robert D. Benford and David A. Snow, "Framing Processes and Social Movements: An Overview and Assessment," *Annual Review of Sociology* 26 (2000): 611–39; Juliane Reinecke and Shahzad Ansari, "Microfoundations of Framing: The Interactional Production of Collective Action Frames in the Occupy Movement," *Academy of Management Journal* 64, no. 2 (2021): 378–408.

76. As Craig Calhoun puts it eloquently in his introduction to the book, "ordinary people can participate, perhaps more than in most of history, but only in elite managed institutions." Caroline W. Lee et al., *Democratizing Inequalities: Dilemmas of the New Public Participation* (New York University Press, 2015), 13, xiii. The problem of self-interest in participatory governance is not necessarily a recent phenomenon. Arnstein's conjugation suggests that participation often appears open-ended but ultimately serves the ends of someone: "je participe, tu participes . . . vous participez, ils profitent." Arnstein, "Ladder of Citizen Participation."

77. This is what institutional theorists like DiMaggio and Powell refer to as mimetic, normative, and coercive isomorphic pressures. DiMaggio and Powell, "Iron Cage Revisited"; Ion Bogdan Vasi, "Thinking Globally, Planning Nationally and Acting Locally: Nested Organizational Fields and the Adoption of Environmental Practices," *Social Forces* 86, no. 1 (2007): 113–36.

78. Christof Brandtner et al., "Enacting Governance Through Strategy: A Comparative Study of Governance Configurations in Sydney and Vienna," *Urban Studies* 54, no. 5 (2016): 1075–91.

79. Theda Skocpol et al., *Bringing the State Back In* (Cambridge University Press, 1985); Peter B. Evans, *Embedded Autonomy: States and Industrial Transformation* (Cambridge University Press, 1995).

80. Michael McQuarrie and Nicole P. Marwell, "The Missing Organizational Dimension in Urban Sociology," *City and Community* 8, no. 3 (2009): 247.

81. McQuarrie and Marwell, "Missing Organizational Dimension in Urban Sociology," 248.

82. Nicole P. Marwell, "Privatizing the Welfare State: Nonprofit Community-Based Organizations as Political Actors," *American Sociological Review* 69, no. 2 (2004): 265–91.

83. Mario L. Small, *Unanticipated Gains: Origins of Network Inequality in Everyday Life* (Oxford University Press, 2009); Sean Safford, "Why the Garden Club Couldn't Save Youngstown: Civic Infrastructure and Mobilization in Economic Crises," Social Working Paper Series on Local Innovation Systems, MIT-IPC-LIS-04-003 (Massachusetts Institute of Technology, Industrial Performance Center, 2004).

84. Safford, "Why the Garden Club Couldn't Save Youngstown," 5; Edward O. Laumann et al., "Community Structure as Interorganizational Linkages," *Annual Review of Sociology* 4, no. 1 (1978): 455–84.

85. McQuarrie and Marwell, "Missing Organizational Dimension in Urban Sociology."

86. Laumann et al., "Community Structure as Interorganizational Linkages"; Renate Mayntz, "New Challenges to Governance Theory," in *Governance as Social and Political Communication*, ed. Henrik Paul Bang (Manchester University

Press, 2003), 27–40; Joseph Galaskiewicz, "An Urban Grants Economy Revisited: Corporate Charitable Contributions in the Twin Cities, 1979–81, 1987–89," *Administrative Science Quarterly* 42 (1997): 445–71.

87. John M. Meyer and Brian Rowan, "Institutionalized Organizations: Formal Structure as Myth and Ceremony," *American Journal of Sociology* 83, no. 2 (1977): 340–63; DiMaggio and Powell, "Iron Cage Revisited"; Mark C. Suchman, "Managing Legitimacy: Strategic and Institutional Approaches," *Academy of Management Review* 20, no. 3 (1995): 571–610.

88. I laid out his perspective in a review article with historian Claire Dunning. We argue that nonprofit organizations are something like an infrastructure to cities, not unlike the streets and bridges that make movement and connections within the city possible. Christof Brandtner and Claire Dunning, "Nonprofits as Urban Infrastructure," in *The Nonprofit Sector: A Research Handbook*, ed. Walter W. Powell and Patricia Bromley, 3rd ed. (Stanford University Press, 2020), 271–91.

89. Chris Ansell and Alison Gash, "Collaborative Governance in Theory and Practice," *Journal of Public Administration Research and Theory* 18, no. 4 (2008): 543–71.

90. Krystal Laryea and Christof Brandtner, "Organizations as Drivers of Social and Systemic Integration: Contradiction and Reconciliation through Loose Demographic Coupling and Community Anchoring," *Research in the Sociology of Organizations* 90 (2024): 177–200; Christof Brandtner et al., "Neighborhood Effects on Integrative Organizational Practices in Five Global Cities," *Nature Cities* 1 (2024): 853–60, https://doi.org/10.1038/s44284-024-00154-1.

91. Donald Kettl, "The Global Revolution in Public Management: Driving Themes, Missing Links," *Journal of Policy Analysis and Management* 16, no. 3 (1997): 446–62; Nicole P. Marwell, "Privatizing the Welfare State: Nonprofit Community-Based Organizations as Political Actors," *American Sociological Review* 69, no. 2 (2004): 265–91; Stephen Osborne, *The New Public Governance?* (Routledge, 2010).

92. Robert J. Sampson et al., "Civil Society Reconsidered: The Durable Nature and Community Structure of Collective Civic Action," *American Journal of Sociology* 111, no. 3 (2005): 673–714.

93. In chapter 5, I will discuss this association between the presence of nonprofits and a place's "civic capacity: to adopt practices ostensibly addressing social and environmental problems.

94. See Hayagreeva Rao and Henrich R. Greve, "Disasters and Community Resilience: Spanish Flu and the Formation of Retail Cooperatives in Norway," *Academy of Management Journal* 61, no. 1 (2018): 5–25; Sunasir Dutta, "Creating in the Crucibles of Nature's Fury: Associational Diversity and Local Social

Entrepreneurship After Natural Disasters in California, 1991–2010," *Administrative Science Quarterly* 62, no. 3 (2017): 443–83.

95. Christopher Marquis et al., "Community Isomorphism and Corporate Social Action," *Academy of Management Review* 32, no. 3 (2007): 8.

96. András Tilcsik and Christopher Marquis, "Punctuated Generosity: How Mega-Events and Natural Disasters Affect Corporate Philanthropy in U.S. Communities." *Administrative Science Quarterly* 58, no. 1 (2013): 111–48, https://doi .org/10.1177/0001839213475800.

97. Robert J. Sampson, *Great American City: Chicago and the Enduring Neighborhood Effect* (University of Chicago Press, 2012).

98. See Jeremy R. Levine, *Constructing Community: Urban Governance, Development, and Inequality in Boston* (Princeton University Press, 2021); Nicole P. Marwell, "Privatizing the Welfare State: Nonprofit Community-Based Organizations as Political Actors." *American Sociological Review* 69, no. 2 (2004): 265–91; Clarence N. Stone, "Urban Regimes and the Capacity to Govern: A Political Economy Approach," *Journal of Urban Affairs* 15, no. 1 (1993): 1–28; Floyd Hunter, *Community Power Structure: A Study of Decision Makers* (University of North Carolina Press, 1953); Robert A. Dahl, *Who Governs? Democracy and Power in an American City* (Yale University Press, 2005); James M. Smith, "Urban Regime Theory," in *The Wiley Blackwell Encyclopedia of Urban and Regional Studies*, ed. Anthony M. Orum (Wiley, 2019), 1–9.

99. Marc Schneiberg and Sarah A. Soule, "Institutionalization as a Contested, Multilevel Process: The Case of Rate Regulation in American Fire Insurance," in *Social Movements and Organization Theory*, ed. Gerald F. Davis et al. (Cambridge University Press, 2010), 122–60; Steven Lukes, *Power: A Radical View* (Palgrave Macmillan, 2005).

100. Dennis R. Young, *If Not for Profit, for What? A Behavioral Theory of the Nonprofit Sector Based on Entrepreneurship* (Lexington, 1983); Laurie E. Paarlberg and Samantha Zuhlke, "Revisiting the Theory of Government Failure in the Face of Heterogeneous Demands," *Perspectives on Public Management and Governance* 2, no. 2 (2019): 103–24.

5. SCALING

1. Katherine A. Trisolini, "All Hands on Deck: Local Governments and the Potential for Bidirectional Climate Change Regulation," *Stanford Law Review* 62 (2009): 698. The high contribution of buildings to energy use was widely discussed on web pages of the US federal government before January 2025. This particular reference stems from a web page of the US Environmental Protection Agency (EPA) which, as of March 2025, shows an error message, but

you can see the Internet Archive's Wayback Machine for a snapshot. In the European Union, the building sector accounts to an estimated 34 percent of energy-related emissions in 2022 according to the European Environmental Agency. US Environmental Protection Agency, "Why Build Green?," https://www.epa.gov/greenbuilding/pubs/whybuild.htm; The Internet Archive, "Why Build Green? (Archival Snapshot)," May 5, 2019, https://web.archive.org/web/20190505181719/https://archive.epa.gov/greenbuilding/web/html/whybuild.html; European Environment Agency, "Greenhouse Gas Emissions from Energy Use in Buildings in Europe," October 31, 2024, https://www.eea.europa.eu/en/analysis/indicators/greenhouse-gas-emissions-from-energy.

2. Trisolini, "All Hands on Deck."

3. John W. Meyer and Brian Rowan, "Institutionalized Organizations: Formal Structure as Myth and Ceremony," *American Journal of Sociology* 83, no. 2 (1977): 340–63; Patricia Bromley et al., "Decoupling Revisited: Common Pressures, Divergent Strategies in the U.S. Nonprofit Sector," *M@n@gement* 15, no. 5 (2012): 468–501.

4. An important additional concern—elaborated in chapter 6—is that the participation in the ecological transition is not always equal, equitable, or just. See Patricia Romero-Lankao et al., "A Framework to Centre Justice in Energy Transition Innovations," *Nature Energy* 8 (2023): 1192–98; Flor Avelino et al., "Just Sustainability Transitions: Politics, Power, and Prefiguration in Transformative Change Toward Justice and Sustainability," *Annual Review of Environment and Resources* 49 (2024): 519–47.

5. Theda Skocpol and Kenneth Finegold, "State Capacity and Economic Intervention in the Early New Deal," *Political Science Quarterly* 97, no. 2 (1982): 255–78; Theda Skocpol, *Diminished Democracy: From Membership to Management in American Civic Life* (University of Oklahoma Press, 2003).

6. For a complementary discussion of civic capacity and its long-term effects, see Hayagreeva Rao and Henrich R. Greve, "Disasters and Community Resilience: Spanish Flu and the Formation of Retail Cooperatives in Norway," *Academy of Management Journal* 61, no. 1 (2018): 5–25.

7. Isaac Martin, "Dawn of the Living Wage: The Diffusion of a Redistributive Municipal Policy," *Urban Affairs Review* 36, no. 4 (2001): 470–96.

8. Bill Gates, *How to Avoid a Climate Disaster: The Solutions We Have and the Breakthroughs We Need* (Vintage, 2021), 42.

9. Dodge Data and Analytics, "Key Trends in the Construction Industry 2015," 2015, https://www.oracle.com/us/assets/key-industry-trends-2773761.pdf. Compound annual growth rate estimated based on a market volume of $120B in 2018. The nonresidential market provides insight into the progression of this growth. Between 2005 and 2008, this market grew from $3 billion to $25 billion. Overall,

nonresidential construction contracted from $212 billion to $154 billion during the economic crisis; green construction continued to grow from $25 billion to $48 billion, increasing its market share from 12 percent to 31 percent. Projections estimate that at $119–134 billion, the current market share is roughly half the nonresidential construction market and continues to grow. Green building is also estimated to have created 1.1 million jobs, 386,000 of which were directly attributed to LEED (adding to some $26.2 billion in wages). See "Green Building Economic Impact Study," Booz Allen Hamilton, 2015, http://go.usgbc .org/2015-Green-Building-Economic-Impact-Study.html.

10. "New York State Green Jobs Study," New York City Labor Market Information Service, CUNY Graduate Center, 2011, https://www.gc.cuny.edu/sites/default /files/2022-01/CUR-Green-Jobs-Study-Chapter-3.pdf.

11. Shuaib Lwasa et al., "Urban Systems and Other Settlements," in *IPCC, 2022: Climate Change 2022: Mitigation of Climate Change. Contribution of Working Group III to the Sixth Assessment Report of the Intergovernmental Panel on Climate Change* (Cambridge University Press, 2022), 861–952.

12. Green buildings are a low-cost climate-change mitigation strategy whose greatest challenge to implementation are weak incentives on the part of potential adopters. Stephen Pacala and Robert Socolow, "Stabilization Wedges: Solving the Climate Problem for the Next 50 Years with Current Technologies," *Science* 305, no. 5686 (2004): 968–72; Marilyn A. Brown et al., *Shrinking the Carbon Footprint of Metropolitan America* (Brookings Institution, 2008); Carbon Mitigation Initiative, "Stabilization Wedges: The Wedges from Efficient Buildings," 2011, https://cmi.princeton.edu/resources/stabilization-wedges/calculations -and-data/efficient-buildings/.

13. Trisolini, "All Hands on Deck," 703.

14. Paul E. Peterson, *City Limits* (University of Chicago Press, 1981).

15. U.S. Green Building Council, *Real Estate and Biodiversity: What You Need to Know* (U.S. Green Building Council, 2023), https://www.usgbc.org/sites/default /files/2023-07/Real-Estate-and-Biodiversity-What-You-Need-to-Know.pdf.

16. GBCI stood first for Green Building Certification Institute and later for Green Business Certification Inc.

17. US Green Building Council, "LEED Credit Library," https://www.usgbc.org /pilotcredits.

18. An evocative image for how small actions of governments and private actors accumulate to meaningful change is the bee swarm model of social change, see Ann Hironaka, *Greening the Globe* (Cambridge University Press, 2014).

19. Tim Bartley, "How Foundations Shape Social Movements: The Construction of an Organizational Field and the Rise of Forest Certification," *Social Problems*

54, no. 3 (2007): 229–55; Tim Bartley et al., *Looking Behind the Label: Global Industries and the Conscientious Consumer* (Indiana University Press, 2015); Tim Bartley and Curtis Child, "Shaming the Corporation: The Social Production of Targets and the Anti-Sweatshop Movement," *American Sociological Review* 79, no. 4 (2014): 653–79, https://doi.org/10.1177/0003122414540653.

20. Magali A. Delmas and David Colgan, *The Green Bundle: Pairing the Market with the Planet* (Stanford University Press, 2018).

21. Jeff Speck, *Walkable City: How Downtown Can Save America, One Step at a Time* (MacMillan, 2013).

22. Brian Barth, "Is LEED Tough Enough for the Climate-Change Era?," *Bloomberg News*, June 5, 2018, https://www.bloomberg.com/news/articles/2018-06-05/reconsidering-leed-buildings-in-the-era-of-climate-change.

23. Rao and Greve, "Disasters and Community Resilience."

24. Slightly adjusted for contemporary context. The original quote is about the Inuit but using dated language.

25. Some of the examples here are drawn from the *American Journal of Sociology* article and other writing that I have done related to this book. See also Christof Brandtner, "How Civic Capacity Gets Urban Social Innovations Started," *Work in Progress* (blog), June 1, 2023, http://www.wipsociology.org/2023/06/01/how-civic-capacity-gets-urban-social-innovations-started.

26. Anil Rupasingha et al., "The Production of Social Capital in US Counties," *Journal of Socio-Economics* 35, no. 1 (2006): 83–101; Raj Chetty et al., "Social Capital I: Measurement and Associations with Economic Mobility," *Nature* 608, no. 7921 (2022): 108–21; Seok-Woo Kwon et al., "Community Social Capital and Entrepreneurship," *American Sociological Review* 78, no. 6 (2013): 980–1008.

27. Thomas Dietz et al., "Climate Change and Society," *Annual Review of Sociology* 46 (2020): 135–58.

28. Paul J. DiMaggio and Walter W. Powell, "The Iron Cage Revisited: Institutional Isomorphism and Collective Rationality in Organizational Fields," *American Sociological Review* 48, no. 2 (1983): 147–60.

29. Showing off materials and techniques is not a coincidence.

30. US Green Building Council, "LEED Professional Credentials," https://www.usgbc.org/credentials.

31. Frank Dobbin, *Inventing Equal Opportunity* (Princeton University Press, 2009); DiMaggio and Powell, "Iron Cage Revisited"; Mitchel Y. Abolafia, *Making Markets: Opportunism and Restraint on Wall Street* (Harvard University Press, 2001).

32. Jeffrey G. York et al., "It's Not Easy Building Green: The Impact of Public Policy, Private Actors, and Regional Logics on Voluntary Standards Adoption," *Academy of Management Journal* 61, no. 4 (2018): 1492–523.

33. On the conception that LEED certifications are driven to a large degree by businesses and environmental entrepreneurs, see York et al., "It's Not Easy Building Green."

34. Daniel C. Matisoff et al., "Performance or Marketing Benefits? The Case of LEED Certification," *Environmental Science and Technology* 48, no. 3 (2014): 2001–7.

35. Trisolini, "All Hands on Deck."

36. Besides a certification fee, construction-related expenses "can increase a project's cost by about 10 to 30 percent." Stephen J. Vamosi, "The True Cost of LEED-Certified Green Buildings," *Heating, Plumbing Air Conditioning Magazine*, January 1, 2011.

37. A 2021 survey showed that the motive of "encouraging sustainable business practices" was the most important reason given besides saving operating costs (66 percent), promoting occupant well-being (80 percent), and increasing worker productivity (59 percent)—mentioned by 76 percent of respondents, and up from 58 percent in the 2016 survey. Dodge Data and Analytics, "World Green Building Trends 2016 Smart Market Report," 2016, https://worldgbc.org/article /world-green-building-trends-2016; Dodge Data and Analytics, "World Green Building Trends 2021 Smart Market Report," 2021, https://www.construction .com/resource/world-green-building-trends-2021.

38. State of Green, "House of Green," https://stateofgreen.com/en/about/about-state -of-green/house-of-green.

39. Christof Brandtner, "Can Cities Be the Source of Scalable Innovations?," *Stanford Social Innovation Review*, March 23, 2023, https://doi.org/10.48558/pwbd-nn15.

40. Pamela S. Tolbert and Lynne G. Zucker, "Institutional Sources of Change in the Formal Structure of Organizations: The Diffusion of Civil Service Reform, 1880–1935," *Administrative Science Quarterly* 28, no. 1 (1983): 22–39.

41. David Knoke, "The Spread of Municipal Reform: Temporal, Spatial, and Social Dynamics," *American Journal of Sociology* 87, no. 6 (1982): 1314–39.

42. Competing certification schemes and rankings sometimes create opportunities for noncompliance among organizations, which we already discussed as one of the reasons why not all organizations act the same, even when they are compliant with institutionalized expectations. Christof Brandtner, "Putting the World in Orders: Plurality in Organizational Evaluation," *Sociological Theory* 35, no. 3 (2017): 200–227.

43. Cox Castle, "CALGreen: A Landmark Law That Mandates Only Modest Change," 2011, https://www.coxcastle.com/publication-calgreen-a-landmark-law-that -mandates-only-modest-change.

44. U.S. Green Building Council, *USGBC Policy Library* (U.S. Green Building Council), https://public-policies.usgbc.org/.

45. Walter W. Powell et al., "Click and Mortar: Organizations on the Web," *Research in Organizational Behavior* 36 (2016): 101–20.

46. The process of distributed adoption through catalysis by well-meaning organizations, legitimation by authorities, and scaling by the bulk of organizations in a place or individuals in an organization provides a general model through which innovative practices can spread in the absence of fiat that I have discussed at length in an article published in the *American Journal of Sociology*. Christof Brandtner, "Green American City: Civic Capacity and the Distributed Adoption of Urban Innovations," *American Journal of Sociology* 128, no. 3 (2022): 627–79.

47. Michael Lounsbury, "Institutional Sources of Practice Variation: Staffing College and University Recycling Programs," *Administrative Science Quarterly* 46, no. 1 (2001): 29–56; Tim Bartley, "How Foundations Shape Social Movements: The Construction of an Organizational Field and the Rise of Forest Certification," *Social Problems* 54, no. 3 (2007): 229–55; Wesley D. Sine and Brandon H. Lee, "Tilting at Windmills? The Environmental Movement and the Emergence of the U.S. Wind Energy Sector," *Administrative Science Quarterly* 54, no. 1 (2009): 123–55; Paul-Brian McInerney, *From Social Movement to Moral Market: How the Circuit Riders Sparked an IT Revolution and Created a Technology Market* (Stanford University Press, 2014).

48. Richard Schragger, *City Power: Urban Governance in a Global Age* (Oxford University Press, 2016); Jeremy R. Levine, "The Privatization of Political Representation," *American Sociological Review* 81, no. 6 (2016): 1251–75.

49. Mark Bevir and Rod A. W. Rhodes, *Governance Stories* (Routledge, 2006); Nicole P. Marwell, "Privatizing the Welfare State: Nonprofit Community-Based Organizations as Political Actors," *American Sociological Review* 69, no. 2 (2004): 265–91; Christof Brandtner et al., "Enacting Governance Through Strategy: A Comparative Study of Governance Configurations in Sydney and Vienna," *Urban Studies* 54, no. 5 (2016): 1075–91. The question of the influence of different interest groups in urban politics has motivated some of the initial debates about coalition-building and community power in urban sociology. For a refreshing discussion of 'who governs,' see Jeremy R. Levine, *Constructing Community: Urban Governance, Development, and Inequality in Boston* (Princeton University Press, 2021).

CONCLUSION

1. On this point, consider Granovetter's germinal insight that what people perceive as "luck" may just be a fortunate constellation of interpersonal networks—in his case, a job seeker knowing people who happen to know that someone is

looking to hire. Many enabling coincidences that empower cities are rendered more likely under certain structural conditions. For instance, a game-changing foundation grant is more likely when the city employs someone who stands a chance of accidentally bumping into a program officer who they went to college with or seeing a call for applications on LinkedIn through a shared acquaintance. Mark Granovetter, *Getting a Job: A Study of Contacts and Careers*, 2nd ed. (University of Chicago Press, 1995).

2. For a discussion of city strategies for strengthening the links between city administrations and civil society, see Christof Brandtner, "Can Cities Be the Source of Scalable Innovations?," *Stanford Social Innovation Review*, March 23, 2023. https://doi.org/10.48558/pwbd-nn15.

3. For this contrast, see Bill Gates, *How to Avoid a Climate Disaster: The Solutions We Have and the Breakthroughs We Need* (Vintage, 2021); ADEME, "Sobriété : un incontournable de la transition écologique," https://infos.ademe .fr/lettre-international-juin-2022/sobriete-un-incontournable-de-la-transition -ecologique.

4. Jacobs challenged the predominant urban planning models of the mid-century United States and advocated for mixed use and denser, walkable neighborhoods. Her championing of communities was much influenced by what she saw from her Greenwich Village apartment on 555 Hudson Street, a street she keenly observed and often used to advance concepts such as the natural surveillance that stem from "eyes on the street" and the spontaneous, choreographed interactions of diverse people in the "sidewalk ballet." Jane Jacobs, " 'The Uses of Sidewalks: Safety': From The Death and Life of Great American Cities (1961)," in *The City Reader*, ed. Richard T. LeGates, Frederic Stout, and Roger W. Caves, 7th ed. (Routledge, 2020), 189–94. For a connection to organizational infrastructure, see Christof Brandtner and Dunning, "Nonprofits as Urban Infrastructure," in *The Nonprofit Sector: A Research Handbook*, ed. Walter W. Powell and Patricia Bromley, 3rd ed. (Stanford University Press, 2020).

5. For discussions about the shifting meaning of change and innovation in the urban context, see Christof Brandtner et al., "Dynamic Persistence of Institutions: Modeling the Historical Endurance of Red Vienna's Public Housing Utopia," *Organization Studies* (2025). https://doi.org/10.1177/01708406251317258; Stephan Leixnering and Markus Höllerer, " 'Remaining the Same or Becoming Another?' Adaptive Resilience Versus Transformative Urban Change," *Urban Studies* 59, no. 6 (2022): 1300–310.

6. Brundtland Commission, *Our Common Future World Commission on Environment and Development* (Oxford University Press, 1987).

7. Maryann P. Feldman and Martin Kenney, *Private Equity and the Demise of the Local: The Loss of Community Economic Power and Autonomy*, Elements in Reinventing Capitalism (Cambridge University Press, 2024).

8. The state policy innovativeness indicates how soon states adopt policies that eventually become widespread and plays a major role in the discussion in chapter 4. See Frederick J. Boehmke and Paul Skinner, "State Policy Innovativeness Revisited," *State Politics and Policy Quarterly* 12, no. 3 (2012): 303–29; Jack L. Walker, "The Diffusion of Innovations Among the American States," *American Political Science Review* 63, no. 3 (1969): 880–99; Frances Berry and William Berry, "State Lottery Adoptions as Policy Innovations: An Event History Analysis," *American Political Science Review* 84, no. 2 (1990): 395–416; Devin Caughey and Christopher Warshaw, "The Dynamics of State Policy Liberalism, 1936–2014," *American Journal of Political Science* 6, no. 4 (2015): 899–913.

9. For a more detailed discussion of the theoretical implications of the findings presented in chapter 4 in particular, see Christof Brandtner and David Suárez, "The Structure of City Action: Institutional Embeddedness and Sustainability Practices in U.S. Cities," *American Review of Public Administration* 51, no. 2 (2021): 121–38.

10. Kent E. Portney, *Taking Sustainable Cities Seriously: Economic Development, the Environment, and Quality of Life in American Cities*, vol. 67 (MIT Press, 2003); Kent E. Portney and Jeffrey M. Berry, "Participation and the Pursuit of Sustainability in U.S. Cities," *Urban Affairs Review* 46, no. 1 (2010): 119–39.

11. John W. Meyer and Brian Rowan, "Institutionalized Organizations: Formal Structure as Myth and Ceremony," *American Journal of Sociology* 83, no. 2 (1977): 340–63; Patricia Bromley et al., "Decoupling Revisited: Common Pressures, Divergent Strategies in the U.S. Nonprofit Sector," *M@n@gement* 15, no. 5 (2012): 468–501; Wesley Longhofer and Andrew Jorgenson, "Decoupling Reconsidered: Does World Society Integration Influence the Relationship between the Environment and Economic Development?," *Social Science Research* 65 (2017): 17–29; Christof Brandtner, "Decoupling Under Scrutiny: Consistency of Managerial Talk and Action in the Age of Nonprofit Accountability," *Nonprofit and Voluntary Sector Quarterly* 50, no. 5 (2021): 1053–78.

12. Chief among them is the import of goods and services from outside the city boundaries, which explains why cities are not as energy efficient as they sometimes claim. Indirect emissions, particularly from the consumption of power produced elsewhere (scope 2 emissions) and resulting from supply chain and transportation systems (scope 3 emissions), are harder to account for than most cities can manage.

13. Tim Hallett, "The Myth Incarnate Recoupling Processes, Turmoil, and Inhabited Institutions in an Urban Elementary School," *American Sociological Review* 75, no. 1 (2010): 52–74; Lars Thøger Christensen et al., "CSR as Aspirational Talk," *Organization* 20, no. 3 (2013): 372–93; Catherine Turco, "Difficult Decoupling: Employee Resistance to the Commercialization of Personal Settings," *American Journal of Sociology* 118, no. 2 (2012): 380–419.

14. For a sociological theorization of the moral stances of urban planners and a review of related literature, see Hillary Angelo and Gianpaolo Baiocchi, who note that "far from being naïve, planners adopt a reflexive, pragmatic morality that allows them to maintain normative commitments even in highly constrained environments." For a more historical one, there are many non-academic studies of the sociology of urban planners and their relationship to political elites, from Chicago's Daniel Burnham (see Carl Smith) to New York's Robert Moses (see Robert Caro). Hillary Angelo and Gianpaolo Baiocchi, "The Moral Work of Participation: Disillusio, Expertise, and Urban Planning Under Neoliberalism," *Qualitative Sociology* 47, no. 3 (2024): 493; Carl Smith, *The Plan of Chicago: Daniel Burnham and the Remaking of the American City* (University of Chicago Press, 2019); Robert A. Caro, *The Power Broker: Robert Moses and the Fall of New York* (Knopf, 1974).

15. Related to the mechanical metaphor or a ratchet that cannot be reversed once it moves up a notch, also note a compelling literature in economics that shows that workers may deliberately limit their output to prevent a high production standard that is difficult to achieve in the long-term. Robert Gibbons, "Incentives and Careers in Organizations," Working Paper (National Bureau of Economic Research, August 1996). Also see Donald Roy, "Quota Restriction and Goldbricking in a Machine Shop," *American Journal of Sociology* 57, no. 5 (1952): 427–42. Examining hesitation to ratcheting up climate goals to unachievable levels is an interesting question in the context of climate goals as well.

16. For deep appreciation for bureaucrats as "visionary agents of change," see Dan Honig, *Mission Driven Bureaucrats: Empowering People to Help Government Do Better* (Oxford University Press, 2024).

17. Organizational scholars have conceptualized such historically endowed capacities as institutional legacies. See Henrich R. Greve and Hayagreeva Rao, "Echoes of the Past: Organizational Foundings as Sources of an Institutional Legacy of Mutualism," *American Journal of Sociology* 118, no. 3 (2012): 635–75; Henrich R. Greve and Hayagreeva Rao, "History and the Present: Institutional Legacies in Communities of Organizations," *Research in Organizational Behavior* 34 (2014): 27–41; Christopher Marquis and Kunyuan Qiao, "History Matters for Organizations: An Integrative Framework for Understanding Influences from the Past," *Academy of Management Review* 50, no. 2 (2024), https://doi.org/10.5465/amr.2022.0238.

18. Christopher M. Rea, "Theorizing Command-and-Commodify Regulation: The Case of Species Conservation Banking in the United States," *Theory and Society* 46 (2017): 21–56.

19. Policies shoring up resistance is an example of negative policy feedback to climate solutions, but political scientist Leah Stokes also makes a strong case for positive policy feedback in which well-designed climate policies create new constituencies for environmental policies. Leah Cardamore Stokes, *Short Circuiting Policy: Interest Groups and the Battle over Clean Energy and Climate Policy in the American States* (Oxford University Press, 2020).

20. Edward O. Laumann et al., "Community Structure as Interorganizational Linkages," *Annual Review of Sociology* 4, no. 1 (1978): 455–84; Edward O. Laumann and David Knoke, *The Organizational State: Social Choice in National Policy Domains* (University of Wisconsin Press, 1987).

21. Organization scholars have recently emphasized the importance of cross-sector partnerships, ecosystems of social innovation, and the orchestrator role of governments; but without noting the important linkages to place and urban politics and economics. Christine M. Beckman et al., "The Social Innovation Trap: Critical Insights into an Emerging Field," *Academy of Management Annals* 17, no. 2 (2023): 684–709, https://doi.org/10.5465/annals.2021.0089.

22. Portney and Berry, "Participation and the Pursuit of Sustainability in U.S. Cities."

23. Patrick Sharkey et al., "Community and the Crime Decline: The Causal Effect of Local Nonprofits on Violent Crime," *American Sociological Review* 82, no. 6 (2017): 1214–40.

24. Robert J. Sampson et al., "Civil Society Reconsidered: The Durable Nature and Community Structure of Collective Civic Action," *American Journal of Sociology* 111, no. 3 (2005): 673–714; Robert J. Sampson, *Great American City: Chicago and the Enduring Neighborhood Effect* (University of Chicago Press, 2012).

25. Jeremy R. Levine, "The Privatization of Political Representation," *American Sociological Review* 81, no. 6 (2016): 1251–75.

26. Social infrastructure is a broad term used to describe the spatial and organizational arrangements that bring together communities, including not only civil society organizations but also libraries, public parks, neighborhood cafés, and other third places. Eric Klinenberg, *Palaces for the People: How Social Infrastructure Can Help Fight Inequality, Polarization, and the Decline of Civic Life* (Crown, 2018); Ramon Oldenburg and Dennis Brissett, "The Third Place," *Qualitative Sociology* 5, no. 4 (1982): 265–84; for a more detailed discussion, see Christof Brandtner et al., "Where Relational Commons Take Place: The City and Its Social Infrastructure as Sites of Commoning," *Journal of Business Ethics* 184, no. 4 (2023): 917–32.

27. Robert E. Park, "The City: Suggestions for the Investigation of Human Behavior in the City Environment," *American Journal of Sociology* 20, no. 5 (1915): 577–612; Robert E. Park and Ernest W. Burgess, *The City* (University of Chicago Press, 1921), 2.

28. As Nicole Marwell and Michael McQuarrie have argued forcefully, "organizational action can affect the distribution of resources, the arrangement of social networks, and the constitution of dispositions and identities." Michael McQuarrie and Nicole P. Marwell, "The Missing Organizational Dimension in Urban Sociology," *City and Community* 8, no. 3 (2009): 247–48.

29. Floyd Hunter, *Community Power Structure: A Study of Decision Makers* (University of North Carolina Press, 1953); Robert A. Dahl, *Who Governs? Democracy and Power in an American City* (Yale University Press, 2005); Harvey Molotch, "The City as a Growth Machine: Toward a Political Economy of Place," *American Journal of Sociology* 82, no. 2 (1976): 309–32.

30. Robert Vargas, *Wounded City: Violent Turf Wars in a Chicago Barrio* (Oxford University Press, 2016); Robert Vargas, "Gangstering Grants: Bringing Power to Collective Efficacy Theory," *City and Community* 18, no. 1 (2019): 369–91.

31. Claire Dunning and I have argued for the urgent need to move from the presence to practices of nonprofits in our 2019 review of the literature on nonprofit organizations as an organizational infrastructure that underpins urban dynamics. Brandtner and Dunning, "Nonprofits as Urban Infrastructure."

32. Dennis R. Young, *If Not for Profit, for What? A Behavioral Theory of the Nonprofit Sector Based on Entrepreneurship* (Lexington, 1983); Elisabeth S. Clemens, *Civic Gifts: Voluntarism and the Making of the American Nation-State* (University of Chicago Press, 2020).

33. Christof Brandtner et al., " Prosperous Places: Processes, Policies, and Practices," in *Industrial and Corporate Change* (2025); Maryann P. Feldman, "The Character of Innovative Places: Entrepreneurial Strategy, Economic Development, and Prosperity," *Small Business Economics* 43, no. 1 (2014): 9–20; Dan Breznitz, *Innovation in Real Places: Strategies for Prosperity in an Unforgiving World* (Oxford University Press, 2021).

34. See James Samuel Coleman, *The Asymmetric Society* (Syracuse University Press, 1982); Charles Perrow, *A Society of Organizations*, vol. 20 (Springer, 1991).

35. Herbert A. Simon, "Organizations and Markets," *Journal of Economic Perspectives* 5, no. 2 (1991): 25–44.

36. Walter W. Powell and Christof Brandtner, "Organizations as Sites and Drivers of Social Action," in *Handbook of Contemporary Sociological Theory*, ed. Seth Abrutyn (Springer, 2016), 269–91, https://doi.org/10.1007/978-3-319-32250-6.

37. Gerald F. Davis, *Managed by the Markets: How Finance Re-Shaped America* (Oxford University Press, 2009); Gerald F. Davis, *Taming Corporate Power in*

the 21st Century, Elements in Reinventing Capitalism (Cambridge University Press, 2022).

38. Patricia Bromley and John W. Meyer, *Hyper-Organization: Global Organizational Expansion* (Oxford University Press, 2015).

39. Evan Schofer et al., "Illiberal Reactions to Higher Education," *Minerva* 60, no. 4 (2022): 509–34; Julia C. Lerch et al., "The Social Foundations of Academic Freedom: Heterogeneous Institutions in World Society, 1960 to 2022," *American Sociological Review* 89, no. 1 (2024): 88–125.

40. Frank Dobbin, *Inventing Equal Opportunity* (Princeton University Press, 2009); Frank Dobbin et al., "Rage Against the Iron Cage: The Varied Effects of Bureaucratic Personnel Reforms on Diversity," *American Sociological Review* 80, no. 5 (2015): 1014–44; John F. Padgett and Walter W. Powell, *The Emergence of Organizations and Markets* (Princeton University Press, 2012).

41. Individuals and groups certainly invigorate and shape organizations, which in the long run affect macrostructures. As sociologists Neil Fligstein and Doug McAdam argue, individuals' socialization and social skills may enable structural change in so-called *fields*. But in reality, many organizations coexist in a place without being in conflict with each other or sharing some collective rationality. Neil Fligstein and Doug McAdam, *A Theory of Fields* (Oxford University Press, 2012); Neil Fligstein, "Social Skill and the Theory of Fields," *Sociological Theory* 19, no. 2 (2001), https://doi.org/10.1111/0735-2751.00132.

42. Sociologists have "brought organizations back into" the study of stratification and inequality by identifying organizational sources of inequality and investigating meaningful efforts to reduce economic disparity in the developing world. James N. Baron and William T. Bielby, "Bringing the Firms Back in: Stratification, Segmentation, and the Organization of Work," *American Sociological Review* 45 (1980): 737–65; Mario L. Small, *Unanticipated Gains: Origins of Network Inequality in Everyday Life* (Oxford University Press, 2009). For example, scholars interested in organizations, occupations, and work show that discrimination and wage disparities can often be traced back to the practices and routines of employers. John M. Amis et al., "The Organizational Reproduction of Inequality," *Academy of Management Annals* 14, 1 (2020): 195–230; David S. Pedulla and Devah Pager, "Race and Networks in the Job Search Process," *American Sociological Review* 84, no. 6 (2019): 983–1012; David S. Pedulla, *Making the Cut: Hiring Decisions, Bias, and the Consequences of Nonstandard, Mismatched, and Precarious Employment* (Princeton University Press, 2020); Adina D. Sterling and Roberto M. Fernandez, "Once in the Door: Gender, Tryouts, and the Initial Salaries of Managers," *Management Science* 64, no. 11 (2018): 5444–60; John W. Meyer and Ronald Jepperson, "The 'Actors'

of Modern Society: The Cultural Construction of Social Agency," *Sociological Theory* 18, no. 1 (2000): 100–120.

43. Coleman, *Foundations of Social Theory*; Meyer and Jepperson, "The 'Actors' of Modern Society: The Cultural Construction of Social Agency."

44. Robert E. Park, "Human Migration and the Marginal Man," *American Journal of Sociology* 33, no. 6 (1928): 881–893.

45. McQuarrie and Marwell, "Missing Organizational Dimension in Urban Sociology"; Sharon Zukin et al., *Global Cities, Local Streets: Everyday Diversity from New York to Shanghai* (Routledge, 2015).

46. For the institutional sources of practice variation, see Michael Lounsbury, "Institutional Sources of Practice Variation: Staffing College and University Recycling Programs," *Administrative Science Quarterly* 46, no. 1 (2001): 29–56.

47. Patricia Bromley, "The Organizational Transformation of Civil Society," in *The Nonprofit Sector: A Research Handbook*, ed. Walter W. Powell and Patricia Bromley, vol. 3 (Stanford University Press, 2020), 123–43; Julia C. Lerch et al., "The Social Foundations of Academic Freedom: Heterogeneous Institutions in World Society, 1960 to 2022," *American Sociological Review* 89, no. 1 (2024): 88–125.

48. Richard Schragger, *City Power: Urban Governance in a Global Age* (Oxford University Press, 2016); Paul E. Peterson, *City Limits* (University of Chicago Press, 1981).

49. These terms are typically used in the management and public administration literatures to refer to problems without simple solutions that require collaboration. See Gerard George et al., "Understanding and Tackling Societal Grand Challenges Through Management Research," *Academy of Management Journal* 59, no. 6 (2016): 1880–95; Ali Aslan Gümüsay et al., "Engaging with Grand Challenges: An Institutional Logics Perspective," *Organization Theory* 1, no. 3 (2020): 1–20; Juliane Reinecke and Shahzad Ansari, "Taming Wicked Problems: The Role of Framing in the Construction of Corporate Social Responsibility," *Journal of Management Studies* 53, no. 3 (2016): 299–329. Although the literature on a slew of related topics may be theoretically disjointed, it has enabled researchers to engage with important problems. See Guillaume Carton et al., "How Not to Turn the Grand Challenges Literature into a Tower of Babel?," *Business and Society* 63, no. 2 (2024): 409–14.

50. These conflicting interests among participants in governance are sometimes referred to as "governance gaps."

51. Such as those begun by Paul E. Peterson and continued by Gerald E. Frug and David J. Barron and Richard Schragger. See Peterson, *City Limits*; Gerald E. Frug and David J. Barron, *City Bound: How States Stifle Urban Innovation* (Cornell University Press, 2013); Schragger, *City Power*; Jon Pierre, "Comparative Urban Governance Uncovering Complex Causalities," *Urban Affairs Review* 40, no. 4

(2005): 446–62; Nicole P. Marwell and Shannon L. Morrissey, "Organizations and the Governance of Urban Poverty," *Annual Review of Sociology* 46 (2020): 233–50; Brandtner et al., "Enacting Governance Through Strategy."

52. Cities and counties also occasionally engage in affirmative litigation, but this is a more recent phenomenon in need of further research, see Amanda Sharkey et al., "Organizational Scarring, Legal Consciousness, and the Diffusion of Local Government Litigation Against Opioid Manufacturers," *American Sociological Review* (2025).

53. There is also room for understanding transnational municipal networks as an evolution in foreign policy. In particular, how city networks can overcome the very real limits to urban development in the Global South deserves greater scrutiny. For a great example of how embeddedness and cohesion shaped the provision of public goods in Johannesburg, South Africa, and Sao Paolo, Brazil, see Benjamin H. Bradlow, *Urban Power: Democracy and Inequality in São Paulo and Johannesburg* (Princeton University Press, 2024).

54. See, for example, Patrick Bergemann and Christof Brandtner, "Territoriality and the Emergence of Norms During the COVID-19 Pandemic," *American Journal of Sociology* 130, no. 5 (2025). https://doi.org/10.1086/733799; Fred Paxton, *Restrained Radicals: Populist Radical Right Parties in Local Government* (Cambridge University Press, 2023).

55. Elisabeth S. Clemens, "From City Club to Nation State: Business Networks in American Political Development," in *Contention and Trust in Cities and States*, ed. Michael Hanagan and Chris Tilly (Springer Netherlands, 2011), 179–98.

56. See Ana Gonzalez, "Making Do: How Cities Pursue Progressive Sustainability Solutions in Light of Empowering or Constraining State Contexts" (master's thesis, University of Chicago, 2023).

57. Tilly never finished *Cities and States in World History*. The introduction to his book was published posthumously by his son Chris Tilly. Michael P. Hanagan and Chris Tilly, *Contention and Trust in Cities and States* (Springer, 2011).

58. Key to his proposed approach is the transformation of a so-called trust network—the relationships through which people generate trust in each other, such as kinship, long-distance trade, local solidarities, and religion. Whereas cities are "storehouses and headquarters for capital" and states attempt to secure a monopoly on coercion, trust networks are the basis of mutual recognition and coordination among places. Tilly argues that integration is a result of trust networks being incorporated in systems of rule—for instance worker representation on corporate boards, as is typical in corporatist countries. Charles Tilly, "Cities, States, and Trust Networks: Chapter 1 of Cities and States in World History," in *Theory and Society: Special Issue in Memory of Charles Tilly, 1929–2008. Cities, States, Trust, and Rule* (Springer, 2011).

59. Manuel Castells, *The Informational City: Information Technology, Economic Restructuring, and the Urban-Regional Process* (Blackwell, 1989).
60. Tilly, "Cities, States, and Trust Networks." 268.

METHODOLOGICAL APPENDIX

1. Christof Brandtner and Patricia Bromley, "The Global Structure of Local Strategy: Institutional Embeddedness and Global Cities' Strategic Urban Governance," *SocArXiv*, 2025, https://doi.org/10.31235/osf.io/ukajc_v1.
2. Christof Brandtner and David Suárez, "The Structure of City Action: Institutional Embeddedness and Sustainability Practices in U.S. Cities," *American Review of Public Administration* 51, no. 2 (2021): 121–38.
3. Christof Brandtner, "Green American City: Civic Capacity and the Distributed Adoption of Urban Innovations," *American Journal of Sociology* 128, no. 3 (2022): 627–79.
4. Ana Gonzalez and Christof Brandtner, "Green in Their Own Way: Pragmatic and Progressive Means for Cities to Overcome Institutional Barriers to Sustainability," *Urban Studies* 61, no. 13 (2024): 2513–30, https://doi.org/10.1177/00420980241239788.

Bibliography

Abolafia, Mitchel Y. *Making Markets: Opportunism and Restraint on Wall Street.* Harvard University Press, 2001.

Acuto, Michele, and Benjamin Leffel. "Understanding the Global Ecosystem of City Networks." *Urban Studies* 58, no. 9 (2021): 1758–74.

Acuto, Michele, Daniel Pejic, Sombol Mokhles, Benjamin Leffel, David Gordon, Ricardo Martinez, Sayel Cortes, and Cathy Oke. "What Three Decades of City Networks Tell Us About City Diplomacy's Potential for Climate Action." *Nature Cities* 1, no. 7 (2024): 451–56.

ADEME. "Sobriété : un incontournable de la transition écologique." https://infos.ademe.fr/lettre-international-juin-2022/sobriete-un-incontournable-de-la-transition-ecologique.

Alderson, Arthur S., and Jason Beckfield. "Power and Position in the World City System." *American Journal of Sociology* 109, no. 4 (2004): 811–51.

Alvarez, José Luis, and Silviya Svejenova. *The Changing C-Suite: Executive Power in Transformation.* Oxford University Press, 2022.

Amis, John M., Johanna Mair, and Kamal A. Munir. "The Organizational Reproduction of Inequality." *Academy of Management Annals* 14, no. 1 (2020): 195–230.

Angelo, Hillary. *How Green Became Good: Urbanized Nature and the Making of Cities and Citizens.* University of Chicago Press, 2021.

Angelo, Hillary, and Gianpaolo Baiocchi. "The Moral Work of Participation: Disillusio, Expertise, and Urban Planning Under Neoliberalism." *Qualitative Sociology* 47, no. 3 (2024): 493–515.

Angelo, Hillary, and David Wachsmuth. "Urbanizing Urban Political Ecology: A Critique of Methodological Cityism." *International Journal of Urban and Regional Research* 39, no. 1 (2015): 16–27.

Angelo, Hillary, and David Wachsmuth. "Why Does Everyone Think Cities Can Save the Planet?" *Urban Studies* 57, no. 11 (2020): 2201–21.

Ansell, Chris, and Alison Gash. "Collaborative Governance in Theory and Practice." *Journal of Public Administration Research and Theory* 18, no. 4 (2008): 543–71.

Anteby, Michel. *Manufacturing Morals: The Values of Silence in Business School Education.* University of Chicago Press, 2013.

Arnstein, Sherry R. "A Ladder of Citizen Participation." *Journal of the American Institute of Planners* 35 (1969): 216–24.

Aronoff, Kate, Alyssa Battistoni, Daniel Aldana Cohen, and Thea Riofrancos. *A Planet to Win: Why We Need a Green New Deal.* Verso, 2019.

Avelino, Flor, Katinka Wijsman, Frank van Steenbergen, Shivant Jhagroe, Julia Wittmayer, Sanne Akerboom, Kristina Bogner, Esther F Jansen, Niki Frantzeskaki, and Agni Kalfagianni. "Just Sustainability Transitions: Politics, Power, and Prefiguration in Transformative Change Toward Justice and Sustainability." *Annual Review of Environment and Resources* 49 (2024): 519–47.

Barber, Benjamin R. *Cool Cities: Urban Sovereignty and the Fix for Global Warming.* Yale University Press, 2017.

Barber, Benjamin R. *If Mayors Ruled the World: Dysfunctional Nations, Rising Cities.* Yale University Press, 2013.

Barkan, Ross. "What Happens to New York's Municipal ID Card Under the Trump Administration?" *New York Magazine*, December 4, 2016. https://nymag.com /intelligencer/2016/12/what-happens-to-idncy-under-the-trump-administration .html.

Baron, James N., and William T. Bielby. "Bringing the Firms Back In: Stratification, Segmentation, and the Organization of Work." *American Sociological Review* 45 (1980): 737–65.

Barron, Laignee. "Here's What the EPA's Website Looks Like After a Year of Climate Change Censorship." *Time Magazine*, March 1, 2018. http://time.com/5075265 /epa-website-climate-change-censorship.

Barth, Brian. "Is LEED Tough Enough for the Climate-Change Era?" *Bloomberg News*, June 5, 2018. https://www.bloomberg.com/news/articles/2018-06-05/reconsidering -leed-buildings-in-the-era-of-climate-change.

Bartley, Tim. "How Foundations Shape Social Movements: The Construction of an Organizational Field and the Rise of Forest Certification." *Social Problems* 54, no. 3 (2007): 229–55.

Battilana, Julie, and Marissa Kimsey. "Should You Agitate, Innovate, or Orchestrate?" *Sanford Social Innovation Review*, September 18, 2017. https://doi.org/10.48558 /3YGB-3M56.

Beckman, Christine M., Jovanna Rosen, Jeimee Estrada-Miller, and Gary Painter. "The Social Innovation Trap: Critical Insights into an Emerging Field." *Academy of Management Annals* 17, no. 2 (2023): 684–709. https://doi.org/10.5465/annals .2021.0089.

Benford, Robert D., and David A. Snow. "Framing Processes and Social Movements: An Overview and Assessment." *Annual Review of Sociology* 26 (2000): 611–39.

Bergemann, Patrick, and Christof Brandtner. "Territoriality and the Emergence of Norms During the COVID-19 Pandemic." *American Journal of Sociology* 130, no. 5 (2025). https://doi.org/10.1086/733799.

Berger, Peter L., and Thomas Luckmann. *The Social Construction of Reality: A Treatise in the Sociology of Knowledge.* Anchor, 1966.

Berry, Christopher R., Anthony Fowler, Tamara Glazer, Samantha Handel-Meyer, and Alec MacMillen. "Evaluating the Effects of Shelter-in-Place Policies During the COVID-19 Pandemic." *Proceedings of the National Academy of Sciences* 118, no. 15 (2021): e2019706118.

Berry, Frances, and William Berry. "State Lottery Adoptions as Policy Innovations: An Event History Analysis." *American Political Science Review* 84, no. 2 (1990): 395–416.

Bevir, Mark, and Rod A. W. Rhodes. *Governance Stories.* Routledge, 2006.

Binder, Amy. "For Love and Money: Organizations' Creative Responses to Multiple Environmental Logics." *Theory and Society* 36, no. 6 (2007): 547–71.

Blau, Peter Michael. "Consultation Among Colleagues." In *The Dynamics of Bureaucracy: Study of Interpersonal Relations in Two Government Agencies.* Rev. ed. University of Chicago Press, 1963.

Boehmke, Frederick J., and Paul Skinner. "State Policy Innovativeness Revisited." *State Politics and Policy Quarterly* 12, no. 3 (2012): 303–29.

Boli, John, and George M. Thomas. "World Culture in the World Polity: A Century of International Non-Governmental Organization." *American Sociological Review* 62, no. 2 (1997): 171–90.

Botelho, Tristan L., Ranjay Gulati, and Olav Sorenson. "The Sociology of Entrepreneurship Revisited." *Annual Review of Sociology* 50 (2024): 341–64.

Bradlow, Benjamin H. *Urban Power: Democracy and Inequality in São Paulo and Johannesburg.* Princeton University Press, 2024.

Brandtner, Christof. "Can Cities Be the Source of Scalable Innovations?" *Stanford Social Innovation Review*, March 23, 2023. https://doi.org/10.48558/pwbd-nn15.

Brandtner, Christof. "Decoupling Under Scrutiny: Consistency of Managerial Talk and Action in the Age of Nonprofit Accountability." *Nonprofit and Voluntary Sector Quarterly* 50, no. 5 (2021): 1053–78.

Brandtner, Christof. "Green American City: Civic Capacity and the Distributed Adoption of Urban Innovations." *American Journal of Sociology* 128, no. 3 (2022): 627–79.

Brandtner, Christof. "How Civic Capacity Gets Urban Social Innovations Started." *Work in Progress* (blog), June 1, 2023. http://www.wipsociology.org/2023/06/01/how-civic-capacity-gets-urban-social-innovations-started.

Brandtner, Christof. "Putting the World in Orders: Plurality in Organizational Evaluation." *Sociological Theory* 35, no. 3 (2017): 200–227.

Brandtner, Christof, Parham Ashur, and Bhargav Srinivasa Desikan. "Dynamic Persistence of Institutions: Modeling the Historical Endurance of Red Vienna's Public Housing Utopia." *Organization Studies* (2025). https://doi.org/10.1177/01708406251317258.

Brandtner, Christof, Luís M. A. Bettencourt, Marc G. Berman, and Andrew J. Stier. "Creatures of the State? Metropolitan Counties Compensated for State Inaction in Initial U.S. Response to COVID-19 Pandemic." *PLOS ONE* 16, no. 2 (2021): e0246249.

Brandtner, Christof, and Patricia Bromley. "Neoliberal Governance, Evaluations, and the Rise of Win–Win Ideology in Corporate Responsibility Discourse, 1960–2010." *Socio-Economic Review* 20, no. 4 (2021): 1933–60.

Brandtner, Christof, and Patricia Bromley. "Neoliberal Governance, Evaluations, and the Rise of Win–Win Ideology in Corporate Responsibility Discourse, 1960–2010." *Socio-Economic Review* 20, no. 4 (2021): 1933–60.

Brandtner, Christof, and Patricia Bromley. "The Global Structure of Local Strategy: Institutional Embeddedness and Global Cities' Strategic Urban Governance." *SocArXiv*, 2025. https://doi.org/10.31235/osf.io/ukajc_v1.

Brandtner, Christof, Gordon C. C. Douglas, and Martin Kornberger. "Where Relational Commons Take Place: The City and Its Social Infrastructure as Sites of Commoning." *Journal of Business Ethics* 184, no. 4 (2023): 917–32.

Brandtner, Christof, and Claire Dunning. "Nonprofits as Urban Infrastructure." In *The Nonprofit Sector: A Research Handbook*, ed. Walter W. Powell and Patricia Bromley. 3rd ed. Stanford University Press, 2020.

Brandtner, Christof, Markus A. Höllerer, Renate E. Meyer, and Martin Kornberger. "Enacting Governance Through Strategy: A Comparative Study of Governance Configurations in Sydney and Vienna." *Urban Studies* 54, no. 5 (2016): 1075–91.

Brandtner, Christof, Krystal Laryea, Gowun Park, Wei Luo, Michael Meyer, David Suárez, Hokyu Hwang, and Walter W. Powell. "Neighborhood Effects on

Integrative Organizational Practices in Five Global Cities." *Nature Cities* 1 (2024): 853–60. https://doi.org/10.1038/s44284-024-00154-1.

Brandtner, Christof, and Walter W. Powell. "Capturing the Civic Lives of Cities: An Organizational, Place-Based Perspective on Civil Society in Global Cities." *Global Perspectives* 3, no. 1 (2022): 36408.

Brandtner, Christof, Walter W. Powell, and Aaron Horvath. "From Iron Cage to Glass House: Repurposing of Bureaucratic Management and the Turn to Openness." *Organization Studies* 45, no. 2 (2024): 193–221.

Brandtner, Christof, Olav Sorenson, and Maryann Feldman. "Prosperous Places: Processes, Policies, and Practices. In *Industrial and Corporate Change.* 2025.

Brandtner, Christof, and David Suárez. "The Structure of City Action: Institutional Embeddedness and Sustainability Practices in U.S. Cities." *American Review of Public Administration* 51, no. 2 (2021): 121–38.

Bratman, Michael E. *Shared Agency: A Planning Theory of Acting Together.* Oxford University Press, 2013.

Brenner, Neil. *New State Spaces: Urban Governance and the Rescaling of Statehood.* Oxford University Press, 2004.

Brenner, Neil. "The Hinterland Urbanised?" *Architectural Design* 86, no. 4 (2016): 118–27.

Brenner, Neil, and Christian Schmid. "The 'Urban Age' in Question." *International Journal of Urban and Regional Research* 38, no. 3 (2014): 731–55.

Breznitz, Dan. *Innovation in Real Places: Strategies for Prosperity in an Unforgiving World.* Oxford University Press, 2021.

Briscoe, Forrest, and Sean Safford. "The Nixon-in-China Effect: Activism, Imitation, and the Institutionalization of Contentious Practices." *Administrative Science Quarterly* 53, no. 3 (2008): 460–91.

Bromley, Patricia. "The Organizational Transformation of Civil Society." In *The Nonprofit Sector: A Research Handbook*, ed. Walter W. Powell and Patricia Bromley. Vol. 3. Stanford University Press, 2020.

Bromley, Patricia, Hokyu Hwang, and Walter W. Powell. "Decoupling Revisited: Common Pressures, Divergent Strategies in the U.S. Nonprofit Sector." *M@n@gement* 15, no. 5 (2012): 468–501.

Bromley, Patricia, and John W. Meyer. *Hyper-Organization: Global Organizational Expansion.* Oxford University Press, 2015.

Bromley, Patricia, and John W. Meyer. "'They Are All Organizations': The Cultural Roots of Blurring Between the Nonprofit, Business, and Government Sectors." *Administration and Society* 49, no. 7 (2014). https://doi.org/10.1177/0095399714548268.

Bromley, Patricia, and Walter W. Powell. "From Smoke and Mirrors to Walking the Talk: Decoupling in the Contemporary World." *Academy of Management Annals* 6, no. 1 (2012): 483–530.

Bromley, Patricia, and Amanda Sharkey. "Casting Call: The Expanding Nature of Actorhood in US Firms, 1960–2010." *Accounting, Organizations and Society* 59 (2017): 3–20.

Brundtland Commission. *Our Common Future: World Commission on Environment and Development.* Oxford University Press, 1987.

Bulkeley, Harriet. *Cities and Climate Change.* Routledge, 2013.

Bulkeley, Harriet, and Vanesa Castán Broto. "Government by Experiment? Global Cities and the Governing of Climate Change." *Transactions of the Institute of British Geographers* 38, no. 3 (2013): 361–75.

Burawoy, Michael. *Manufacturing Consent: Changes in the Labor Process Under Monopoly Capitalism.* University of Chicago Press, 2012.

Bush, Sarah Sunn, and Jennifer Hadden. "Density and Decline in the Founding of International NGOs in the United States." *International Studies Quarterly* 63, no. 4 (2019): 1133–46.

Callenbach, Ernest. *Ecotopia.* Banyan Tree, 1975.

Campbell, John L. *Institutional Change and Globalization.* Princeton University Press, 2004.

Carbon Mitigation Initiative. "Stabilization Wedges: The Wedges from Efficient Buildings," 2011. https://cmi.princeton.edu/resources/stabilization-wedges/calculations-and-data/efficient-buildings/.

Caro, Robert A. *The Power Broker: Robert Moses and the Fall of New York.* Knopf, 1974.

Carton, Guillaume, Julia Parigot, and Thomas Roulet. "How Not to Turn the Grand Challenges Literature into a Tower of Babel?" *Business and Society* 63, no. 2 (2024): 409–14.

Castells, Manuel. *The Informational City: Information Technology, Economic Restructuring, and the Urban-Regional Process.* Blackwell, 1989.

Caughey, Devin, and Christopher Warshaw. "The Dynamics of State Policy Liberalism, 1936–2014." *American Journal of Political Science* 6, no. 4 (2015): 899–913.

C40 Cities. "Global Network of Mayors React to President Trump Withdrawing the US from the Paris Climate Agreement." Press release, January 21, 2025. https://www.c40.org/news/global-network-of-mayors-react-to-president-trump-paris-agreement/.

Chetty, Raj, Matthew O Jackson, Theresa Kuchler, Johannes Stroebel, Nathaniel Hendren, Robert B Fluegge, Sara Gong, Federico Gonzalez, Armelle Grondin, and Matthew Jacob. "Social Capital I: Measurement and Associations with Economic Mobility." *Nature* 608, no. 7921 (2022): 108–21.

Christensen, Lars Thøger, Mette Morsing, and Ole Thyssen. "CSR as Aspirational Talk." *Organization* 20, no. 3 (2013): 372–93.

Clemens, Elisabeth S. *Civic Gifts: Voluntarism and the Making of the American Nation-State.* University of Chicago Press, 2020.

Clemens, Elisabeth S. "From City Club to Nation State: Business Networks in American Political Development." In *Contention and Trust in Cities and States*, ed. Michael Hanagan and Chris Tilly. Springer Netherlands, 2011.

Cole, Wade M. "World Polity or World Society? Delineating the Statist and Societal Dimensions of the Global Institutional System." *International Sociology* 32, no. 1 (2017): 86–104.

Coleman, James. *The Asymmetric Society*. Syracuse University Press, 1982.

Coleman, James. *Foundations of Social Theory*. Harvard University Press, 1990.

Cox Castle. "CALGreen: A Landmark Law That Mandates Only Modest Change," 2011. https://www.coxcastle.com/publication-calgreen-a-landmark-law-that-mandates -only-modest-change.

Czarniawska, Barbara. *A Tale of Three Cities: Or the Glocalization of City Management*. Oxford University Press, 2002.

Czarniawska, Barbara, and Guje Sevón, eds. *Translating Organizational Change*. Walter de Gruyter, 1996.

Dacin, M. Tina, Tammar Zilber, Mélodie Cartel, and Ewald Kibler. "Navigating Place: Extending Perspectives on Place in Organization Studies." *Organization Studies* 45, no. 8 (2024): 1191–212.

Dahl, Robert A. *Who Governs? Democracy and Power in an American City*. Yale University Press, 2005.

Darwin, Charles. *On the Origin of Species by Means of Natural Selection, or the Preservation of Favoured Races in the Struggle for Life*. Murray, 1859.

Davis, Gerald F. *Managed by the Markets: How Finance Re-Shaped America*. Oxford University Press, 2009.

Davis, Gerald F. *Taming Corporate Power in the 21st Century*. Elements in Reinventing Capitalism. Cambridge University Press, 2022.

De Graauw, Els. *Making Immigrant Rights Real: Nonprofits and the Politics of Integration in San Francisco*. Cornell University Press, 2016.

Delmas, Magali A., and David Colgan. *The Green Bundle: Pairing the Market with the Planet*. Stanford University Press, 2018.

Deslatte, Aaron. "Implementing City Sustainability: Overcoming Administrative Silos to Achieve Functional Collective Action, by Rachel M. Krause and Christopher Hawkins." *Journal of Urban Affairs* 44, no. 2 (2022): 278–80.

DeVore, Chuck. "Texas Town's Environmental Narcissism Makes Al Gore Happy While Sticking Its Citizens with the Bill." *Fox News*, January 29, 2019. https:// www.foxnews.com/opinion/texas-towns-environmental-narcissism-makes-al -gore-happy-while-sticking-its-citizens-with-the-bill.

Dietz, Thomas, Rachael L. Shwom, and Cameron T. Whitley. "Climate Change and Society." *Annual Review of Sociology* 46 (2020): 135–58.

DiMaggio, Paul J., and Walter W. Powell. "The Iron Cage Revisited: Institutional Isomorphism and Collective Rationality in Organizational Fields." *American Sociological Review* 48, no. 2 (1983): 147–60.

The District of Columbia. *Sustainable DC Plan*, 2013. https://sustainable.dc.gov/sites /default/files/dc/sites/sustainable/page_content/attachments/SustainableDCPlan _web.pdf.

Dobbin, Frank. *Forging Industrial Policy: The United States, Britain, and France in the Railway Age.* Cambridge University Press, 1994.

Dobbin, Frank. *Inventing Equal Opportunity.* Princeton University Press, 2009.

Dobbin, Frank, Daniel Schrage, and Alexandra Kalev. "Rage Against the Iron Cage: The Varied Effects of Bureaucratic Personnel Reforms on Diversity." *American Sociological Review* 80, no. 5 (2015): 1014–44.

Dodge Data and Analytics. "Key Trends in the Construction Industry 2015." Dodge Data and Analytics, 2015. https://www.oracle.com/us/assets/key-industry-trends -2773761.pdf.

Dodge Data and Analytics. "World Green Building Trends 2016 Smart Market Report." Dodge Data and Analytics, 2016. https://worldgbc.org/article/world-green-building -trends-2016.

Dodge Data and Analytics. "World Green Building Trends 2021 Smart Market Report." Dodge Data and Analytics, 2021. https://www.construction.com/resource /world-green-building-trends-2021.

Drori, Gili S., Yong Suk Jang, and John W. Meyer. "Sources of Rationalized Governance: Cross-National Longitudinal Analyses, 1985–2002." *Administrative Science Quarterly* 51, no. 2 (2006): 205–29.

Drori, Gili S., John W. Meyer, and Hokyu Hwang. "Global Organization: Rationalization and Actorhood as Dominant Scripts." *Research in the Sociology of Organizations* 27 (2009): 17–43.

Drori, Gili S., John W. Meyer, and Hokyu Hwang. *Globalization and Organization: World Society and Organizational Change.* Oxford University Press, 2006.

Drori, Gili S., John W. Meyer, Francisco O. Ramirez, and Evan Schofer. *Science in the Modern World Polity: Institutionalization and Globalization.* Stanford University Press, 2003.

DuPuis, Nicole, Trevor Langan, Christiana McFarland, Angelina Panettieri, and Brooks Rainwater. "City Rights in an Era of Preemption: A State-by-State Analysis." National League of Cities, 2018. https://www.nlc.org/resource/city-rights-in-an-era-of -preemption-a-state-by-state-analysis.

Dutta, Sunasir. "Creating in the Crucibles of Nature's Fury: Associational Diversity and Local Social Entrepreneurship After Natural Disasters in California, 1991–2010." *Administrative Science Quarterly* 62, no. 3 (2017): 443–83.

Espeland, Wendy N., and Michael Sauder. *Engines of Anxiety: Academic Rankings, Reputation, and Accountability.* Russell Sage, 2016.

Espeland, Wendy N., and Mitchell L. Stevens. "A Sociology of Quantification." *European Journal of Sociology* 49, no. 3 (2008): 401–36.European Environment Agency. "Greenhouse Gas Emissions from Energy Use in Buildings in Europe," October 31, 2024. https://www.eea.europa.eu/en/analysis/indicators /greenhouse-gas-emissions-from-energy.

Feiock, Richard C., and Hee Soun Jang. "Nonprofits as Local Government Service Contractors." *Public Administration Review* 69, no. 4 (2009): 668–80.

Feiock, Richard, Rachel M. Krause, Christopher Hawkings, and Cali Curley. "The Integrated City Sustainability Database." *Urban Affairs Review* 50, no. 4 (2014): 577–89.

Feldman, Maryann P. "The Character of Innovative Places: Entrepreneurial Strategy, Economic Development, and Prosperity." *Small Business Economics* 43, no. 1 (2014): 9–20.

Feldman, Maryann P. *The Geography of Innovation.* Vol. 2. Springer Science and Business Media, 2013.

Feldman, Maryann P., and Martin Kenney. *Private Equity and the Demise of the Local: The Loss of Community Economic Power and Autonomy.* Elements in Reinventing Capitalism. Cambridge University Press, 2024.

Fisher, Dana R., and William R. Freudenburg. "Ecological Modernization and Its Critics: Assessing the Past and Looking Toward the Future." *Society and Natural Resources* 14, no. 8 (2001): 701–9.

Fleming, Lee, King Charles III, and Adam I. Juda. "Small Worlds and Regional Innovation." *Organization Science* 18, no. 6 (2007): 938–54.

Fligstein, Neil. "Social Skill and the Theory of Fields." *Sociological Theory* 19, no. 2 (2001). https://doi.org/10.1111/0735-2751.00132.

Fligstein, Neil, and Doug McAdam. *A Theory of Fields.* Oxford University Press, 2012.

Florida, Richard. "The Economic Geography of Talent." *Annals of the Association of American Geographers* 92, no. 4 (2002): 743–55.

Florida, Richard. *The Rise of the Creative Class.* Basic Books, 2002.

Florida, Richard. *Who's Your City? How the Creative Economy Is Making Where to Live the Most Important Decision of Your Life.* Vintage Canada, 2010.

Ford, Norman D. *America's 50 Best Cities in Which to Live, Work, and Retire.* Harian, 1958.

Frank, David John, Ann Hironaka, and Evan Schofer. "The Nation-State and the Natural Environment over the Twentieth Century." *American Sociological Review* 65, no. 1 (2000): 96–116.

Fredericks, Rosalind. *Garbage Citizenship: Vital Infrastructures of Labor in Dakar, Senegal.* Duke University Press, 2018.

Frickel, Scott, and James R. Elliott. *Sites Unseen: Uncovering Hidden Hazards in American Cities*. Russell Sage, 2018.

Frug, Gerald E., and David J. Barron. *City Bound: How States Stifle Urban Innovation*. Cornell University Press, 2013.

Furnari, Santi. "Interstitial Spaces: Microinteraction Settings and the Genesis of New Practices between Institutional Fields." *Academy of Management Review* 39, no. 4 (2014): 439–62.

Galaskiewicz, Joseph. "Interorganizational Relations." *Annual Review of Sociology* 11 (1985): 281–304.

Galaskiewicz, Joseph. "An Urban Grants Economy Revisited: Corporate Charitable Contributions in the Twin Cities, 1979–81, 1987–89." *Administrative Science Quarterly* 42 (1997): 445–71.

Galaskiewicz, Joseph, and Stanley Wasserman. "Mimetic Processes Within an Interorganizational Field: An Empirical Test." *Administrative Science Quarterly* 34, no. 3 (1989): 454–79.

Garrido, Marco Z. *The Patchwork City: Class, Space, and Politics in Metro Manila*. University of Chicago Press, 2019.

Gates, Bill. *How to Avoid a Climate Disaster: The Solutions We Have and the Breakthroughs We Need*. Vintage, 2021.

GaWC. "Globalization and World Cities." https://gawc.lboro.ac.uk.

Gehl Architects. "Home—Gehl." https://www.gehlpeople.com.

Gemeente Rotterdam. *Rotterdam Resilience Strategy: Ready for the 21st Century*, 2016. https://resilientcitiesnetwork.org/downloadable_resources/Network/Rotterdam-Resilience-Strategy-English.pdf.

Gensler. "Cities and Urban Design." https://www.gensler.com/expertise/cities-urban-design.

George, Gerard, Jennifer Howard-Grenville, Aparna Joshi, and Laszlo Tihanyi. "Understanding and Tackling Societal Grand Challenges through Management Research." *Academy of Management Journal* 59, no. 6 (2016): 1880–95.

Gibbons, Robert. "Incentives and Careers in Organizations." Working Paper. National Bureau of Economic Research, August 1996.

Gibbons, Robert. "March-ing Toward Organizational Economics." *Industrial and Corporate Change* 29, no. 1 (2020): 89–94.

Giffinger, Rudolf, Gudrun Haindlmaier, and Hans Kramar. "The Role of Rankings in Growing City Competition." *Urban Research and Practice* 3, no. 3 (2010): 299–312.

Glaeser, Edward. *Triumph of the City*. Penguin, 2011.

Gonzalez, Ana. "Making Do: How Cities Pursue Progressive Sustainability Solutions in Light of Empowering or Constraining State Contexts." Master's thesis, University of Chicago, 2023.

Gonzalez, Ana, and Christof Brandtner. "Green in Their Own Way: Pragmatic and Progressive Means for Cities to Overcome Institutional Barriers to Sustainability." *Urban Studies* 61, no. 13 (2024): 2513–30. https://doi.org/10.1177/00420980241239788.

Gounaridis, Dimitrios, and Joshua P. Newell. "The Social Anatomy of Climate Change Denial in the United States." *Scientific Reports* 14, no. 1 (2024): 2097.

Granovetter, Mark. "Economic Action and Social Structure: The Problem of Embeddedness." *American Journal of Sociology* 91, no. 3 (1985): 481–510.

Granovetter, Mark. *Getting a Job: A Study of Contacts and Careers.* 2nd ed. University of Chicago Press, 1995.

Granovetter, Mark. *Society and Economy: Framework and Principles.* Harvard University Press, 2017.

Grant, Don, Andrew Jorgenson, and Wesley Longhofer. *Super Polluters: Tackling the World's Largest Sites of Climate-Disrupting Emissions.* Columbia University Press, 2020.

Green, Amy. "With Gov. Scott and Legislature in Denial, Tiny Town Adapts on Its Own to Climate Change." *The Miami Herald*, July 15, 2018. https://www.miamiherald.com/news/state/florida/article214355019.html.

"Green Building Economic Impact Study." Booz Allen Hamilton, 2015. http://go.usgbc.org/2015-Green-Building-Economic-Impact-Study.html.

Greenwood, Royston. *Patterns of Management in Local Government.* Martin Robertson, 1980.

Greve, Henrich R., and Hayagreeva Rao. "Echoes of the Past: Organizational Foundings as Sources of an Institutional Legacy of Mutualism." *American Journal of Sociology* 118, no. 3 (2012): 635–75.

Greve, Henrich R., and Hayagreeva Rao. "History and the Present: Institutional Legacies in Communities of Organizations." *Research in Organizational Behavior* 34 (2014): 27–41.

Grimmelikhuijsen, Stephan G., and Mary K. Feeney. "Developing and Testing an Integrative Framework for Open Government Adoption in Local Governments." *Public Administration Review* 77, no. 4 (2017): 579–90.

Grimmer, Justin, Margaret E. Roberts, and Brandon M. Stewart. *Text as Data: A New Framework for Machine Learning and the Social Sciences.* Princeton University Press, 2022.

Gümüsay, Ali Aslan, Laura Claus, and John Amis. "Engaging with Grand Challenges: An Institutional Logics Perspective." *Organization Theory* 1, no. 3 (2020): 1–20.

Hall, Peter. *Cities of Tomorrow: An Intellectual History of Urban Planning and Design Since 1880.* Wiley-Blackwell, 2014.

Hallett, Tim. "The Myth Incarnate Recoupling Processes, Turmoil, and Inhabited Institutions in an Urban Elementary School." *American Sociological Review* 75, no. 1 (2010): 52–74.

Hanagan, Michael P., and Chris Tilly. *Contention and Trust in Cities and States.* Springer, 2011.

Harvey, David. *A Brief History of Neoliberalism.* Oxford University Press, 2007.

Harvey, David. *Rebel Cities: From the Right to the City to the Urban Revolution.* Verso, 2012. http://www.worldcat.org/title/rebel-cities-from-the-right-to-the-city -to-the-urban-revolution/oclc/943969433.

Hawkins, Christopher V., Rachel M. Krause, Richard C. Feiock, and Cali Curley. "Making Meaningful Commitments: Accounting for Variation in Cities' Investments of Staff and Fiscal Resources to Sustainability." *Urban Studies* 53, no. 9 (2016): 1902–24.

Hedström, Peter. *Dissecting the Social: Social Mechanisms and the Principles of Analytical Sociology.* Cambridge University Press, 2005.

Hironaka, Ann. *Greening the Globe.* Cambridge University Press, 2014.

Hoffmann, Matthew J. *Climate Governance at the Crossroads: Experimenting with a Global Response After Kyoto.* Oxford University Press, 2011.

Honig, Dan. *Mission Driven Bureaucrats: Empowering People to Help Government Do Better.* Oxford University Press, 2024.

Hood, Christopher. "A Public Management for All Seasons." *Public Administration* 69, no. 1 (1991): 3–19.

Howard, Ebenezer. *Garden Cities of To-Morrow.* 11th ed. MIT Press, 2001.

Hsu, Tiffany. "He Wanted to Unclog Cities. Now He's 'Public Enemy No. 1.' " *The New York Times,* March 28, 2023. https://www.nytimes.com/2023/03/28/technology /carlos-moreno-15-minute-cities-conspiracy-theories.html.

Hunter, Floyd. *Community Power Structure: A Study of Decision Makers.* University of North Carolina Press, 1953.

Hwang, Hokyu. "Planning Development: Globalization and the Shifting Locus of Planning." In *Globalization and Organization: World Society and Organizational Change,* ed. Gili S. Drori, John W. Meyer, and Hokyu Hwang. Oxford University Press, 2006.

Hwang, Hokyu, and Jeannette A Colyvas. "Ontology, Levels of Society, and Degrees of Generality: Theorizing Actors as Abstractions in Institutional Theory." *Academy of Management Review* 45, no. 3 (2020): 570–95.

Hwang, Hokyu, and Walter W. Powell. "The Rationalization of Charity: The Influences of Professionalism in the Nonprofit Sector." *Administrative Science Quarterly* 54, no. 2 (2009): 268–98.

Ingram, Paul, Jeffrey Robinson, and Marc L. Busch. "The Intergovernmental Network of World Trade: IGO Connectedness, Governance, and Embeddedness." *American Journal of Sociology* 111, no. 3 (2005): 824–58.

International City/County Management Association (ICMA). "ICMA Survey Research: 2015 Local Government Sustainability Practices Survey Report." ICMA,

2016. https://icma.org/documents/icma-survey-research-2015-local-government
-sustainability-practices-survey-report.

International City/County Management Association (ICMA). "ICMA Survey Research: 2018 Municipal Form of Government Survey." ICMA, July 2, 2019. https://icma.org/2018-municipal-fog-survey.

Jacobs, Jane. "'The Uses of Sidewalks: Safety': From The Death and Life of Great American Cities (1961)." In *The City Reader*, ed. Richard T. LeGates, Frederic Stout, and Roger W. Caves, 7th ed. Routledge, 2020.

Jepperson, Ronald, and John W. Meyer. "Multiple Levels of Analysis and the Limitations of Methodological Individualisms." *Sociological Theory* 29, no. 1 (2011): 54–73.

Jessop, Bob. "The Entrepreneurial City: Re-Imaging Localities, Redesigning Economic Governance, or Restructuring Capital." In *Transforming Cities: Contested Governance and New Spatial Divisions*, ed. Nick Jewson and Susanne MacGregor, vol. 46. Routledge, 1997.

Jessop, Bob, and Ngai-Ling Sum. "An Entrepreneurial City in Action: Hong Kong's Emerging Strategies in and for (Inter) Urban Competition." *Urban Studies* 37, no. 12 (2000): 2287–313.

Johnston, Sadhu Aufochs, Steven S. Nicholas, and Julia Parzen. *The Guide to Greening Cities*. Springer, 2013.

Kadushin, Charles. "Friendship Among the French Financial Elite." *American Sociological Review* 60, no. 2 (1995): 202–21.

Kane, Joseph W., Adie Turner, Caroline George, and Jamal Russell Black. "Not According to Plan: Exploring Gaps in City Climate Planning and the Need for Regional Action." Brookings Institution, 2022. https://www.brookings.edu/articles/not
-according-to-plan-exploring-gaps-in-city-climate-planning-and-the-need-for
-regional-action.

Kass, Jonathon. "How Copenhagen Can Inspire Bay Area Cities to Go Big on Bikes." SPUR Urban Center, August 31, 2022. https://www.spur.org/news/2022-08-31/how
-copenhagen-can-inspire-bay-area-cities-go-big-bikes.

Katz, Bruce, and Jennifer Bradley. *The Metropolitan Revolution*. Brookings Institution, 2014.

Kettl, Donald. "The Global Revolution in Public Management: Driving Themes, Missing Links." *Journal of Policy Analysis and Management* 16, no. 3 (1997): 446–62.

Khurana, Rakesh. *From Higher Aims to Hired Hands: The Social Transformation of American Business Schools and the Unfulfilled Promise of Management as a Profession*. Princeton University Press, 2010.

King, Brayden G., Teppo Felin, and David A. Whetten. "Perspective-Finding the Organization in Organizational Theory: A Meta-Theory of the Organization as a Social Actor." *Organization Science* 21, no. 1 (2010): 290–305.

Klinenberg, Eric. "The Key to Surviving Climate Change? Build Tight-Knit Communities." *Wired*, October 25, 2016. https://www.wired.com/2016/10/klinenberg-transforming-communities-to-survive-climate-change.

Klinenberg, Eric. *Heat Wave: A Social Autopsy of Disaster in Chicago.* 2nd ed. University of Chicago Press, 2015.

Klinenberg, Eric. *Palaces for the People: How Social Infrastructure Can Help Fight Inequality, Polarization, and the Decline of Civic Life.* Crown, 2018.

Knoke, David. "The Spread of Municipal Reform: Temporal, Spatial, and Social Dynamics." *American Journal of Sociology* 87, no. 6 (1982): 1314–39.

Knox, Paul L., and Peter J. Taylor. *World Cities in a World-System.* Cambridge University Press, 1995.

Kornberger, Martin. "Governing the City: From Planning to Urban Strategy." *Theory, Culture, and Society* 29, no. 2 (2012): 84–106.

Kornberger, Martin, and Chris Carter. "Manufacturing Competition: How Accounting Practices Shape Strategy Making in Cities." *Accounting, Auditing, and Accountability Journal* 23, no. 3 (2010): 325–49.

Kornberger, Martin, and Stewart Clegg. "Strategy as Performative Practice: The Case of Sydney 2030." *Strategic Organization* 9, no. 2 (2011): 136–62.

Kornberger, Martin, Renate E. Meyer, Christof Brandtner, and Markus A. Höllerer. "When Bureaucracy Meets the Crowd: Studying 'Open Government' in the Vienna City Administration." *Organization Studies* 38, no. 2 (2017): 179–200.

Kornberger, Martin, Renate E. Meyer, and Markus A. Höllerer. "Exploring the Long-Term Effect of Strategy Work: The Case of Sustainable Sydney 2030." *Urban Studies* 58, no. 16 (2021): 3316–34.

Koshar, Rudy. German Travel Cultures. Berg, 2000.

Koslov, Liz. "The Case for Retreat." *Public Culture* 28, no. 2 (2016): 359–87.

Krause, Rachel M. "An Assessment of the Greenhouse Gas Reducing Activities Being Implemented in US Cities." *Local Environment* 16, no. 2 (2011): 193–211.

Krause, Rachel M., and Christopher Hawkins. *Implementing City Sustainability: Overcoming Administrative Silos to Achieve Functional Collective Action.* Temple University Press, 2021.

Krinsky, John, and Maud Simonet. *Who Cleans the Park? Public Work and Urban Governance in New York City.* University of Chicago Press, 2017.

Krücken, Georg, and Frank Meier. "Turning the University into an Organizational Actor." In *Globalization and Organization: World Society and Organizational Change*, ed, Gili S. Drori, John W. Meyer, and Hokyu Hwang. Oxford University Press, 2006.

Kuala Lumpur City Hall. *Kuala Lumpur Climate Action Plan 2050*, 2021. https://aipalync.org/storage/documents/main/kuala-lumpur-climate-action-plan-2025_1713867229.pdf.

Kunzig, Robert. "The World's Most Improbable Green City." *National Geographic*, April 4, 2017. https://www.nationalgeographic.com/environment/article/dubai -ecological-footprint-sustainable-urban-city.

Kwon, Seok-Woo, Colleen Heflin, and Martin Ruef. "Community Social Capital and Entrepreneurship." *American Sociological Review* 78, no. 6 (2013): 980–1008.

Lamont, Michèle. "Toward a Comparative Sociology of Valuation and Evaluation." *Annual Review of Sociology* 38, no. 1 (2012): 201–21.

Laryea, Krystal, and Christof Brandtner. "Organizations as Drivers of Social and Systemic Integration: Contradiction and Reconciliation Through Loose Demographic Coupling and Community Anchoring." *Research in the Sociology of Organizations* 90 (2024): 177–200.

Laumann, Edward O., Joseph Galaskiewicz, and Peter V. Marsden. "Community Structure as Interorganizational Linkages." *Annual Review of Sociology* 4, no. 1 (1978): 455–84.

Laumann, Edward O., and David Knoke. *The Organizational State: Social Choice in National Policy Domains*. University of Wisconsin Press, 1987.

Lee, Caroline W., Michael McQuarrie, and Edward T. Walker. *Democratizing Inequalities: Dilemmas of the New Public Participation*. New York University Press, 2015.

Lee, Neil. *Innovation for the Masses: How to Share the Benefits of the High-Tech Economy*. University of California Press, 2024.

Lefebvre, Henri. *The Urban Revolution*. University of Minnesota Press, 2003.

Leixnering, Stephan, and Markus Höllerer. "'Remaining the Same or Becoming Another?' Adaptive Resilience versus Transformative Urban Change." *Urban Studies* 59, no. 6 (2022): 1300–1310.

Lerch, Julia C., David John Frank, and Evan Schofer. "The Social Foundations of Academic Freedom: Heterogeneous Institutions in World Society, 1960 to 2022." *American Sociological Review* 89, no. 1 (2024): 88–125.

Levine, Jeremy R. *Constructing Community: Urban Governance, Development, and Inequality in Boston*. Princeton University Press, 2021.

Levine, Jeremy R. "The Paradox of Community Power: Cultural Processes and Elite Authority in Participatory Governance." *Social Forces* 95, no. 3 (2017): 1155–79.

Levine, Jeremy R. "The Privatization of Political Representation." *American Sociological Review* 81, no. 6 (2016): 1251–75.

Lin, Jean Yen-chun. *A Spark in the Smokestacks: Environmental Organizing in Beijing Middle-Class Communities*. Columbia University Press, 2023.

Lipset, Seymour Martin. *American Exceptionalism: A Double-Edged Sword*. Norton, 1996.

Longhofer, Wesley, and Andrew Jorgenson. "Decoupling Reconsidered: Does World Society Integration Influence the Relationship Between the Environment and Economic Development?" *Social Science Research* 65 (2017): 17–29.

Longhofer, Wesley, and Evan Schofer. "National and Global Origins of Environmental Association." *American Sociological Review* 75, no. 4 (2010): 505–33.

Lounsbury, Michael. "Institutional Sources of Practice Variation: Staffing College and University Recycling Programs." *Administrative Science Quarterly* 46, no. 1 (2001): 29–56.

Lounsbury, Michael, Marc Ventresca, and Paul M. Hirsch. "Social Movements, Field Frames and Industry Emergence: A Cultural–Political Perspective on US Recycling." *Socio-Economic Review* 1, no. 1 (2003): 71–104.

Lukes, Steven. *Power: A Radical View*. Palgrave Macmillan, 2005.

Lwasa, Shuaib, Karen C. Seto, Xuemei Bai, et al. "Urban Systems and Other Settlements." *IPCC, 2022: Climate Change 2022: Mitigation of Climate Change. Contribution of Working Group III to the Sixth Assessment Report of the Intergovernmental Panel on Climate Change.* Cambridge University Press, 2022.

Macaulay, Stewart. "Non-Contractual Relations in Business: A Preliminary Study." In *The Sociology of Economic Life*, ed. Mark Granovetter and Richard Swedberg. Routledge, 2018.

Manduca, Robert, and Robert J. Sampson. "Punishing and Toxic Neighborhood Environments Independently Predict the Intergenerational Social Mobility of Black and White Children." *Proceedings of the National Academy of Sciences* 116, no. 16 (2019): 7772–77.

March, James G. "The Business Firm as a Political Coalition." *Journal of Politics* 24, no. 4 (1962): 662–78.

March, James G., and Herbert A. Simon. *Organizations*. Wiley, 1993.

Marquis, Christopher, and Julie Battilana. "Acting Globally but Thinking Locally? The Enduring Influence of Local Communities on Organizations." *Research in Organizational Behavior* 29 (2009): 283–302.

Marquis, Christopher, Gerald F. Davis, and Mary Ann Glynn. "Golfing Alone? Corporations, Elites, and Nonprofit Growth in 100 American Communities." *Organization Science* 24, no. 1 (2013): 39–57.

Marquis, Christopher, Mary Ann Glynn, and Gerald F. Davis. "Community Isomorphism and Corporate Social Action." *Academy of Management Review* 32, no. 3 (2007): 925–45.

Marquis, Christopher, Michael Lounsbury, and Royston Greenwood. "Introduction: Community as an Institutional Order and a Type of Organizing." In *Communities and Organizations*, Research in the Sociology of Organizations, ed. Christopher Marquis, Michael Lounsbury, and Royston Greenwood, vol. 33. Emerald Group, 2011.

Marquis, Christopher, and Kunyuan Qiao. "History Matters for Organizations: An Integrative Framework for Understanding Influences from the Past." *Academy of Management Review* 50, no. 2 (2024). https://doi.org/10.5465/amr.2022.0238.

Martin, Isaac. "Dawn of the Living Wage: The Diffusion of a Redistributive Municipal Policy." *Urban Affairs Review* 36, no. 4 (2001): 470–96.

Martin, John Levi. *The Explanation of Social Action*. Oxford University Press, 2011.

Marwell, Nicole P. "Privatizing the Welfare State: Nonprofit Community-Based Organizations as Political Actors." *American Sociological Review* 69, no. 2 (2004): 265–91.

Marwell, Nicole P., and Shannon L. Morrissey. "Organizations and the Governance of Urban Poverty." *Annual Review of Sociology* 46 (2020): 233–50.

Matisoff, Daniel C., Douglas S. Noonan, and Anna M. Mazzolini. "Performance or Marketing Benefits? The Case of LEED Certification." *Environmental Science and Technology* 48, no. 3 (2014): 2001–7.

Mayntz, Renate. "New Challenges to Governance Theory." In *Governance as Social and Political Communication*, ed. Henrik Paul Bang. Manchester University Press, 2003.

Mazzucato, Mariana. *Mission Economy: A Moonshot Guide to Changing Capitalism*. Penguin UK, 2021.

McAdam, Doug. *Political Process and the Development of Black Insurgency, 1930–1970*. University of Chicago Press, 2010.

McInerney, Paul-Brian. *From Social Movement to Moral Market: How the Circuit Riders Sparked an IT Revolution and Created a Technology Market*. Stanford University Press, 2014.

McQuarrie, Michael, and Nicole P. Marwell. "The Missing Organizational Dimension in Urban Sociology." *City and Community* 8, no. 3 (2009): 247–68.

McTarnaghan, Sara, Jorge Morales-Burnett, and Rebecca Marx. *Urban Resilience: From Global Vision to Local Practice*. Urban Institute, 2022.

Meyer, John W., John Boli, George M. Thomas, and Francisco O. Ramirez. "World Society and the Nation-State." *American Journal of Sociology* 103, no. 1 (1997): 144–81.

Meyer, John W., and Ronald Jepperson. "The 'Actors' of Modern Society: The Cultural Construction of Social Agency." *Sociological Theory* 18, no. 1 (2000): 100–120.

Meyer, John W., and Brian Rowan. "Institutionalized Organizations: Formal Structure as Myth and Ceremony." *American Journal of Sociology* 83, no. 2 (1977): 340–63.

Meyer, John W., and W. Richard Scott. *Organizational Environments: Ritual and Rationality*. SAGE, 1983.

Millard-Ball, Adam. "The Limits to Planning: Causal Impacts of City Climate Action Plans." *Journal of Planning Education and Research* 33, no. 1 (2013): 5–19.

Molotch, Harvey. "The City as a Growth Machine: Toward a Political Economy of Place." *American Journal of Sociology* 82, no. 2 (1976): 309–32.

Molotch, Harvey, William Freudenburg, and Krista E. Paulsen. "History Repeats Itself, but How? City Character, Urban Tradition, and the Accomplishment of Place." *American Sociological Review* 65, no. 6 (2000): 791–823.

Morgan, Kevin. "The Learning Region: Institutions, Innovation and Regional Renewal." *Regional Studies* 31, no. 5 (2010): 491–503.

Mullinix, Nayiri. "Nearly 15 Percent of Americans Deny Climate Change Is Real, AI Study Finds." Michigan News: University of Michigan, February 14, 2024. https://news .umich.edu/nearly-15-of-americans-deny-climate-change-is-real-ai-study-finds.

Mumford, Lewis. "What Is a City?" *Architectural Record* 82, no. 5 (1937): 59–62.

Naumovska, Ivana, Vibha Gaba, and Henrich R Greve. "The Diffusion of Differences: A Review and Reorientation of 20 Years of Diffusion Research." *Academy of Management Annals* 15, no. 2 (2021): 377–405.

"New York State Green Jobs Study." New York City Labor Market Information Service, CUNY Graduate Center, 2011. https://www.gc.cuny.edu/sites/default /files/2022-01/CUR-Green-Jobs-Study-Chapter-3.pdf.

Norris, Davon. "The Illusion of Transparency: The Political Double Standard in City Credit Ratings." *Socio-Economic Review* 21, no. 2 (2023): 1125–50.

NYC Mayor's Office of Climate & Environmental Justice. *PlaNYC: Getting Sustainability Done*, 2023. https://www.nyc.gov/content/climate/pages/reports-and -publications/planyc.

Oldenburg, Ramon, and Dennis Brissett. "The Third Place." *Qualitative Sociology* 5, no. 4 (1982): 265–84.

Osborne, Stephen. *The New Public Governance?* Routledge, 2010.

"Our Cities, Ourselves." *Nature Cities* 1, no. 1 (2024): 1–1. https://doi.org/10.1038 /s44284-023-00030-4.

Paarlberg, Laurie E., and Samantha Zuhlke. "Revisiting the Theory of Government Failure in the Face of Heterogeneous Demands." *Perspectives on Public Management and Governance* 2, no. 2 (2019): 103–24.

Pacala, Stephen, and Robert Socolow. "Stabilization Wedges: Solving the Climate Problem for the Next 50 Years with Current Technologies." *Science* 305, no. 5686 (2004): 968–72.

Paddison, Ronan. "City Marketing, Image Reconstruction and Urban Regeneration." *Urban Studies* 30, no. 2 (1993): 339–49.

Padgett, John F., and Christopher K. Ansell. "Robust Action and the Rise of the Medici, 1400–1434." *American Journal of Sociology* 98, no. 6 (1993): 1259–319.

Padgett, John F., and Paul D. McLean. "Organizational Invention and Elite Transformation: The Birth of Partnership Systems in Renaissance Florence." *American Journal of Sociology* 111, no. 5 (2006): 1463–1568.

Padgett, John F., and Walter W. Powell. *The Emergence of Organizations and Markets*. Princeton University Press, 2012.

"Paris Climate Action Plan: Plan Climat 2024–2030." C40. https://www
.c4oknowledgehub.org/s/article/In-Paris-we-re-implementing-accelerated-actions
-now-to-halve-our-emissions-by-2030-and-achieve-carbon-neutrality-by-2050

Park, Robert E. "The City: Suggestions for the Investigation of Human Behavior in the City Environment." *American Journal of Sociology* 20, no. 5 (1915): 577–612.

Park, Robert E. "Human Migration and the Marginal Man." *American Journal of Sociology* 33, no. 6 (1928): 881–893.

Park, Robert E. "Succession, an Ecological Concept." *American Sociological Review* 1, no. 2 (1936): 171–79.

Park, Robert E., and Ernest W. Burgess. *The City*. University of Chicago Press, 1921.

Paxton, Fred. *Restrained Radicals: Populist Radical Right Parties in Local Government*. Cambridge University Press, 2023.

Pedersen, Jesper Strandgaard, and Frank Dobbin. "The Social Invention of Collective Actors on the Rise of the Organization." *American Behavioral Scientist* 40, no. 4 (1997): 431–43.

Pedulla, David S. *Making the Cut: Hiring Decisions, Bias, and the Consequences of Nonstandard, Mismatched, and Precarious Employment*. Princeton University Press, 2020.

Pedulla, David S., and Devah Pager. "Race and Networks in the Job Search Process." *American Sociological Review* 84, no. 6 (2019): 983–1012.

Perrow, Charles. "Organisations and Global Warming." In *Routledge Handbook of Climate Change and Society*, ed. Steven Brechin and Seungyun Lee. Routledge, 2010.

Perrow, Charles. *A Society of Organizations*. Vol. 20. Springer, 1991.

Peters, B. Guy, and Jon Pierre. "Developments in Intergovernmental Relations: Towards Multi-Level Governance." *Policy and Politics* 29, no. 2 (2001): 131–36.

Peters, B. Guy, and Jon Pierre. "Governance Without Government? Rethinking Public Administration." *Journal of Public Administration Research and Theory* 8, no. 2 (1998): 223–43.

Peterson, Paul E. *City Limits*. University of Chicago Press, 1981.

Pierre, Jon. "Comparative Urban Governance Uncovering Complex Causalities." *Urban Affairs Review* 40, no. 4 (2005): 446–62.

Podolny, Joel M. "Networks as the Pipes and Prisms of the Market." *American Journal of Sociology* 107, no. 1 (2001): 33–60.

Pollitt, Christopher, and Geert Bouckaert. *Public Management Reform: A Comparative Analysis: Into the Age of Austerity*. 4th ed. Oxford University Press, 2017.

Portney, Kent E. *Taking Sustainable Cities Seriously: Economic Development, the Environment, and Quality of Life in American Cities*. Vol. 67. MIT Press, 2003.

Portney, Kent E., and Jeffrey M. Berry. "Participation and the Pursuit of Sustainability in U.S. Cities." *Urban Affairs Review* 46, no. 1 (2010): 119–39.

Powell, Walter W., and Christof Brandtner. "Organizations as Sites and Drivers of Social Action." In *Handbook of Contemporary Sociological Theory*, ed. Seth Abrutyn. Springer, 2016. https://doi.org/10.1007/978-3-319-32250-6.

Powell, Walter W., and Paul DiMaggio, eds. *The New Institutionalism in Organizational Analysis*. University of Chicago Press, 1991.

Powell, Walter W., Aaron Horvath, and Christof Brandtner. "Click and Mortar: Organizations on the Web." *Research in Organizational Behavior* 36 (2016): 101–20.

Rao, Hayagreeva, and Henrich R. Greve. "Disasters and Community Resilience: Spanish Flu and the Formation of Retail Cooperatives in Norway." *Academy of Management Journal* 61, no. 1 (2018): 5–25.

Rao, Hayagreeva, Philippe Monin, and Rodolphe Durand. "Institutional Change in Toque Ville: Nouvelle Cuisine as an Identity Movement in French Gastronomy." *American Journal of Sociology* 108, no. 4 (2003): 795–843.

Rea, Christopher M. "Theorizing Command-and-Commodify Regulation: The Case of Species Conservation Banking in the United States." *Theory and Society* 46 (2017): 21–56.

Rea, Christopher M., and Scott Frickel. "The Environmental State: Nature and the Politics of Environmental Protection." *Sociological Theory* 41, no. 3 (2023): 255–81.

Reinecke, Juliane, and Shahzad Ansari. "Microfoundations of Framing: The Interactional Production of Collective Action Frames in the Occupy Movement." *Academy of Management Journal* 64, no. 2 (2021): 378–408.

Reinecke, Juliane, and Shahzad Ansari. "Taming Wicked Problems: The Role of Framing in the Construction of Corporate Social Responsibility." *Journal of Management Studies* 53, no. 3 (2016): 299–329.

Ren, Xuefei. *Governing the Urban in China and India: Land Grabs, Slum Clearance, and the War on Air Pollution*. Princeton University Press, 2020.

Resilient Cities Network. "Homepage." https://resilientcitiesnetwork.org.

Rhodes, Rod A. W. "Understanding Governance: Ten Years On." *Organization Studies* 28, no. 8 (2007): 1243–64.

Ritchie, Hannah, Veronika Samborska, and Max Roser. "Urbanization." *Our World in Data*, February 21, 2024. https://ourworldindata.org/urbanization.

Roberts, Kale. "Three Decades of Sustainability: ICLEI at 30 Enters Next 'Decade of Local Action.'" ICLEI USA, September 4, 2020. https://icleiusa.org/iclei-at-30.

Rogers, Everett. *Diffusion of Innovations*. 5th ed. Free Press, 2003.

Romero-Lankao, Patricia, Nicole Rosner, Christof Brandtner, Christopher Rea, Adolfo Mejia-Montero, Francesca Pilo, Fedor Dokshin, Vanesa Castan-Broto, Sarah Burch, and Scott Schnur. "A Framework to Centre Justice in Energy Transition Innovations." *Nature Energy* 8 (2023): 1192–98. https://doi.org/10.1038/s41560-023-01351-3.

Rosenzweig, Cynthia, William Solecki, Stephen A. Hammer, and Shagun Mehrotra. "Cities Lead the Way in Climate–Change Action." *Nature* 467, no. 7318 (2010): 909–11.

Rothwell, Jonathan, José Lobo, Deborah Strumsky, and Mark Muro. *Patenting Prosperity: Invention and Economic Performance in the United States and Its Metropolitan Areas.* Brookings Institution, 2013. https://www.brookings.edu/wp-content/uploads/2016/06/patenting-prosperity-rothwell.pdf.

Roy, Donald. "Quota Restriction and Goldbricking in a Machine Shop." *American Journal of Sociology* 57, no. 5 (1952): 427–42.

Ruddick, Sue, Linda Peake, Gökbörü S. Tanyildiz, and Darren Patrick. "Planetary Urbanization: An Urban Theory for Our Time?" *Environment and Planning D: Society and Space* 36, no. 3 (2018): 387–404.

Rupasingha, Anil, Stephan J Goetz, and David Freshwater. "The Production of Social Capital in US Counties." *Journal of Socio-Economics* 35, no. 1 (2006): 83–101.

Safford, Sean. "Why the Garden Club Couldn't Save Youngstown: Civic Infrastructure and Mobilization in Economic Crises." Social Working Paper Series on Local Innovation Systems, MIT-IPC-LIS-04-003. Massachusetts Institute of Technology, Industrial Performance Center, 2004.

Sahlin-Andersson, Kerstin, and Lars Engwall. *The Expansion of Management Knowledge: Carriers, Flows, and Sources.* Stanford University Press, 2002.

Sampson, Robert J. *Great American City: Chicago and the Enduring Neighborhood Effect.* University of Chicago Press, 2012.

Sampson, Robert J. "Neighbourhood Effects and Beyond: Explaining the Paradoxes of Inequality in the Changing American Metropolis." *Urban Studies* 56, no. 1 (2019): 3–32.

Sampson, Robert J., Doug McAdam, Heather MacIndoe, and Simón W. Weffer-Elizondo. "Civil Society Reconsidered: The Durable Nature and Community Structure of Collective Civic Action." *American Journal of Sociology* 111, no. 3 (2005): 673–714.

Sassen, Saskia. *Cities in a World Economy.* 4th ed. Pine Forge, 2012.

Sassen, Saskia. *The Global City: New York, London, Tokyo.* Princeton University Press, 2001.

Saxenian, AnnaLee. "Inside-Out: Regional Networks and Industrial Adaptation in Silicon Valley and Route 128." *Cityscape: A Journal of Policy Development and Research* 2, no. 2 (1994).

Scharpf, Fritz W. "Community and Autonomy: Multi-Level Policy-Making in the European Union." *Journal of European Public Policy* 1, no. 2 (1994): 219–42.

Schneiberg, Marc, and Elisabeth S. Clemens. "The Typical Tools for the Job: Research Strategies in Institutional Analysis." *Sociological Theory* 24, no. 3 (2006): 195–227.

Schneiberg, Marc, and Sarah A. Soule. "Institutionalization as a Contested, Multilevel Process: The Case of Rate Regulation in American Fire Insurance." In *Social*

Movements and Organization Theory, ed. Gerald F. Davis, Doug McAdam, W. Richard Scott, and Mayer N. Zald. Cambridge University Press, 2010.

Schofer, Evan, and Ann Hironaka. "The Effects of World Society on Environmental Protection Outcomes." *Social Forces* 84, no. 1 (2005): 25–47.

Schofer, Evan, Julia C. Lerch, and John W. Meyer. "Illiberal Reactions to Higher Education." *Minerva* 60, no. 4 (December 1, 2022): 509–34.

Schragger, Richard. *City Power: Urban Governance in a Global Age*. Oxford University Press, 2016.

Selznick, Philip. *TVA and the Grass Roots: A Study of Politics and Organization*. Vol. 3. University of California Press, 1949.

Sharkey, Amanda, and Patricia Bromley. "Can Ratings Have Indirect Effects? Evidence from the Organizational Response to Peers' Environmental Ratings." *American Sociological Review* 80, no. 1 (2014): 63–91.

Sharkey, Amanda, Kathryne M. Young, Christof Brandtner, and Patrick Bergemann. "Organizational Scarring, Legal Consciousness, and the Diffusion of Local Government Litigation Against Opioid Manufacturers." *American Sociological Review*. (2025).

Sharkey, Patrick, Gerard Torrats-Espinosa, and Delaram Takyar. "Community and the Crime Decline: The Causal Effect of Local Nonprofits on Violent Crime." *American Sociological Review* 82, no. 6 (2017): 1214–40.

Shi, Linda, Eric Chu, and Jessica Debats. "Explaining Progress in Climate Adaptation Planning across 156 US Municipalities." *Journal of the American Planning Association* 81, no. 3 (2015): 191–202.

Shils, Edward. "The Sociology of Robert E. Park." *The American Sociologist* 27, no. 4 (1996): 88–106.

Simon, Herbert A. *Administrative Behavior*. 4th ed. Simon and Schuster, 2013.

Simon, Herbert A. "Organizations and Markets." *Journal of Economic Perspectives* 5, no. 2 (1991): 25–44.

Sine, Wesley D., and Brandon H. Lee. "Tilting at Windmills? The Environmental Movement and the Emergence of the U.S. Wind Energy Sector." *Administrative Science Quarterly* 54, no. 1 (2009): 123–55.

Singapore Ministry of National Development. "About the Project." https://www.mnd.gov.sg/tianjinecocity/who-we-are.

Singh, Chandni, Mythili Madhavan, Jasmitha Arvind, and Amir Bazaz. "Climate Change Adaptation in Indian Cities: A Review of Existing Actions and Spaces for Triple Wins." *Urban Climate* 36 (2021): 100783.

Skocpol, Theda. *Diminished Democracy: From Membership to Management in American Civic Life*. University of Oklahoma Press, 2003.

Skocpol, Theda, Peter Evans, and Dietrich Rueschemeyer. *Bringing the State Back In*. Cambridge University Press, 1985.

Skocpol, Theda, and Kenneth Finegold. "State Capacity and Economic Intervention in the Early New Deal." *Political Science Quarterly* 97, no. 2 (1982): 255–78.

Small, Mario L. *Unanticipated Gains: Origins of Network Inequality in Everyday Life*. Oxford University Press, 2009.

Smith, Carl. *The Plan of Chicago: Daniel Burnham and the Remaking of the American City*. University of Chicago Press, 2019.

Smith, James M. "Urban Regime Theory." In *The Wiley Blackwell Encyclopedia of Urban and Regional Studies*, ed. Anthony M. Orum. Wiley, 2019.

Sorenson, Olav, and Pino G Audia. "The Social Structure of Entrepreneurial Activity: Geographic Concentration of Footwear Production in the United States, 1940–1989." *American Journal of Sociology* 106, no. 2 (2000): 424–62.

Speck, Jeff. *Walkable City: How Downtown Can Save America, One Step at a Time*. MacMillan, 2013.

Spicer, Jason, Tamara Kay, and Marshall Ganz. "Social Entrepreneurship as Field Encroachment: How a Neoliberal Social Movement Constructed a New Field." *Socio-Economic Review* 17, no. 1 (2019): 195–227.

State of Green. "House of Green." https://stateofgreen.com/en/about/about-state-of-green/house-of-green.

Steil, Justin Peter, and Ion Bogdan Vasi. "The New Immigration Contestation: Social Movements and Local Immigration Policy Making in the United States, 2000–2011." *American Journal of Sociology* 119, no. 4 (2014): 1104–55.

Sterling, Adina D., and Roberto M. Fernandez. "Once in the Door: Gender, Tryouts, and the Initial Salaries of Managers." *Management Science* 64, no. 11 (2018): 5444–60.

Stinchcombe, Arthur L. "Social Structure and Organizations." In *Handbook of Organizations*, ed. James G. March. Rand McNally, 1965.

Stokes, Leah Cardamore. *Short Circuiting Policy: Interest Groups and the Battle over Clean Energy and Climate Policy in the American States*. Oxford University Press, 2020.

Stone, Clarence N. "Urban Regimes and the Capacity to Govern: A Political Economy Approach." *Journal of Urban Affairs* 15, no. 1 (1993): 1–28.

Storper, Michael. *Keys to the City*. Princeton University Press, 2013.

Strang, David, and John W. Meyer. "Institutional Conditions for Diffusion." *Theory and Society* 22, no. 4 (1993): 487–511.

Strang, David, and Sarah A. Soule. "Diffusion in Organizations and Social Movements: From Hybrid Corn to Poison Pills." *Annual Review of Sociology* 24 (1998): 265–90.

Strang, David, and Nancy Brandon Tuma. "Spatial and Temporal Heterogeneity in Diffusion." *American Journal of Sociology* 99, no. 3 (1993): 614–39.

Suchman, Mark C. "Managing Legitimacy: Strategic and Institutional Approaches." *Academy of Management Review* 20, no. 3 (1995): 571–610.

Taylor, Peter J. "Comment: Parallel Paths to Understanding Global Intercity Relations." *American Journal of Sociology* 112, no. 3 (2006): 881–94.

The City of Amsterdam. *New Amsterdam Climate Roadmap: Amsterdam Climate Neutral 2050*, 2020. https://assets.amsterdam.nl/publish/pages/943415/roadmap_amsterdam_climate_neutral_2050_2.pdf.

The City of Brussels. *City of Brussels Climate Plan*, 2022. https://www.brussels.be/sites/default/files/bxl/221130%20%20Plan%20Climat%20Version%20finale%20EN.pdf.

The City of Hiroshima. *International Peace Culture City: Connecting Hiroshima to the Future (6th Hiroshima City Basic Plan)*, 2020. https://www.city.hiroshima.lg.jp/_res/projects/default_project/_page_/001/009/818/220813_339228_misc.pdf.

The City of New York. *PlaNYC: A Greener, Greater New York*, 2007. https://www.nyc.gov/html/planyc/downloads/pdf/publications/full_report_2007.pdf.

The City of Palo Alto. 2022 Sustainability and Climate Action Plan, 2023. https://www.paloalto.gov/files/assets/public/v/1/sustainability/reports/2022-scap-report_final.pdf

The Economist. "California Leads Subnational Efforts to Curb Climate Change." September 15, 2018. https://www.economist.com/international/2018/09/15/california-leads-subnational-efforts-to-curb-climate-change.

The Internet Archive. "Why Build Green? (Archival Snapshot)," May 5, 2019. https://web.archive.org/web/20190505181719/https://archive.epa.gov/greenbuilding/web/html/whybuild.html.

Thompson, James D. *Organizations in Action: Social Science Bases of Administrative Theory*. Transaction, 1967.

Tilcsik, András, and Christopher Marquis. "Punctuated Generosity: How Mega-Events and Natural Disasters Affect Corporate Philanthropy in U.S. Communities." *Administrative Science Quarterly* 58, no. 1 (2013): 111–48. https://doi.org/10.1177/0001839213475800.

Tilly, Charles. "Cities, States, and Trust Networks: Chapter 1 of Cities and States in World History." In *Theory and Society: Special Issue in Memory of Charles Tilly, 1929–2008. Cities, States, Trust, and Rule*, ed. Michael P Hanagan and Chris Tilly. Springer, 2011.

Tolbert, Pamela S., and Lynne G. Zucker. "Institutional Sources of Change in the Formal Structure of Organizations: The Diffusion of Civil Service Reform, 1880–1935." *Administrative Science Quarterly* 28, no. 1 (1983): 22–39.

Tozer, Laura, Harriet Bulkeley, and Linjun Xie. "Transnational Governance and the Urban Politics of Nature-Based Solutions for Climate Change." *Global Environmental Politics* 22, no. 3 (2022): 81–103.

Trisolini, Katherine A. "All Hands on Deck: Local Governments and the Potential for Bidirectional Climate Change Regulation." *Stanford Law Review* 62 (2009): 669–722.

Turco, Catherine. "Difficult Decoupling: Employee Resistance to the Commercialization of Personal Settings." *American Journal of Sociology* 118, no. 2 (2012): 380–419.

Ulpiani, Giulia, Nadja Vetters, Giulia Melica, and Paolo Bertoldi. "Towards the First Cohort of Climate-Neutral Cities: Expected Impact, Current Gaps, and Next Steps to Take to Establish Evidence-Based Zero-Emission Urban Futures." *Sustainable Cities and Society* 95 (2023): 104572.

United Cities and Local Governments (UCLG). "The Lampedusa Charter for Dignified Human Mobility and Territorial Solidarity." UCLG, n.d. https://www.uclg.org/sites/default/files/lampedusa-_carta-eng.pdf.

Urban Sustainability Directors Network. "History." https://www.usdn.org/history.html.

US Centers for Disease Control and Prevention. "NCHS Urban-Rural Classification Scheme for Counties." https://www.cdc.gov/nchs/data-analysis-tools/urban-rural.html.

US Environmental Protection Agency. "Homepage," January 19, 2017. https://19january2017snapshot.epa.gov.US Environmental Protection Agency. "Why Build Green?" https://www.epa.gov/greenbuilding/pubs/whybuild.htm.

US Green Building Council. "LEED Credit Library." https://www.usgbc.org/pilotcredits.

US Green Building Council. "LEED Professional Credentials." https://www.usgbc.org/credentials.

U.S. Green Building Council. *USGBC Policy Library.* https://public-policies.usgbc.org/.

U.S. Green Building Council. *Real Estate and Biodiversity: What You Need to Know.* U.S. Green Building Council, 2023. https://www.usgbc.org/sites/default/files/2023-07/Real-Estate-and-Biodiversity-What-You-Need-to-Know.pdf.

Vamosi, Stephen J. "The True Cost of LEED-Certified Green Buildings." *Heating, Plumbing Air Conditioning Magazine*, 2011.

Vargas, Robert. "Gangstering Grants: Bringing Power to Collective Efficacy Theory." *City and Community* 18, no. 1 (2019): 369–91.

Vargas, Robert. *Wounded City: Violent Turf Wars in a Chicago Barrio.* Oxford University Press, 2016.

Vasi, Ion Bogdan. "Thinking Globally, Planning Nationally and Acting Locally: Nested Organizational Fields and the Adoption of Environmental Practices." *Social Forces* 86, no. 1 (2007): 113–36.

Vasi, Ion Bogdan, and David Strang. "Civil Liberty in America: The Diffusion of Municipal Bill of Rights Resolutions After the Passage of the USA PATRIOT Act." *American Journal of Sociology* 114, no. 6 (2009): 1716–64.

Vedres, Balázs, and David Stark. "Structural Folds: Generative Disruption in Overlapping Groups." *American Journal of Sociology* 115, no. 4 (2010): 1150–90.

"Vienna in Figures 2022." Statistics Vienna, 2022. https://www.wien.gv.at/statistik /pdf/viennainfigures-2022.pdf.

Vox. "A French City Announced It Would Serve Meatless School Lunches. The Backlash Was Swift." April 1, 2021. https://www.vox.com/future-perfect/22360062 /meat-vegetarian-vegan-lyon-france-culture-identity.

Wachsmuth, David. "City as Ideology: Reconciling the Explosion of the City Form with the Tenacity of the City Concept." *Environment and Planning D: Society and Space* 32, no. 1 (2014): 75–90.

Wachsmuth, David, and Hillary Angelo. "Green and Gray: New Ideologies of Nature in Urban Sustainability Policy." *Annals of the American Association of Geographers* 108, no. 4 (2018): 1038–56.

Walker, Edward T., and Andrew Malmuth. "The Natural Gas Industry, the Republican Party, and State Preemption of Local Building Decarbonization." *npj Climate Action* 3, no. 1 (2024): 98.

Walker, Jack L. "The Diffusion of Innovations Among the American States." *American Political Science Review* 63, no. 3 (1969): 880–99.

Warner, Mildred, and Amir Hefetz. "Managing Markets for Public Service: The Role of Mixed Public-Private Delivery of City Services." *Public Administration Review* 68, no. 1 (2008): 155–66.

Weber, Max. *Economy and Society*. University of California Press, 1978.

Whittington, Kjersten B., Jason Owen-Smith, and Walter W. Powell. "Networks, Propinquity, and Innovation in Knowledge-Intensive Industries." *Administrative Science Quarterly* 54, no. 1 (2009): 90–122.

York, Jeffrey G., Siddharth Vedula, and Michael J. Lenox. "It's Not Easy Building Green: The Impact of Public Policy, Private Actors, and Regional Logics on Voluntary Standards Adoption." *Academy of Management Journal* 61, no. 4 (2018): 1492–523.

Young, Dennis R. *If Not for Profit, for What? A Behavioral Theory of the Nonprofit Sector Based on Entrepreneurship*. Lexington, 1983.

Zhang, Lin, Yanliu Lin, Pieter Hooimeijer, and Stan Geertman. "Heterogeneity of Public Participation in Urban Redevelopment in Chinese Cities: Beijing versus Guangzhou." *Urban Studies* 57, no. 9 (2020): 1903–19.

Zukin, Sharon. "Reconstructing the Authenticity of Place." *Theory and Society* 40, no. 2 (2011): 161–65.

Zukin, Sharon, Philip Kasinitz, and Xiangming Chen. *Global Cities, Local Streets: Everyday Diversity from New York to Shanghai*. Routledge, 2015.

Index

Access Living, 174
actorhood, 11–12, 61; Meyer and Jepperson on, 59, 236n31; of organizations, 59; through strategic plans, 56–60
Acuto, Michele, 251n36
Adams, Eric, 36
adaptation, 98, 118–19, 191
ADEME. *See Agence de la Transition Écologique*
administrative adoption, 183, *184*
administrative city action, 2, 19, 221; attribution emphasizing, 12; backlash to, 109; civic capacity influencing, 163, 183, 187; distributed city action relationship with, 149, 196; for green construction, 154; variation in, 150
Africa, strategic plans in, 30
Agence de la Transition Écologique (ADEME), 257n32
agency, 235n26, 235n28

Agenda 21, 68
aggregation, city action through, 11, 12, 23, 138
AI. *See* artificial intelligence
Alvarez, José Luis, 67
American Cities Climate Challenge, 126, 186
American Planning Association, 70
Amsterdam (Netherlands), 37, 54
Angelo, Hillary, 126; on green and gray sustainability, 256n25; *How Green Became Good* by, 232n8; on urban planning, 272n14
Anna Karenina (Tolstoy), 35
anti-immigrant actions, of US federal government, 250n27
AP. *See* LEED Accredited Professionals
architecture companies, green construction specialization of, 169–70
Arizona State University (ASU), 104
Arnstein, Sherry, 253n2, 262n76

artificial intelligence (AI), 124–25
Arup (London, UK), 247n6
ASU. *See* Arizona State University
Asymmetric Society (Coleman), 202
Athens (Greece), 38
attribution, city action through, 11–12, 22, 138, 183
Australia, Sydney, 25, 55
Austria. *See* Vienna
Automata (Caballero) (2025), 229n2
Automaton (city pod design) (2025), *xii*

Baiocchi, Gianpaolo, 272n14
Barber, Benjamin, 260n58
Barron, David, 119, 120
Battilana, Julie, 10
B Corps certification, 157
bee swarm model of social change (Hironaka), 266n18
Behind the Green Mask (conspiracy theory pamphlet), 123
Beijing (China), 245n50
Berger, Peter, 14
Berkowitz, Michael, 39
Blasio, Bill de, 36
Bloomberg, Michael, 25, 26
Bloomberg Center for Cities, at Harvard University, 68
Bloomberg Philanthropies, 72, 221
BREEAM. *See* Building Research Establishment Environmental Assessment Method
Bretton Woods treaties, IMF and World Bank established from, 91
Briscoe, Forest, 128
British Independent, 229n2
Bromley, Patricia, 12, 220; coding guideline of, 42; on hyper-organization, 59, 202; on decoupling, 52–53

Brookings Institution, 241n9
Brundtland Report (1987), 68, 192
Brussels (Belgium), 37–38
building codes, 154
Building Research Establishment Environmental Assessment Method (BREEAM), 178
built environment: energy use of, 147, 152, 264n1; from ICMA Sustainability Survey, 106; IPCC on, 153; retrofitting of, 66. *See also* green construction; Leadership in Energy and Environmental Design
Bulkeley, Harriet, 231n2
bumble-bees, 237n48
Burawoy, Michael, 244n44
Burnham Plan, of Chicago, 57

Caballero, Oscar M, *xii*, 229n2
CALGreen, 180
Calhoun, Craig, 262n76
California (US), 136; CSOs of, 62, 102–3; Energy Code Title 24, 147; Kiva in, 160; Oakland, 256n24; San Mateo Public Library in, 146, 147, 174, 176. *See also* Palo Alto
California Academy of Sciences, 174
Callenbach, Ernest, ix
Campbell, John, 67
carbon emissions: city climate action on, 231n4; LEED credits for, 155–56
Carbon Neutral Cities Alliance: collaboration of, 84; of USDN, 81
carbon neutrality: CPH 2025 strategic plan for, 26, 37, 38, 65, 102; for Dubai, 50; Hong Kong plan for, 240n5
cast of actors, 140; in governance, 41, 202, 205; of organizational infrastructure, 138

catalysis, of distributed adoption, 150–51. *See also* green construction catalysis

cats, red clover relationship with, 237n48

CDC. *See* Centers for Disease Control and Prevention

Census Bureau, 20

Census Bureau, US, 160

Centers for Disease Control and Prevention (CDC), 257n34

C40 cities. *See* City Climate Leadership Group

Chesapeake Bay Foundation, 176

Chicago (US): Burnham Plan of, 57; civic capacity influencing, 162; collaboration of, 80; in competition, 78–79; Global Covenant of Mayors influencing, 84–85; Green Permit Program of, 181; municipal ID in, 250n27; peer cities with, 113–14; strategic plan of, 38–39; University of Chicago, 116

Chicago Center for Green Technology, 177

chief resilience officers (CROs), 71, 189; collaboration of, 83–84; on mitigation and adaptation, 191; of New York City, 25, 26

chief sustainability officers (CSOs), 189; of California, 62, 102–3; of Chicago, 134–35, 261n75; on ICLEI, 123; on built environment, 152; on climate goals, 85, 256n22; on collaboration, 99; on competition, 78–80; on public participation, 134; reactive approach of, 114–15.

childcare centers, 9

China, 31, 128; national planning of, 58, 245n50; resource-conserving city goal of, 50–51; Shanghai, 113–14; strategic plans in, 27, 30, 35, 240n5; Tianjin in, 190

China Sustainable Energy Program, 70

Cincinnati (Ohio, US), 182

cities. *See specific topics*

Cities and States in World History (Tilly), 277n57

city action, 7, 226, 231n5; actorhood influencing, 236n31; through aggregation, 11, 12, 23, 138; through attribution, 11–12, 22, 138, 183; Brookings Institution on, 241n9; civic capacity relationship with, 148, *148*, 204, *204*; through cultural lens, 67; discretion of, 117–18; through distributed adoption, 184; dual embeddedness model of, 16–17, *17*, 18–20, 188; environmental sociology on, 237n45; ICLEI relationship with, 144; inequality of, 208; intercity associations shaping, 93–95, *95*, 96–99, 188; political economy on, 14, 19; public participation in, 134; social structure and, 11–12; staff capacity for, 189; state relationship with, 115–20, *121*; variation in, 113

city administrations, 16, *17*; building codes under, 154; city climate action of, 19, 21–22, 110, 151; civil servants in, 198; civil society organizations and, 199; dual embeddedness of, 110; green construction legitimization of, 180–81; innovation of, 115; intercity association relationship with, 127–28, 225; local government distinguished from, 238n52; organizational charts of, 233n16; principal component analysis of, 109; professional associations

city administrations (*continued*)
targeting, 199; silos within,
49; strategic plans of, 28, 46;
sustainability practices of, 20, 23,
130, 149; of Vienna, 25, 27, 241n8
city climate action, xi, *3*, 6–7, 67, 188;
on carbon emissions, 231n4; city
action distinguished from, 231n5;
of city administrations, 19, 21–22,
110, 151; civic capacity relationship
with, 187; data sources on, *218*;
dual embeddedness influencing,
110; global phenomenon of, 28;
institutional superstructure
relationship with, 18–19, 24, 63, *63*;
Millard-Ball on, 109; nonprofits
influencing, 198; organizational
infrastructure relationship with,
18–19, 24, 137–38; organizational
lens on, 2–3, 8–11, 29, 185; of Palo
Alto, 101–3; public participation
in, 137; rational models defied by,
255n18; social construction of,
239n56; sociocultural perspective
on, 14; as urban innovation, 190–
97; US, 4, 5, 110–12, *112*, 113–15;
variation in, 29, 64, 104, 115, 144,
149. *See also* city learning; city
planning; strategic plans
city climate inaction, 231n2
City Climate Leadership Group (C40),
2, 40, 70, 92, 185, 221; collaboration
of, 80, 82–83, 84; economic status of,
96–97; institutional influence of, 122;
membership to, 89, 199; strategic
plans influenced by, 46, 85, 96–97
city development, 13
city innovativeness index, 194
CityKey, 250n27. *See also* municipal
identification

city learning, 21, 151; intercity
associations facilitating, 143; on
municipal identification, 250n27;
through town halls, 189
City Limits (Peterson), 119
city managers, 225; apolitical stance of,
130–31; professional associations
targeting, 199; US mayors compared
to, 129–31, 132–33, *133*, 144, 187
city networks. *See* intercity associations
city planning, 21
city power: bifurcation of, 205–9, *209*,
210; distributed, 208
civic action, 10, 140
civic capacity, 23, 64, 145;
administrative and distributed
city action influenced by, 163, 183,
187; city action relationship with,
148, *148*, 204, *204*; city climate
action relationship with, 187; green
construction relationship with,
160, 161, *161*, 162–63, 182–83, 184,
187; investments in, 189; nonprofit
relationship with, 19, 140, 141, 160,
187, 263n93; social infrastructures
empowered by, 199; state capacity
relationship with, 150, 151–52
Civic Life of Cities Lab, 59, 140
Civil Rights Act, 116
civil servants, 53, 198
civil service reform, US, 14
civil society, 15, 143, 149, 182, 187, 199;
leaders in 5, 18, 190
civil society organizations, 15, 16–17,
188, 252n41; city administrations
and, 199; within organizational
infrastructure, 197. *See also* civic
capacity; nonprofits
Cleveland (Ohio, US): civic capacity of,
162; strategic plan changes of, 42

climate change, 143; disasters, 6; "grand challenge" and "wicked problem" of, 206, 276n49; master frame of, 27; NYC Office for Climate and Environmental Justice on, 36; political contentions around, 129; strategic plans on, 43, 47, 47, 48; US federal government erasure of, 118. *See also* city climate action

climate goals, 85, 196, 256n22, 272n15

Climate Mayors, 127

Climate Protection Plan, Paris (2007), 31

Close, Philippe, 37–38

coding: categories of, 226; guideline, 42; of strategic plans, 41–42, 47, 47, 48, 242n13

Coleman, James, 202, 236n33

collaboration: of C40 cities, 80, 82–83, 84; competition relationship with, 99; expertise gaps bridged through, 82–83; of intercity associations, 96, 98, 251nn34–36; with peer cities, 79–85; significance of, 85–87, 87, 88–90, 90, 91–92; social interactions routinized with, 83–84; standard setting with, 84–85; strategic plans influenced by, 63, 63, 64–66; of USDN, 80–81, 82, 83, 84, 92

collective efficacy, 198

communities, 9; city action relationship with, 17; organization relationship with, 10–11; place-based, 203, 204; sustainable development goal on, 91, 247n8

competition: Chicago CSO on, 78–80; collaboration relationship with, 99; culture and, 81–82; within economic order, 74–79; between green building certifications, 268n42;

significance of, 85–87, 87, 88–90, 90, 91–92; strategic plans influenced by, 52, 63, 63, 64–66, 78

comprehensive plan, strategic plan compared to, 48–49

concrete manufacturing, 152

confirmation bias, 142

confounders, 109

construction industry, 152–53, 166, 265n9

COP21. *See* United Nations Climate Change Conference

Copenhagen (Denmark), 122; C40 membership of, 89; CPH 2025 strategic plan of, 26, 37, 38, 65, 102; Danish Industry Foundation in, 177; governance tensions with, 117; ranking of, 78

Copenhagen: Solutions for Sustainable Cities, 26, 65

corporate citizens, 202; coding guidelines from, 42; corporation role as, 239n53

corporate demography, 201

corporate social responsibility (CSR), 51, 141, 142, 195

corporations, 16; corporate citizen role of, 239n53; Marquis on, 141–42; rationalized practices of, 59

council-manager systems, US: ICMA advocating for, 129–30; rationalized practices of, 131; sustainability practices of, 132, 133

COVID-19 pandemic, US response to, 116–17

critical urban theorists, 50, 245n49

Cronbach's alpha, 254n12

CROs. *See* chief resilience officers

CSOs. *See* chief sustainability officers

CSR. *See* corporate social responsibility

cultural superstructure, symbiotic
substructure relationship with,
15–16, 238n50
culture, 238n50; of books, 68, 69;
competition and, 81–82; global
scripts influencing, 63, 63, 64–69,
69, 70–71, 71, 72–74; of higher
education, 67–68; of INGOs, 70,
71; of 100 Resilient Cities, 71–72;
organizations shaped by, 12;
significance of, 85–87, 87, 88–90, 90,
91–92
Czarniawska, Barbara, 41

Dakar (Senegal), 239n54
Dallas (Texas, US), 181
Danish Industry Foundation, 177
Darwin, Charles, 237n48
data sources, 218–19
data visualization, 221
decoupling, 51, 52–53, 58
democracy, direct, 137
Democrat vote share, sustainability
practice relationship with, 132–33, 133
Denmark. See Copenhagen
Department of Defense, US, 179
Department of Energy, US, 147–48
Department of Housing and Urban
Development, US, 118, 222, 253n2
Department of Transportation, US, 70
Deslatte, Aaron, 256n24
Detroit (Michigan, US), 73
diffusion, 242n14; of green building
certifications, 164, 165, 166, 166,
167; of green buildings, 159–61, 161,
162–64; S-curve of, 164, 166, 166,
167; of strategic plans, 28, 32–33, 34,
86, 220; of urban innovation, 23
Dillon's rule (1868), 107, 254n9. See also
Home Rule

diplomatic corps, 210
distributed adoption, 183, 184, 269n46;
catalysis of, 150–51; city action
through, 184; legitimization and
scaling of, 151
distributed city action, 2, 12, 19, 221;
administrative city action relationship
with, 149, 196; backlash to, 209; civic
capacity influencing, 163, 183, 187;
green construction as, 152–55
Dobbin, Frank, 111
Dodge Data and Analytics, 173, 268n37
Doorninck, Marieke van, 37
DPR Architecture, 170
Drori, Gili, 131
dual embeddedness, 5, 13; between-
city inequalities explained by, 193;
city action model, 16–17, 17, 18–20,
188; of city administrations and
city climate action, 110; within
organizational infrastructure and
institutional superstructure, 16, 17;
strategic plans influenced by, 94;
sustainability practice correlation
with, 112
Dubai (United Arab Emirates), 50
Dunning, Claire, 10, 160, 263n88,
274n31
Durkheim, Émile, 13–14, 86

East Palo Alto (California, US), 253n3
Eco-City Tianjin, 190
ecological modernization, 237n45
ecological transition, 265n4
Economic Intelligence Unit, 76
economic interests, 13, 135, 136–37
economic order: competition within,
74–79; within intercity associations,
93–94, 248n16; rankings of, 75–76
economic production, 237n45, 272n15

economic status: of C40 cities, 96–97; strategic plan diffusion based on, 33, 34

Economist, 76

Economist Impact, Resilient Cities Index by, 76–77

Economy and Society (Weber), 15

Ecotopia (Callenbach), ix

education: backlash against, 202; for executives, 68; nonprofit organizations in, 171; relationship between city action and, 43, 59, 67, 73, 124, 129, 151, 198, 256n24

Eigenvalue, 254n12

Elementary Forms (Durkheim), 13–14

elitism, 248n14

embeddedness: of organizations, 250n33; rationalized practice relationship with, 131. *See also* dual embeddedness

Energy Code Title 24 (California), 147

EnergyStar accreditations, 102, 178

energy use: of built environment, 147, 152, 264n1; LEED, 172–73

Engel, Friedrich, 13, 86

environmental, social, and governance reporting (ESG), 157

environmental footprint, 6

environmental justice, 7, 193

Environmental Protection Agency (EPA), 116, 158, 258n38, 264n1

environmental sociology, 237n45

EPA. *See* Environmental Protection Agency

ESG. *See* environmental, social, and governance reporting

ethnography, 227

European Commission, information and communications technology strategy of, 69–70

European Environmental Agency, 264n1

European Innovation Partnership for Smart Cities and Communities, 69–70

evaluative devices, 75–76

event-history models, 94

expertise gaps, collaboration bridging, 82–83

explanatory variables, of institutional superstructure and organizational infrastructure, *218*

Fair Trade, 157

federal government, US: anti-immigrant actions of, 250n27; on built environment energy use, 264n1; climate change erasure by, 118; COVID-19 pandemic response failure of, 116; green construction of, 179–80

Feiock, Richard, 131, 255n16, 255n18

field mice, 237n48

fifteen-minute city, 123–24

firms, Gibbons on, 235n27

Fish and Wildlife Service (FWS), 116

Fligstein, Neil, 275n41

floating signifiers, 239n56

Florida, Richard, 25

Florida (US): adaptation action area in, 118–19; Richard, 25

Floyd, George, 42, 192–93

France: *Agence de la Transition Écologique* in, 257n32; Climate Protection Plan of Paris, 31; Green Party in, 259n53

Frank, David, 73

Fredericks, Rosalind, 239n54

Fridays for Future, 65

Friend, Gil, 102–3

Frug, Gerald, 119, 120
FWS. *See* Fish and Wildlife Service

Galaskiewicz, Joseph, 139, 234n25
Garden City of To-Morrow (Howard), ix
Gates, Bill, 152, 190
GaWC. *See* Globalization and World
Cities
Gehl Architects, 69, 135
General Services Administration
(GSA): Green Globes certifications
of, 178, 180; LEED certifications of,
179
Gensler (San Francisco, US), 170, 247n6
German Ruhrtal, urban greening in,
232n8
GHGs. *See* greenhouse gas emissions
Gibbons, Bob, 235n27
Glaeser, Ed, 81
Global Covenant of Mayors, 2, 84–85,
127
globalization, 86
Globalization and World Cities
(GaWC), *34*, 75, 88, 220
Global North, 33
Global Parliament of Mayors, 127,
260n58
Global South, 193, 227, 277n53
Gonzalez, Ana, 114, 124, 222, 226
Google, 202; Sidewalk Labs, 69; smart
city design by, 190
governance, 203, 226, 244n42, 255n17;
cast of actors in, 41, 202, 205;
competing levels of, 117–18, 119, 207;
gaps, 276n50; higher education on,
67–68; isomorphism of, 119–20,
261n74; market perspectives
dominating, 197; multilevel, 73;
organizational infrastructure
influencing, 140; organizations

influencing, 139. *See also* city
administrations; council-manager
systems, US; US federal government
grand challenge, of climate change,
206, 276n49
Granovetter, Mark, 269n1
gray sustainability, 256n25
Great Recession, green construction
during, 153, 166, 173
green building certifications, 155–59,
196; competition between, 268n42;
diffusion of, 164, *165*, 166, *166*, 167;
public policy on, 178, 181, 182; US
Department of Energy on, 147–48.
See also Leadership in Energy and
Environmental Design
green buildings: diffusion of, 159–61,
161, 162–64; for San Mateo Public
Library, 147, 174, 176
green construction, 220–21;
architecture company specialization
in, 169–70; of city administrations,
180–81; civic capacity relationship
with, 160, 161, *161*, 162–63, 182–83,
184, 187; costs of, 173, 268n36; as
distributed city action, 152–55; by
EPA, 158; during Great Recession,
153, 166, 173; institutionalization
of, 183; jobs, 265n9; legitimization
of, 167, 178–82; market for,
152–53; mitigation with, 266n12;
for office buildings, 167–68, 170;
organizational infrastructure
shaping, 148, *148*, 149; of US
federal government, 179–80. *See*
also Leadership in Energy and
Environmental Design
green construction catalysis, 167;
with nonprofits, 174–75, *175*, 176;
through proof of concept, 171–72,

172, 173–75, *175*, 176–78; USGBC on, 171, 174

Green Globes, 178, 180

greenhouse gas emissions (GHGs), 8; Global Covenant of Mayors pledge reducing, 2; ICLEI inventory of, 121; Palo Alto on, 101, 253n3; scopes of, 255n13, 271n12; strategic plans on, 52; from urbanization, 231n2

Green Party (France), 259n53

Green Permit Program, of Chicago, 181

green sustainability, 256n25

greenwashing, of LEED, 157–58

Greenwich Village (New York City), 270n4

GSA. *See* General Services Administration

Guide to Greening Cities, The (Johnson), 81

Hannan, Michael, 94

hard power, soft power relationship with, 206

Harvard University, 68, 98

Hawkins, Christopher, 256n24

Herron, Ron, ix, *xii*, 185, 229n2

Hewlett Foundation (Palo Alto, California), 176, 221

higher education, on urban management and governance, 67–68

Hironaka, Ann, 73

Hiroshima (Japan), 39–40, 44

HKU. *See* University of Hong Kong

HOK Group, 170

Home Rule, 254n9

Hong Kong (China), 27, 240n5

Howard, Ebenezer, ix

How Green Became Good (Angelo), 232n8

How to Avoid a Climate Disaster (Gates), 152, 190

Hsu, Angel, 231n4

human ecology, 15

humble-bees, 237n48

Hurricane Sandy, 25, 135, 247n10

hyper-organization, 59, 202

IBM. *See* International Business Machines Corporation

ICLEI. *See* Local Governments for Sustainability

ICMA. *See* International City and County Management Association

ICT. *See* information and communications technology

If Mayors Ruled the World (Barber), 260n58

Illinois Green Alliance, 177, 222

IMF. *See* International Monetary Fund

immigration laws, US, 115

implementation: gap, 195; science, 9; of strategic plans and sustainability strategies, 194–95

India, 31; Mumbai and Surat in, 241n10; strategic plans in, 30, 35

inequality: between cities, 193; of city action, 208; from organizations, 275n42; in urban sustainability, 192, 194

information and communications technology (ICT), 69–70

INGOs. *See* international nongovernmental organizations

innovation: of city administrations, 115; LEED credits for, 156–57; public policy, 271n8. *See also* urban innovation

institutional influences, of intercity associations, 122–23

institutionalization, 14; of green
  construction, 183; of LEED, 167–68,
  173, 182
institutional superstructure, 3, 17,
  87, 149, 187; city climate action
  relationship with, 18–19, 24, 63,
  63; dual embeddedness within, 16,
  17; evolution of, 67; explanatory
  variables related to, *218*; intercity
  associations included in, 185;
  organizational infrastructure
  relationship with, 23, 100, 104,
  104, 137–43, 164, 185, *186*, 188,
  221; strategic plans influenced
  by, 22, 60–61, 85–86, 93. *See also*
  collaboration; competition; culture
institutional theory, 204–5
institution nexus, 55
intercity associations, 20, 89, *90*, 92;
  bifurcation of, 210; city action
  shaped by, 93–95, *95*, 96–99, 188;
  city administration relationship
  with, 127–28, 225; city learning
  facilitated by, 143; collaboration
  of, 96, 98, 251nn34–36; economic
  order within, 93–94, 248n16; GaWC
  models of, 88; institutional influence
  of, 122–23; within institutional
  superstructure, 185; interview
  protocol for, 224; isolate, 90–91;
  New York City affiliations with, 66;
  position within, 93–95, *95*, 96–97,
  99–100, 220, 251n36; ranking
  relationship with, 126; regional,
  125–26; soft power of, 207; strategic
  plans influenced by, 85, 93–95,
  95, 99, 186, 252n43; sustainability
  practices increased with, 186–87;
  transnational, 277n53. *See also* City
  Climate Leadership Group; 100

Resilient Cities; UN-Habitat; Urban
  Sustainability Directors Network
interest groups, 269n49
Intergovernmental Panel on Climate
  Change (IPCC), 38, 153
International Business Machines
  Corporation (IBM), 69, 76
International City and County
  Management Association (ICMA),
  20, 70, 104; simple ordinary least
  squares regressions used in, 220;
  Sustainability Survey of, 105, 106,
  108, 220, 254n11
International Code Council, 154
International Monetary Fund (IMF), 91
international nongovernmental
  organizations (INGOs), 247n8;
  culture of, 70, *71*; strategic plans
  influenced by, 93
International Organization for
  Standardization (ISO), 76
Internet Archive, 30–31, 264n1
interorganizational associations, *17*
interpersonal networks, 269n1
interviews, 20, 21; city and state
  characteristics in, *228*; geographic
  regions of, 227, 239n55; guideline for,
  222, 225; protocol for, 222, 223–24;
  respondents for, 221–22, 225, 227;
  with USGBC, 159
IPCC. *See* Intergovernmental Panel on
  Climate Change
IRB. *See* Stanford Institutional Review
  Board for Human Subjects
ISO. *See* International Organization for
  Standardization
isolate intercity associations, 90–91
isomorphism, 41, 119–20, 261n74,
  262n77
isopraxis, 41

Jacobs, Jane, 270n4
Jepperson, Ron, 59, 236n31
Jessop, Bob, 74
Jim Crow laws, 116
Johnson, Sadhu Aufochs, 81
just transition, 7, 193–94

KCAP Architects and Planners
   (Rotterdam, Netherlands), 247n6
Keene, Jim, 102
Kiva (San Francisco, California), 160
Kline, Kendra, 222
Klinenberg, Eric, 6, 232n9
Kornberger, Martin, 55
Krause, Rachel M., 255n16, 256n24
Krinsky, John, 239n54
Kuala Lumpur (Malaysia), 39
Kyoto Protocol (1992), 1, 27

labor, organizational infrastructure of,
   239n54
Lampedusa Charter, 115
land use, from ICMA Sustainability
   Survey, 106
Latin America, strategic plans in, 30
Laumann, Ed, 139
Leadership in Energy and
   Environmental Design (LEED), 160,
   265n9; brand recognition of, 168–69;
   civic capacity relationship with, 148;
   credits for, 155–57; energy use of,
   172–73; Gold certification, 147, 148,
   156, 180; greenwashing of, 157–58;
   institutionalization of, 167–68, 173,
   182; Platinum certification, 148, 156,
   176, 177; public policy and, 181, 182;
   registrations, 171, 172, 175; scaling
   of, 167; sectors adopting, 175, 175;
   Silver certification of, 156, 159; in US
   cities, 158–59, 161, 161, 162, 164, 165,

166, 166, 167, 175; US Secretary of
   the Navy on, 179–80. See also green
   building certifications; US Green
   Building Council
League of Nations, Covenant of, 91
LEED Accredited Professionals (AP),
   169, 171, 183
LEED Cities, 179
LEED Home, 171
LEED Project Directory, 221
LEED Public Policy Library, 221
LEED Volume, 170
Leffel, Benjamin, 251n36
legitimization: of distributed adoption,
   151; of green construction, 167,
   178–82
L'Enfant, Pierre Charles, 57
Levine, Jeremy R., 143, 239n56
Line (horizontal skyscraper) (Saudi
   Arabia), 190
livability rankings, 76
Local Governments for Sustainability
   (ICLEI), 70, 76, 97, 185, 221; city
   action relationship with, 144;
   CSO on, 123; on fiscal savings and
   sustainability, 135; GHG inventory
   of, 121; membership of, 88, 199; for
   smaller and midsize cities, 113, 122;
   sustainability practice relationship
   with, 122, 123; US council-manager
   system advocated for by, 129–30;
   World Congress of, 82
Luckmann, Thomas, 14
Lucknow (India), 31

Malmuth, Andrew, 258n42
Mansueto Institute of Urban
   Innovation, at University of
   Chicago, 116
Marquis, Christopher, 10, 141–42

Marwell, Nicole, 138–39, 234n21; on nonprofits, 140, 143; on organizational action, 274n28

Marx, Karl, 13

master frame: of climate change, 27; of strategic plans, 240n6, 242n20

Matthew effect, 129

mayors: Close, 37–38; Global Covenant of Mayors, 2, 84–85, 127; Global Parliament of Mayors, 127, 260n58; of Kuala Lumpur, 39; strategic plans announced by, 33

mayors, US, 145; city climate action led by, 5; city managers compared to, 129–31, 132–33, *133*, 144, 187; intercity associations supported by, 127–28; of NYC, 25, 26, 36; Paris Agreement reaffirmed by, 1–2, 4; Republican, 128, 129

Mazzucato, Mariana, 54

McAdam, Doug, 275n41

McKinsey, 135

McQuarrie, Michael, 138–39, 234n21, 274n28

Medici, Cosimo di, 11

Mercer, 27, 76, 77, 240n4

meso-level organizational societies, 202–4, *204*, 205

Meta, 103, 202

methodology, 240n1; data sources and variables, *218–19*; for strategic plans, 30–31, 217, 220

Metropolis, 97, 252n43

Meyer, John, 14; on actorhood, 59, 236n31; on decoupling, 51, 58; on hyper-organization, 202; on world polity, 252n41

Miami Herald, 118–19

Millard-Ball, Adam, 109

Minnesota, Twin Cities of, 234n25

Mitchell Park Community Center (Palo Alto, California), 102

mitigation, 191, 266n12

Molotch, Harvey, 248n14

moonshots, 54

Morenos, Carlos, 123–24

motivations, 111

multiplexity, 250n33

Mumbai (India), 241n10

Mumford, Lewis, x

municipal identification (municipal ID), 115, 250n27. *See also CityKey*

National Center on Charitable Statistics, 20

National Geographic, 50

National League of Cities (NLC), 88, 221

National Oceanic and Atmospheric Administration (NOAA), 116

National Public Radio (NPR), 160

National Resource Defense Council, 145

Nature Cities, 91

Nature of the City, The (Weber), 15

Naval facilities, US, 180

New York City (US): C40 membership of, 89; global scripts influencing, 73; Greenwich Village, 270n4; intercity association affiliations with, 66; mayors of, 25, 26, 36; municipal ID in, 250n27; Office for Climate and Environmental Justice, 36; Staten Island, 119; sustainability planning of, 25; 30 Rockefeller Plaza in, 72, 92. *See also* OneNYC; PlaNYC

Nixon-in-China effect, 128

NLC. *See* National League of Cities

NOAA. *See* National Oceanic and Atmospheric Administration

nonprofits, 10, 16–17; city climate action influenced by, 198; civic capacity relationship with, 19, 140, 141, 160, 187, 263n93; Dunning on, 263n88, 274n31; environmental, 163, 198; green construction catalysis with, 174–75, *175*, 176; infrastructure compared to, 263n88; politics influenced by, 143; rationalized practices of, 59; sustainability practice relationship with, 140–41, 142, 145; urban outcomes impacted by, 200–201; urban sustainability influenced by, 198–99

NPR. *See* National Public Radio

NVivo, 226

Oakland (California, US), 256n24

Obama, Barack, 4, 133, 179, 232n9

office buildings, green construction for, 167–68, 170

Office for Climate and Environmental Justice, NYC, 36

Ohio (US): Cincinnati, 182; Cleveland, 42, 162

oil crisis (1973), 168

OLS. *See* simple ordinary least squares regressions

100 Resilient Cities (100RC), 39, 42, 185, 221; city size range within, 113; criteria for, 127, 199; funding from, 186; media effort of, 72; on resilience, 205; of Rockefeller Foundation, 70, 71–72, 247n10; strategic plans and, 46, 51–52, 85; technical assistance offered by, 70; 30 Rockefeller Plaza center of, 72, 92. *See also* Resilient Cities Network

OneNYC, 36; comprehensive lens of, 49; public participation in, 134

open systems, in organizational studies, xi, xiii, 230n7

organizational action, 274n28

organizational infrastructure, 3, 145, 187, 198; adoption and, 150–52; cast of actors within, 138; city climate action relationship with, 18–19, 24, 137–38; civil society organizations within, 197; dual embeddedness within, 16, *17*; enabling, 18; explanatory variables related to, *218*; governance influenced by, 140; green construction shaped by, 148, *148*, 149; institutional superstructure relationship with, 23, 100, 104, *104*, 137–43, 164, 185, *186*, 188, 221; of labor, 239n54; of nonprofits, 274n31; research opportunities on, 202; sustainability strategies shaped by, 22; urban innovation impacted by, 194; variation in, 201. *See also* dual embeddedness

organizational learning, 80

organizational lens, on city climate action, 2–3, 8–11, 29, 185

organizational sociology, place-based, 203

organizational studies, 273n21; open systems in, xi, xiii, 230n7; urban studies interdisciplinary connection with, x, xi, 9, 10

organizational theory: on decoupling, 51, 52; meso-level, 203

organizational thinking, 59

organization of cities, 197–202

organizations, 9, 234n25, 238n51, 275n41; actorhood of, 59; Coleman on, 236n33; community relationship with, 10–11; cultural norms and expectations shaping, 12;

organizations (*continued*)
    embeddedness of, 250n33; inequality originating from, 275n42; roles of, 244n42; urban scholars neglecting, 234n21; urban sociology on, 138–39, 234n21
Organizations in Action (Thompson), x
Ottawa (Canada) (2003), 31

Padgett, John, 10–11, 246n2
Palantir, 103
Palo Alto (California, US), 66, 253nn3–4; city climate action of, 101–3; Hewlett Foundation in, 176, 221; town hall in, 101–3, 134, 136
Paradox of Community Power, The (Levine), 239n56
Paris (France), 31
Paris Agreement (2016), 144, 154; city commitments to, 231n3, 256n22; strategic plans contributing to, 39; US withdrawal from, 1–2, 4, 62, 72
Park, Robert, 86, 204, 237n47; on symbiotic and cultural societies, 238n50; on symbiotic substructure and cultural superstructure, 15–16, 238n50; on web of life, 200
partisanship, 144, 187, 207–8
peer cities, 65–66, 98; city size relationship with, 113–14; collaboration with, 79–85
Pentagon, 180
performance indicators, of strategic plans, 45
Perrow, Charles, 202
Peterson, Paul, 119, 154
Philip Merrill Environmental Center, 176
place-based communities, 203, 204
PlaNYC, 25, 26–27, 31, 40, 253n44; Organics program, 38; public

participation in, 134; 2023 update of, 36–37. *See also* OneNYC
Poisson models, 221
policy think tanks, 241n9
political economy, 13, 86; on city action, 14, 19; on competition, 74
politics: city manager stance on, 130–31; around climate change, 129; coalitions in, 74; nonprofits influencing, 143
population, 6
Portney, Kent, 110
Porto Alegre (Brazil), 89
positionality, 226
Powell, Woody, 52–53, 246n2
PR. *See* public relations
principal component analysis, of city administrations, 109
private sector, 15
professional associations, 18, 22–23; agendas of, 98; city managers and administrators targeted by, 199; strategic plan relationship with, 94, 96, 97
protests, for Floyd, 42, 192
public housing, in Vienna, 27, 239n56
public participation: Arnstein critiquing, 253n2, 262n76; in city action, 134; city administrations involving, 151; in city climate action, 137; in strategic plans, 135–36
public policy, 197; on green building certifications, 178, 181, 182; green construction legitimization through, 178–82; ICMA Sustainability Survey targets for, 106; innovation, 271n8; LEED Public Policy Library, 221; negative and positive feedback in, 273n19
public relations (PR), 51

qualitative analysis, 221–27, *228*
quantitative datasets, 20, 21

racism, institutional, 193
railway systems, Dobbin analyzing, 111
Rainforest Alliance, 157
rankings, 249n19; of economic
  order, 75–76; intercity association
  relationship with, 126; of Vienna,
  77–78
rationalized practices, 22; of city
  climate commitments, 149;
  embeddedness relationship with,
  131; of rhetorical commitments, 64;
  of strategic plans, 58, 59, 60; Suárez
  on, 105; of US council-manager
  systems, 131; variation of, 73
rational models, city climate action
  defying, 255n18
receptor cites, 73
recognition programs, from ICMA
  Sustainability Survey, 106
recycling, from ICMA Sustainability
  Survey, 106
red clover, cats influencing, 237n48
regulation, state and national, 116, 180
Renaissance Florence, 10–11
renewable energy, 153; strategic plan of
  Washington, D.C. on, 52; in Texas,
  119
Republican mayors, US, 128, 129
resilience: 100 Resilient Cities on, 205;
  in strategic plans, 48; sustainability
  compared to, 191
Resilience by Design, 50, 221
Resilient Cities Index, by *Economist
  Impact*, 76–77
Resilient Cities Network, 39, 72, 221.
  See also 100RC
retrofitting, of built environment, 66

Rio Conference on Sustainable
  Development (1992), 68, 70, 73
Rockefeller Foundation, 48, 70, 71–72,
  221, 247n10
Rogers, Everett, 32
Rotterdam (Netherlands): KCAP
  Architects and Planners in, 247n6;
  strategic plan of, 39
Rowan, Brian, 14, 51, 58
Royal Danish Academy, 78

Safford, Sean, 128, 139
Saint Paul (Minnesota, US), 136
same-sex partner benefits, 128
sanctuary cities, 115
San Francisco (US): capital plan, 195;
  Gensler in, 170, 247n6; Kiva in, 160;
  strategic plan of, 38
San Francisco Bay Area Planning
  and Urban Research Association
  (SPUR), 65, 199
San Mateo Public Library (San Mateo,
  California), 146, 147, 174, 176
scaling, 21; of distributed adoption, 151;
  of green construction, 167–70
Schneiberg, Marc, 138
Schofer, Evan, 73
S-curve, of diffusion, 164, 166, *166*, 167
SDGs. *See* sustainable development
  goals
Secretary of the Navy, US, 179–80
Senegal, Dakar, 239n54
SGR. *See* Smart, Green, Resilient
  strategy
Shanghai (China), 113–14
Sharkey, Amanda, 12, 42
Sharkey, Patrick, 200
shelter-in-place policies, 258n35
Shi, Linda, 53
Shils, Edward, 238n49

Sidewalk Labs, 69
Siemens, 76
Sierra Club, 145
Simon, Herbert, 202
Simonet, Maud, 239n54
simple ordinary least squares
    regressions (OLS), 220
Singapore: resource-conserving city
    goal of, 50–51; World City Summit
    in, 82
Singh, Chandni, 241n10
Sister City Network, 92
Small, Mario, 9–10
small cities: ICLEI for, 113, 122;
    Paris Agreement commitment of,
    256n22
Smart, Green, Resilient strategy (SGR),
    240n5
social action, 11, 12, 99, 235n26
social capital, civic capacity relationship
    with, 163
social construction, xi, xiii, 239n56
social infrastructures, 199, 273n26
social interactions, collaboration
    routinizing, 83–84
social norms, 13
social structure, city action and, 11–16
social theory, meso-foundations
    of, 204
Society of Organizations, A (Perrow),
    202
sociocultural perspective, on city
    climate action, 14
soft power, 196, 197; hard power
    relationship with, 206; of intercity
    associations, 207
Soule, Sarah, 138
Southern Sustainability Director's
    Network (SSDN), 221; expertise gaps

bridged by, 82; safe space created
    by, 126
SPUR. *See* San Francisco Bay Area
    Planning and Urban Research
    Association
SSDN. *See* Southern Sustainability
    Director's Network
stakeholders, in strategic plans, 54, 55
Stanford Institutional Review Board for
    Human Subjects (IRB), 222
Stanford Social Innovation Review, 177
Stanford University, 59, 103, 140, 160,
    253n4
STAR Communities, 179
state capacity, civic capacity
    relationship with, 150, 151–52
State Department, US, 4
Staten Island (New York City, US), 119
states, 209, *209*; city action relationship
    with, 115–20, *121*; interview sample
    characteristics of, *228*; preemption
    of, 119, 144, 258n42; regulation of,
    116, 180; USGBC on, 179; variation
    within, 120, *121*
steel manufacturing, 152
STEP 2000, of Vienna, 31
Stokes, Leah, 273n19
Strang, David, 94
strategic plans, 28, 32, 36, 244n44;
    absence of, 33, 35; actorhood
    through, 56–60; of Amsterdam,
    37, 54; C40 cities influencing,
    46, 85, 96–97; change tracked
    in, 46–47, *47*, 48; of Chicago,
    38–39; in China, 27, 30, 35, 240n5;
    on climate change, 43, 47, *47*,
    48; coding of, 41–42, 47, *47*, 48,
    242n13; collaboration influencing,
    63, *63*, 64–66; competition

influencing, 52, 63, *63*, 64–66, 78; of Copenhagen, 26, 37, 38, 65, 102; diffusion of, 28, 32–33, *34*, 86, 220; equity in, 191–92; of Hiroshima, 39–40, 44; implementation of, 194–95; institutional superstructure influencing, 22, 60–61, 85–86, 93; intercity associations influencing, 85, 93–95, *95*, 99, 186, 252n43; interview protocol for, 223–24; master frame of, 240n6, 242n20; methodology for, 30–31, 217, 220; 100 Resilient Cities and, 46, 51–52, 85; private decisions influenced by, 245n48; professionals driving, 53–54; PR stunts, 51; public participation in, 135–36; purposive, 29; rank mobility through, 78; rationalized practice of, 58, 59, 60; SDGs linked to, 40, 45, 58, 244n45; sustainability practice relationship with, 109; of Sydney, 25, 55; thematic emphases of, 43–44, *44*; time frame of, 42–43; tools of, 45, *45*, 46; trustworthiness of, 48–56; turn toward, 240n6, 241n7; variation in, 40, 46, 66, 205; of Vienna, 27, 28, 38, 54–55, 58; of Washington, DC, 48–49, 52. *See also* city climate action; culture; PlaNYC; sustainability strategies

strategy scholars, 245n49

Suárez, David, 105, 122, 142, 220

Surat (India), 241n10

sustainability, 226; Brundtland report defining, 68, 192; economic interests for, 135, 136–37; green and gray, 256n25; New York City planning for, 25; resilience compared to, 191. *See also* urban sustainability

sustainability practices: of city administrations, 20, 23, 130, 149; CSR relationship with, 142; Democrat vote share relationship with, 132–33, *133*; endogeneity of, 110; ICLEI relationship with, 122, *123*; from ICMA Sustainability Survey, 105, 106, 108, 220, 254n11; intercity associations increasing, 186–87; nonprofit relationship with, 140–41, 142, 145; self-reported, 108; strategic plan relationship with, 109; in US, 107–8, 112, *112*, 113, 132, *133*; of US council-manager systems, 132, *133*

sustainability strategies, 23; implementation of, 194–95; organizational infrastructure shaping, 22; of US cities, 108

Sustainability Survey, ICMA, 105, 106, 108, 220, 254n11

sustainable development goals (SDGs), 206, 233n16; strategic plans linked to, 40, 45, 58, 244n45; Sustainable Cities and Communities goal of, 91, 247n8; Vienna influenced by, 196–97

Sustainable San Mateo, 199

Svejenova, Silviya, 67

Sydney (Australia), 25, 55

symbiotic substructure, cultural superstructure relationship with, 15–16, 238n50

Taylor, Peter, 75

technological solutions, 7, 190

term frequency and inversed document frequency (TF-IDF), 243n29

Tesseract, 243n29

Texas (US): Dallas, 181; renewable
    energy in, 119
text as data, 243n28
TF-IDF. *See* term frequency and
    inversed document frequency
30 Rockefeller Plaza (New York City),
    72, 92
Thompson, James, x
Tianjin (Eco-City) (China), 190
Tilcsik, András, 141–42
Tilly, Charles, 209, 210; *Cities and
    States in World History* by, 277n57;
    on trust networks, 277n58
Tokio Marine Group, 77
Tolbert, Pam, 177
Tolstoy, Leo, 35
Toronto (Canada): Chicago peer city
    with, 113; smart city design by
    Google in, 190
town halls: city learning through, 189;
    in Palo Alto, 101–3, 134, 136
transportation: from ICMA
    Sustainability Survey, 106; in
    strategic plans, 43
Treaty of Versailles, 91
Trump, Donald, 1, 170
Trump administration, 207
trust network, 277n58
Tuma, Nancy, 94
Twin Cities, of Minnesota,
    234n25

UCLG. *See* United Cities and Local
    Governments
UN. *See* United Nations
UN-Habitat, 76, 90; C40 cities
    compared to, 97; World Urban
    Forum of, 82, 96
United Cities and Local Governments
    (UCLG), 115

United Nations (UN): charter of, 91;
    conspiracy theory on, 123. *See also*
    sustainable development goals
United Nations Climate Change
    Conference (COP21) (2015), 4
United Nations World Urbanization
    Prospects, 6
United States (US): city action in,
    115–20, *121*; city climate action, 4,
    5, 110–12, *112*, 113–15; civil service
    reform in, 14; COVID-19 pandemic
    response of, 116–17; Detroit,
    73; green building certification
    diffusion in, 164, *165*, 166, *166*,
    167; green building diffusion in,
    159–61, *161*, 162–64; immigration
    laws, 115; intercity associations
    in, 125–26; LEED in, 158–59, 161,
    161, 162, 164, *165*, 166, *166*, 167,
    175; Paris Agreement withdrawal
    of, 1–2, 4, 62, 72; Saint Paul, 136;
    sustainability practices in, 107–8,
    112, *112*, 113, 132, *133*; sustainability
    surveyed across, 104–10. *See also*
    California; Chicago; council-
    manager systems, US; Florida;
    mayors, US; New York City; Ohio;
    Palo Alto; San Francisco; Texas;
    Washington, D.C.
University of Chicago, Mansueto
    Institute of Urban Innovation
    at, 116
University of Hong Kong (HKU), 27
UN Population, 217
urban drama, x
urban grant economy, 139
urban greening, in German Ruhrtal,
    232n8
*Urban Imperative Towards Competitive
    Cities, The* (Glaeser) (2015), 81

urban innovation, 24, 198; city climate action as, 190–97; diffusion of, 23; variation in, 194
Urban Institute, 83, 127, 222
urbanization, GHGs from, 231n2
urban management: in books, 68, 69; consultants in, 69; higher education on, 67–68
urban planning: Angelo and Baiocchi on, 272n14; Jacobs on, 270n4; strategic plans compared to, 48
urban revolution, 115, 119
urban-rural divide, 117
urban scholars, organizations neglected by, 234n21
urban social movements, 17
urban sociology, 50; interest groups influencing, 269n49; on organizations, 138–39, 234n21
urban studies, organizational studies interdisciplinary connection with, x, xi, 9, 10
urban sustainability, 231n2; distribution of, 220; green sustainability and gray sustainability of, 256n25; inequality in, 192, 194; nonprofits influencing, 198–99; partisan issue of, 144; variation in, 256n24
Urban Sustainability Directors Network (USDN), 221; collaboration facilitated by, 80–81, 82, 83, 84, 92; institutional influence of, 122. *See also* Southern Sustainability Directors Network
urban theory frameworks, 143
US. *See* United States
USDN. *See* Urban Sustainability Directors Network

US Green Building Council (USGBC), 20, 22, 147, 221; on demonstration projects, 176–77; on green construction catalysis, 171, 174; interviews with, 159; public policy and, 178, 181; on states, 179. *See also* Leadership in Energy and Environmental Design

Vancouver (Canada), 38
Vargas, Robert, 200–201
variables, for quantitative studies, *218–19*
Vienna (Austria), 57, 107; city administration of, 25, 27, 241n8; in competition, 79; public housing in, 27, 239n56; ranking of, 77–78; SDGs influencing, 196–97; STEP 2000 of, 31; strategic plan of, 27, 28, 38, 54–55, 58

Wachsmuth, David, 126, 256n25
Walker, Edward T., 258n42
Walking City on the Ocean (Herron) (1966), ix, *xii*, 185, 229n2
Washington, DC (US): city climate action of, 4; LEED certifications in, 158–59, 175; L'Enfant planning, 57; strategic plan of, 48–49, 52
water, from ICMA Sustainability Survey, 106
Weber, Max, 16, 86, 237n47; on capitalistic and bureaucratized society, 238n49; *The Nature of the City* and *Economy and Society* by, 15
We're Still In campaign, 1–2, 62
WhatsApp, collaboration through, 84

White House Rose Garden, 1
wicked problem, of climate change,
206, 276n49
Wired, Klinenberg in, 232n9
World Bank, 135; Bretton Woods
treaties establishing, 91; *The Urban
Imperative Towards Competitive
Cities* funded by, 81
World City Summit (Singapore), 82
World Health Organization, 73

world polity, 252n37, 252n41
world society, 252n41
World Urban Forum, of UN-Habitat,
82, 96
Wounded City (Vargas), 200–201

Yearbook of International
Organizations (2022), 70, 247n8

Zucker, Lynne, 177

GPSR Authorized Representative: Easy Access System Europe, Mustamäe tee
50, 10621 Tallinn, Estonia, gpsr.requests@easproject.com

www.ingramcontent.com/pod-product-compliance
Lightning Source LLC
Chambersburg PA
CBHW022136020426
42334CB00015B/913